Serendipity in Science

Research Lab hallway at opening of Niskayuna, NY GE Research Center, 1950. Left front to back: Dr. V. J. Schaefer; Dr. B. Vonnegut; Dr. J. D. Cobine; Dr. K. B. Blodgett; Mr. A. H. Young; Dr. J. R. C. Brown; Dr. P. D. Johnson. Right front to back: Dr. W. R. Whitney; Dr. C. J. Gallagher; Dr. A. W. Hull; Dr. S. Dushman; Dr. J. M. Lafferty; Dr. S. Roberts; Dr. J. S. Prener; Mr. J. Martens, Dr. L. R. Koller; Mr. E. Finley; Mr. R. W. Larson. Courtesy of Schenectady Museum.

Serendipity in Science

Twenty Years at Langmuir University

An Autobiography
by
Vincent J. Schaefer, ScD

Compiled and Edited by Don Rittner

Square Circle Press
Voorheesville, New York

Serendipity In Science: Twenty Years at Langmuir University
An Autobiography by Vincent J. Schaefer, ScD

Published by
Square Circle Press LLC
137 Ketcham Road
Voorheesville, NY 12186
www.squarecirclepress.com

First American paperback edition, 2013. Revised with corrections, 2016.
Printed and bound in the United States of America on acid-free, durable paper.
ISBN 13: 978-0-9856926-3-6
ISBN 10: 0-9856926-3-4
Library of Congress Control Number: 2013933876

Compiled and Edited by Don Rittner.

Original transcription by Michael Sullivan.

Foreword by Michael Sullivan.

Preface by Don Rittner.

A Brief Biographical Sketch of Vincent J. Schaefer by Duncan Blanchard.

Publisher's Acknowledgments
Cover ©2013 by Square Circle Press; design by Richard Vang. Illustrations are credited within the text.

The acknowledgments of the Editor appear elsewhere in this book.

A portion of the proceeds from this book will be used for a college scholarship and grants program for students in the field of atmospheric science.

Contents

V Remembering Vince

Chapter 13 Vince's Colleagues Recollect

Chapter 14 Vincent Schaefer's Serendipitous World

Foreword

His formal education cut short by the need to help support his family during the Great Depression, Vincent Schaefer spent a lifetime educating himself. As part of his 'informal' education he spent twenty years working as a laboratory assistant, and later colleague, to Nobel Prize winner Dr. Irving Langmuir. While at "the Labs" (General Electric Laboratories) he was involved in the basic research that led to the development and refinement of color TV, sonar, cloud seeding, and many other technological advances.

As a child growing up I didn't appreciate much of this, but I knew that my grandfather, Vincent J. Schaefer, was very special.

He was a tall and imposing figure, but his warm smile and a laugh that cascaded gently through three octaves put everyone instantly at ease. He had an infectious curiosity about practically everything and seemed to always have three or four fascinating projects going on. If an idea occurred to him, he would head down to his workshop in the basement and set about creating a prototype. He was always finding patterns, and experimenting to see what explanation might be lurking behind them.

Some of my fondest memories are of time spent with him. As we walked in the woods he would point out interesting rock formations, deformed trees, or patterns in ice, and ask me to try to figure out what had caused them. Most often he knew the answer, but when he didn't he was quick to say, "Wouldn't it be great to be the one who figured it out!?" When I was around ten years old he taught me how to capture snowflakes, and preserve their structure as a plastic replica on a glass slide using a process he devised. Several times he recreated for us that magical moment where he discovered the use of dry ice to seed supercooled water vapor to make it snow in a freezer in his cellar.

Working on this book brought me closer to the memory of my grandfather, and helped me to know him in a much more personal way than I had before. I hope that others who read it will be similarly touched by the account of his approach to lifelong learning.

Michael Sullivan

Preface

If you are lucky in life you will meet someone along the way that will have a profound effect on the way you look at the world. I was such a lucky lad when I was nineteen and a freshman at the University at Albany in New York State. Lou Ismay, who was the head of the university's Environmental Forum, decided it was time for me to meet Vince Schaefer, the man who invented cloud seeding. Lou enlisted the aid of another friend, Al Hulstrunk, who then arranged for me to go to the Schenectady Airport where Vince had set up his famous freezer for demonstrating cloud seeding.

It didn't take long to like Vince. He was an outgoing, very friendly man who wanted to share his knowledge with you. That was the beginning of a friendship and mentoring for me for the next twenty-five years. While Vince was famous for cloud seeding, he was so much more than that. He loved archeology, natural history, hiking, and everything else. I shared those interests as well and that is why Vince was such a good mentor. He loved to share and I was eager to learn.

When Vince was starting the Dutch Barn Preservation Society years ago, we had long talks about what it should do but also about the Dutch contributions to the American way of life that at the time was poorly understood. Vince of course owned one of the most classic Dutch barns– the Teller-Schermerhorn barn on his property, along with a Dutch-styled home that he built by hand with his brothers in the Town of Rotterdam off Schermerhorn Road, a location which had a historic home or two already located there.

During my early career as city archeologist for Albany I would often consult with Vince, since as a youth and throughout adulthood he had not only located and excavated many local sites, he also created the Van Epps-Hartley Chapter of the New York State Archeological Association.

When I began championing the preservation of the Albany Pine Bush, Vince was right there to help me since he studied the area in the 1950s. When I needed to learn about the geology of the region, Vince was happy to take me around and show me features and landmarks including several caves, Vroman's Nose, drumlins, Thacher Park, and of course, the Pine Bush.

As you will find from reading his autobiography, Vince was a keen observer and writer. He often told me to look for the obvious in that nature's solutions to how it functions was often very simple and logical and that we humans tend to try to make it too complex. Vince's lab director Roger Cheng proved that observation by making major discoveries in the formation of acid rain, the freezing of a water drop, and evaporation of seawater simply by studying three drops of water. Roger, a native of China, came to America with his sights set on working specifically for Vince and spent the next twenty years running his lab. Vince encouraged his lab director to do his own experiments when he had down time, something I doubt happens very much in today's labs.

Throughout my career I have tried to emulate Vince's penchant for sharing information. As you will read in his autobiography, Vince gives much credit to his own mentor, Irving Langmuir, but it was Vince's inquisitiveness and perseverance that separated him from so many others. His relationships with others, including fellow scientists, was remarkable and devoid of the jealousies and competitiveness so prevalent today.

Toward the end of his life Vince had an offer from a publishing company to write his story. He agreed, and every morning for over a year he brought his handwritten text to his secretary at the ASRC, getting a typed manuscript back each afternoon. Once completed, he submitted the manuscript to the publisher. Weeks later Vince received a letter from the publisher explaining that the manuscript was a true "biography," but that they wanted a book with more of a broad educational twist. Vince felt too exhausted to take up the publisher's challenge and dropped the idea. However, he continued to correct the manuscript, and declared it "complete and finished" just a few months before his death.

The main text, written in his own words, often speaks to the present time in which he wrote. The pieces that comprise the "memories" section at the end of this published version were solicited from Vince's colleagues, friends, and family by his daughter Kathie after he passed away, and thus they also often speak in the present tense, or speak directly to Vince's family. These were edited for consistency and, except for "published" pieces, were given titles to distinguish them from each other.

As the editor, I tried very hard to retain the main text as Vince had written it, but some editing was required. For example, commas were sometimes inserted to elucidate some of his longer sentences, and a few spelling changes were made, mainly for consistency. One specific word that I discussed with the publisher was "archaeology." Vince, even from an early age, considered himself to be a modern, North American practitioner of this science, and therefore preferred "archeology" (without the second "a"). Out of respect for Vince, the modern spelling is used throughout the book. But despite these few changes and edits, this book remains as Vince wrote it.

It is my hope that his writings will inspire many more to open their eyes to the mysteries of the world and not be afraid to go on that path to explore and understand them. More importantly, while on that journey try to be a mentor to a young mind who shares that same curiosity. Vince would approve.

When Vince passed in 1993, his friend Everett Rau summed up what everyone knew about the man: "Vince was outstanding to many. He was like a straight tree in a forest of people. To a stranger walking the forest this tree was strong, thoughtful, sharing, a giver, a doer. What joy just to stand and listen and learn and be encouraged to think and find answers for the benefit of others."

Don Rittner, Historian
City and County of Schenectady

Acknowledgments

I would like to acknowledge Jim Schaefer and the entire Schaefer family for allowing me the privilege of publishing Vince's autobiography. Thanks to Mike Sullivan for taking the time to arrange and transcribe Vince's copious notes about his life and getting it into a digital format. Special thanks to Duncan Blanchard, a surviving member of Project Cirrus, for his biographical sketch of Vince, and also to Roger Chang for photos of Vince and his colleagues. Chrissie Reilly, CECOM LCMC Staff Historian from the Aberdeen Proving Ground, Maryland, supplied the photos of Project Cirrus and the H-1 Hurricane Seeding conducted in 1948. Chris Hunter, Archivist at the Schenectady Museum (now the Museum of Innovation and Science), was instrumental in supplying many of the Project Cirrus photos, as well as those of Vince and his colleagues at GE. Special thanks to Susan Holland for proofreading and to Richard Vang at Square Circle Press for bringing this project to the public.

I have included footnotes where I felt it appropriate to update or expand on something Vince mentioned.

Don Rittner, Editor

A Biographical Sketch of Vincent J. Schaefer

History was about to be made. On the afternoon of the 13th of November, 1946, a small, single-engine plane approached a large cloud east of Schenectady. The altitude was 14,000 feet, the air temperature -20°C. Inside the cramped quarters of the aircraft, Vincent Schaefer was giving last minute instructions to the pilot on the experiment they were about to carry out within the cloud. As they flew into the cloud, Schaefer dropped about three pounds of dry ice. Later he wrote in his notebook, "I was thrilled to see long streamers of snow falling from the base of the cloud." He shouted to the pilot to swing around, and "we passed through a mass of glistening snow crystals." They then went west of the cloud, and "we observed draperies of snow which seemed to hang for 2 to 3,000 feet below us, and saw the cloud drying up rapidly."

No wonder Schaefer was excited. What he had accomplished that day was the first successful demonstration that a natural supercooled cloud could sometimes be converted into a cloud of ice crystals.

Vincent Schaefer was not a meteorologist by training. For that matter, he had no formal training in any branch of science. But a first-class mind, unsurpassed powers of observation, and an insatiable desire to understand natural phenomena enabled him to chart a unique trail that led from one scientific adventure to another. Although he never knew in what area of natural science his next discovery might come, and he certainly had no inkling a few years before that he would discover the world's first practical technique of cloud seeding, those who knew him were not surprised at what he did on that day in November, 1946. His total communion with nature seemed almost to decree it.

At the age of sixteen, with financial hardships pressing in on the family, he dropped out of high school and went to work for the General Electric Company and completed the rigorous four-year course for apprentice toolmakers. He worked elsewhere briefly but eventually became an instrument maker at the General Electric Research Laboratory in Schenectady, NY.

At the research laboratory, the Nobel Laureate Irving Langmuir noticed that this young instrument maker had a keen mind that could pose interesting scientific questions, and an amazing talent to build simple devices with which the questions could be answered. In 1931, the year before Langmuir won the Nobel Prize in chemistry, he asked Schaefer to become his research assistant and work with him in the laboratory. Thus began a scientific adventure that during the next twenty years was to blaze a trail from one successful experiment to the next. It was a marvelous scientific symbiosis in which the older Langmuir would make theoretical predictions for the latest scientific puzzle, and the younger Schaefer would approach the answer with deceptively simple experiments. By letting theory and experiment both feed back into the other, they made rapid progress in research. During the 1930s they published papers on a variety of topics in surface chemistry.

Shortly before World War II they were asked by the government to design a filter for gas masks to trap toxic smokes. In working on this project they took advantage of serendipity, which in Langmuir's definition was "the art of profiting from unexpected occurrences." When they finished the toxic smoke project, they became involved in making more efficient smoke generators to screen military operations at sea. With Langmuir supplying the theory, and Schaefer the elegantly simple experiments (at one point an oil can, some oils, and a heater that can be bought in any hardware store), they quickly developed smoke generators vastly more efficient than those used by the armed forces.

From smoke generators, they turned to precipitation static on aircraft. Pilots flying through snow in the Aleutians reported that the impact of snow on the plane produced unwanted static and loss of radio contact. The government asked Langmuir and Schaefer to look into this problem.

They decided to work atop Mount Washington in the wintertime. There, in the clouds and strong winds, Schaefer exposed several metal surfaces to the clouds at temperatures of -10^{0} to -30^{0} Celsius; but instead of ice crystals striking them, to his surprise the surfaces rapidly became covered with ice. The clouds were composed of supercooled water droplets! Why weren't there snowflakes in the clouds? This was a question that so excited Schaefer that he began a series of experiments that culminated in his classic cold box experiment.

He devised an experiment that was beautiful in its simplicity. He took an ordinary home freezer, lined it with black velvet to provide a dark background, breathed into it to produce a supercooled cloud, and looked to see if there were any ice crystals in the beam of a microscope lamp. That's all there was to it, simple yet elegant.

Though the temperature in his cold box was far below 0^{0} Celsius, he seldom saw any ice crystals. He spent weeks trying an endless variety of materials that he sprinkled into a supercooled cloud. Nothing worked until serendipity played a role. On a hot, humid day in July, 1946, the air in the cold box was not as cold as usual. Schaefer, in a desperate attempt to cool the air, placed a block of dry ice in the cold box. When he breathed into the cold box to produce a cloud, the water droplets disappeared and in their place were millions of tiny ice crystals that danced and sparkled in the light beam as though alive. At last, Schaefer had found the trigger by which he could convert supercooled droplets into ice crystals.

In the spring of 1947 the government-sponsored Project Cirrus was organized. During the five years the project was in operation, numerous seeding missions were carried out, countless discoveries were made, many instruments were developed, and a basis for a practical seeding technology was established. Besides Langmuir and Schaefer, Bernard Vonnegut, Raymond Falconer, and Duncan Blanchard were also members of Project Cirrus.

In 1954, Schaefer left General Electric to become Director of Research of the Munitalp Foundation (spell it backwards). During his four years with Munitalp he initiated many research programs, including one on orographic and noctilucent clouds with the International Institute of Meteorology in Stockholm.

In the early 1960s Schaefer was the prime mover behind the formation of the Atmospheric Sciences Research Center at the University at Albany. In large part because of the great admiration they held for Vincent Schaefer, both for the man and the scientist, Vonnegut, Falconer, and Blanchard (three members of the group who had been privileged to work with Schaefer during the exciting years at Project Cirrus, and who had shared with him the exhilaration of being at "Langmuir University"), joined him once again as members of his fledgling research center.

A tribute to Vincent Schaefer must include more than just an account of his research activities. He was never an ivory-tower scientist. To Schaefer the communication to others of the excitement and joy of research and discovery was as much a part of the adventure of learning as breathing is of living. He was convinced that the fun of learning is often stifled by the oppressive atmosphere of the classroom, and that the ability of young people to discover the world around them is enhanced by taking them out into the field and letting them learn by doing. To put these ideas into action, he established in 1969 what was to become the Natural Sciences Institute. Over the next ten years some 500 high-school students from all over the United States took part in these adventures of learning.

Schaefer taught that money and sophisticated scientific equipment do not guarantee the discovery of new truths, and that simple, well-designed experiments are more often the gateway to new discoveries. He put this very eloquently in an article in the 12 December 1959 issue of Saturday Review:

> "Before we can challenge the young, we older people of America need to change many currently popular attitudes. We must accept the idea that money does not purchase new ideas. Expensive equipment and campus-like surroundings for research laboratories are no guarantee of effective spending of research dollars. The idea that the string-paperclip-sealing wax scientist is gone forever and that he has been replaced by the 'team' is a dangerously misleading philosophy. New ideas come from brains– generally in the singular. Our finest heritage is still freedom of opportunity for the individual."

Over the years many honors and awards were bestowed on Vincent Schaefer. In the long list one will find three honorary degrees. In 1948 he was awarded the Doctor of Science degree from the University of Notre Dame, in 1975 the degree of Doctor of Humane Letters from Siena College, and in 1983 the degree of Doctor of Humane Letters from York University in Toronto. In 1953 he received the Robert M. Losey Award from the Institute of Aeronautical Sciences, and in 1957 the American Meteorological Society presented him with the award for Outstanding Contributions to Applied Meteorology. In 1976 he was the first recipient of the Vincent Schaefer Award, presented by the Weather Modification Society, and was given a Special Citation by the American Meteorological Society.

Vincent Schaefer died on July 25, 1993. In the words of an old Hebrew proverb, we should: "Say not in grief he is no more, but live in thankfulness that he was."

Most of this information was taken from my article, "Vincent J. Schaefer: A Tribute," that appeared in the ASRC Report of 1975-76.

I want to end with a few words about my interaction with Vince Schaefer. He more than anyone else made it possible for me to become a scientist. In 1947, just a few months after I was hired by the General Electric Company in Schenectady, Vince welcomed me aboard Project Cirrus, where I spent a happy two years learning what it was like to become a scientist. In 1968, when I was at the Woods Hole Oceanographic Institution, he invited me to become a member of his Atmospheric Sciences Research Center. I accepted the invitation, and so did Ray Falconer and Bernard Vonnegut. The old Project Cirrus gang was together once again. I stayed at the ASRC until I retired in 1989.

Duncan Blanchard

Serendipity in Science

Vince cooking his meal over a campfire with a friend (date unknown). Courtesy of Jim Schaefer.

Vince, Eleanor Nichols and Ray Hagedorn find a turtle, 1932. Courtesy of Jim Schaefer.

I

The Formative Years

Chapter 1
Dreams and Action for the Future

Introduction: The Purpose for This Book

I have had a fascinating life! Its course and development could only happen in America, and I am not sure the happenings which I will tell you about could occur at the present time, though I think they could. The presence of the computer in hiring practices and decision-making has so modified the procedures, rules and regulations, that it might be a barrier to an individual's progress in areas that I traveled without hindrance.

The subtle regimentation that has been programmed into so many aspects of our freedom to act– in academia, the laboratories, big business, and in our everyday life as a citizen – with its codes, restrictions, and limitations, makes me wonder.

However, I have described the way the world has treated me during what was probably the most important twenty-year period of my life, starting when I was 28 years old, single and living at home. Under present standards it might appear that I started with "two strikes" against me. Having to leave high school at the age of 16 when only a sophomore, and then aspiring to move into a region where a doctorate nowadays is a minimum requirement, I managed to overcome obstacles to upward progress that might be insurmountable today.

Although during many of my early years I didn't recognize it, I apparently was fortunate to have discovered the serendipitous method of conducting my activities. In retrospect, I now realize that this method was brought to the attention of the world of science by an old friend of mine, Dr. Willis Whitney. He was extolling its advantages about 1917 or earlier. It has had a profound effect on my life since he discussed it frequently with me. This method, as I now understand it, is the need to develop a streak of innate wisdom, wherein the practitioner prepares the mind with a wealth of information on a wide variety of subjects, and is ready and able to "cash in" on this knowledge at appropriate times. Such action may involve meeting new people or getting a glimpse of some new area of activity of an intellectual or physical nature. These unexpected experiences and new vistas should be seized with enthusiasm, without thinking too deeply about the consequences that might develop by such action. This free-wheeling method of operation is not easily carried out! In fact, the freedom to act in this manner is not commonplace, but it is vital to the success of serendipity.

In a bestselling book, *In Search of Excellence*, Tom Peters has described highly successful organizations that emulate the old GE Research Laboratory developed by "Doc" Whitney shortly after the turn of the century. Throughout Whitney's directorship of the Laboratory, it was a place where intellectual freedom reigned. The same management philosophy was followed by his successor, Dr. William D. Coolidge, and continued for several years after the end of World War II as Dr. Guy Suits succeeded Coolidge. However, by the early fifties, the Laboratory had grown so large that economic pressures and new thoughts on business management began to change the "free" atmosphere. To my dismay, I began to feel that the Lab was no longer an exciting place to be. Thus, in 1952, after Langmuir had retired and local managers in the Laboratory began to organize teams of researchers having assigned projects, I became greatly intrigued with a new opportunity offered to me by the Director of the Munitalp Foundation. After discussing their offer with Dr. Whitney, Langmuir, Suits, Albert Hull and with Katy Blodgett, I decided to leave the Laboratory and accept the post of Director of Research of the Foundation. The next six years were an exciting and fulfilling adventure, but that's another story! Those years were a continuing series of serendipitous events, culminating in a highly satisfying career in academia.

If these essays give a few young people (even one) a glimpse of the way in which one can be prepared to meet the opportunities encountered in life, I will be glad.

Life is an exciting adventure and one that for each individual is only available for a limited period of time. If part of that time is wasted, that loss cannot be recovered. The way of life that I have experienced has been an exciting adventure. I hope one day you will be able to say the same!

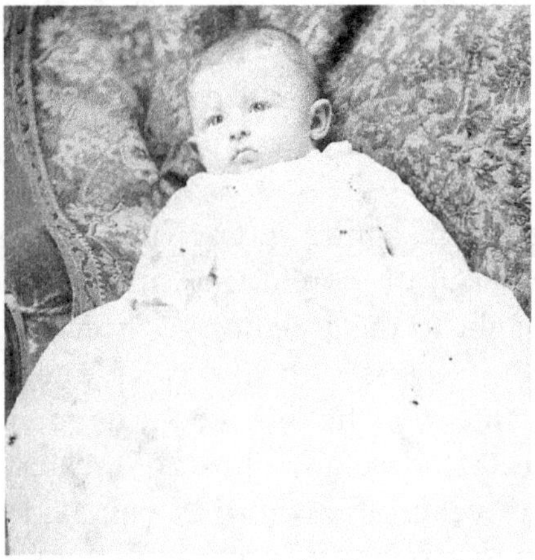

Vincent Schaefer as a baby. Born on July 4, 1906. Courtesy of Jim Schaefer.

Vincent as a three year old on his stoop in 1909. Courtesy of Jim Schaefer.

A Brief Autobiography

1906-1912

I was born on Elm Street in Schenectady on July 4, 1906. We moved to Stanley Street and then, when I was two or three years old, to Cherry Street in Bellevue. About five years later we moved to Arthur Street. On Cherry Street, we rented a flat in a two-family house from John Walsh, and on Arthur Street from John Grupe, the florist. These houses still exist and now (1979) and look better than when we lived there!

1912-1918

When I was six years old, I entered Fulton School on Eleanor Street and, subsequently, Bellevue School on Broadway– a two-story wooden building. When it burned, I continued in Euclid Avenue School, finishing sixth grade there. During this period, I had a marvelous time exploring the environs of Bellevue, which were a paradise for young people: swimming at Campbell's Pond; fishing there and at Bullhead Pond, the Erie Canal, Poentic Kill and Gritzback's Pond; exploring Tippecanoe; sleighing and skiing at Durkee's Hill, the Tramp Path, the Devil's Bend; skating at Fuller's Pond; berry hunting at the Race Track, on the river flats, and Lock 8. At this time (1917), I became a Lone Scout and became involved in the publication of a magazine on archeology.

Vincent around 8-10 years old, wearing his Golden Eagle Jacket. Courtesy of Jim Schaefer.

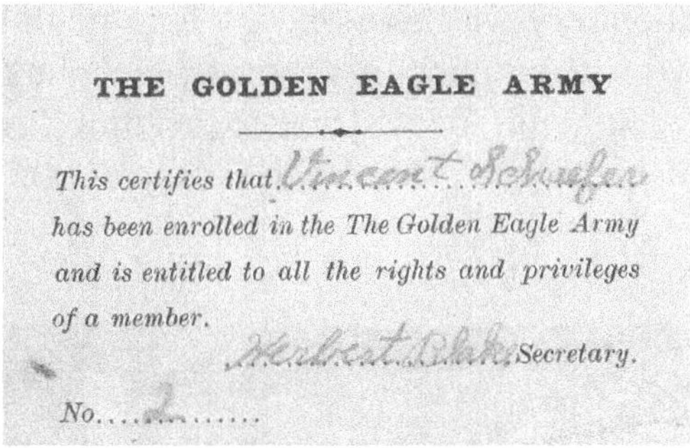

THE GOLDEN EAGLE ARMY

This certifies that Vincent Schaefer
has been enrolled in the The Golden Eagle Army and is entitled to all the rights and privileges of a member.

Herbert Blake *Secretary.*

No. 2

Vincent's Golden Eagle Army card.

Membership Certificate

No 36 This Certifies that

Lone Scout Vincent Schaefer
is a member of
COMRADES TRIBE L. S. A.

Signed Howard Finger
Captain

ACLS
UCLS

Vincent's Comrades Tribe card.

1918-1922

From Bellevue School, I went to Van Corlaer on Guilderland Avenue. There I played first violin in the school orchestra, and upon graduation was selected to present the Class Gift to the principal. Despite an effort to memorize my presentation speech, I completely forgot the words and probably gave the shortest presentation speech on record, when I handed the gift to Principal Edgar Palmer with the single word, "Here!" In 1920, after we had moved to 2637 Augustine Avenue, in what was then called "The French Settlement," I entered Schenectady High School on Nott Terrace in downtown Schenectady.

Vincent about 19 years old. Courtesy of Jim Schaefer.

1922-1927

The late spring of 1922, my parents told me that it would be best if I left school to go to work to eliminate the cost of schooling, and to bring in some money to supplement my Dad's income. Upon the advice of Uncle Joe (Holtslag), I entered the General Electric Apprentice Course to learn the machinists' trade. Preceding the start of the course, I obtained the job of drill press operator in Building 16 at General Electric. As my four-year apprenticeship drew to a close, my intensive effort to be assigned to the Machine Shop of the General Electric Research Laboratory was successful, so that my last month of apprenticeship was completed there. I then became a journeyman and spent a year in the Laboratory Machine Shop.

1927-1932

Meanwhile, I decided to leave General Electric and join the Davey Institute of Tree Surgery in order to satisfy a desire to work outdoors and to do some traveling. After some years with Davey, I left and did some private tree work. During this period, while working for Ed Cushing of the Cushing Stone Company, I formed the Mohawk Valley Hiking Club. I nearly went into the plant nursery business in 1929 with Charlie Mix of Schoharie, but followed the advice of Robert Palmer, Superintendent of General Electric Research Laboratory, and returned to the General Electric Laboratory Machine Shop, where he offered me a job as a model maker. There I worked for a number of scientists, including Irving Langmuir and Katherine Blodgett. In 1932, Langmuir invited me to become his laboratory assistant, and in 1933, I was successful in achieving my transfer to Room 408 in Building 5 at the main plant in Schenectady.

1932-1938

When I moved into Dr. Langmuir's laboratory, I was issued my first Laboratory Notebook. This (and subsequent ones) proved a good record of the variety of projects I worked on during my career in the General Electric Research Laboratory. It was an exciting and rewarding experience. In 1932, I formed the Van Epps-Hartley Chapter of the New York State Archeological Association and, shortly afterward, the Schenectady Wintersports Club. At General Electric I was privileged to have a close rela-

Vince (right) with pals in the Tory Hole Cave at Thacher Park in the Helderbergs. Courtesy of Jim Schaefer.

tionship with Dr. Langmuir, as well as Dr. Willis R. Whitney, Dr. Katherine Blodgett, Dr. Coolidge, Dr. Hull, and many other members of the Laboratory. Initially, I worked directly with Dr. Langmuir, taking directions from him. Gradually he permitted me, and eventually encouraged me, to carry out my own research studies. During this period, I also trained a number of persons in surface chemistry techniques. These individuals included Dr. Harry Sobotka of the Mount Sinai Hospital (cholesterol monolayers), Dr. William Davis of the Eli Lilly Research Lab (insulin film), Dr. David Waugh of Massachusetts Institute of Technology (protein films), and Dr. Eliot Porter of Harvard Medical (surface electric potential of monolayers). Toward the end of the period, my salary and activities reflected a change from Laboratory Assistant to Research Associate in the General Electric Research Laboratory. I was given complete freedom to develop my own programs, though I continued to work closely with Dr. Langmuir on most of his physical research projects until, and even after, he retired.

1938-1943

As I was given the privilege of conducting my own research, I found a great many things in surface chemistry that were intriguing. Just prior to World War II, Dr. Langmuir and I became very much involved in the study of protein monolayers. I also worked with Dr. David Harker on electron microscope techniques; with Dr. Herbert Uhlig on corrosion of metals and crystalline meteorites; with Dr. Charles Bachman on TV tube brightness; with Dr. Lewis Koller on color TV; with Dr. Paul Zemani on micrometeorites; and with Ted Rich on submicroscopic particles, etc. With World War II approaching, Dr. Langmuir, being on some national advisory committees, arranged for us to become involved successively in gas mask filtration of smokes, submarine detection with binaural sound, the properties of fine fibers, smoke, the production of replicas, the formation of artificial fog, and more.

1943-1948

In 1943, our research efforts shifted from fogs to precipitation static. This took us to the summit of Mount Washington in the White Mountains of New Hampshire, where we studied snow static, then supercooled clouds, aircraft wing icing, ice nuclei, and other aspects of cloud physics. The Mount Washington studies led to Whiteface Mountain for helicopter blade icing tests, etc. In the summer of 1946, Langmuir became much excited over my discovery of an effective cloud seeding technique using dry ice in supercooled clouds. After a field demonstration I made over Greylock Mountain in Massachusetts on November 13, 1946, he sought and obtained federal support for further research in cloud modification. This led to the development of Project Cirrus. I was designated General Electric's coordinator for research recommendations, with Vic Fraenckel handling logistics, contracts, etc., for the four or five airplanes assigned to the project.

1948-1953

In 1947, Harry Gisborne, Chief of Fire Research at Region #1 of the United States Forest Service located in Missoula, Montana, had visited me. He wondered whether cloud seeding could affect the pattern of lightning in the thunderstorms of the Northern Rockies. Out of this initial contact came a series of events that in later years led to the initiation and development of Project Skyfire. It also got me involved in a long and rewarding association with United States Forest Service. Our Project Cirrus activities reached a climax in 1950 with a series of field experiments in New Mexico, in cooperation with the New Mexico Institute of Mining and Technology. Shortly afterward, we lost our key pilots to the Korean War activities. Rather than go through the frustration of training a new crew, we recommended that Project Cirrus be terminated. After preparing some comprehensive reports on the laboratory, field and flight activities of the project, we ended operations and took stock of our options.

At about this time, I was approached by Vernon Crudge of New York City, on behalf of the Trustees of the Munitalp Foundation, to determine if I had ideas that might be followed to encourage the development of basic research in meteorology. I put together a set of plans that embodied many of the unfinished projects that we had encountered in Project Cirrus that involved a number of the outstanding young scientists we had met during our program of research activities. To my surprise and pleasure, my suggestions were endorsed. Arrangements were made with Research Laboratory officials so I could spend part time in developing a program for Munitalp. Within the year, I was asked to spend full time on these programs. This meant leaving the Research Laboratory and the General Electric Company. After considerable soul searching, I decided to leave General Electric, although such a move would cause me to lose most of the pension benefits accrued from more than 25 years of continuous services (with five earlier years not counted.)

1953-1958

Upon leaving General Electric, I found it necessary to develop a completely new work plan. Acting as Director of Research, it became necessary for me to go to New York City at least once a month, and to undertake considerable traveling to various parts of the United States and to other countries. This worked out fine from my standpoint. The major fieldwork involved studies in northwestern Montana and northern Arizona, where I was accompanied by my wife Lois, and our children, Sue, Kathie, and Jim. In addition to these projects, we developed programs with Jack Barrows, the successor of Harry Gisborne in the Forest Service; with Frank Ludlum in England and Scandinavia (noctilucent clouds); Carl Gustaf-Rossby in Stockholm, along with his graduate student, Erik Erikkson (atmospheric chemistry); Herbert Riehl of the University of Chicago (long range forecasting); Phil Church of the University of Washington, with his graduate student, Donald Fuquay (connective clouds); Joanne Malkus at Illinois Institute of Technology and later Woods Hole; Walter O. Roberts of the High Altitude Observatory (solar-terrestrial relationships); Ted Fujita of the University of Chicago (mesometeorology); and quite a few others. I assembled a very small staff– Alice Klopfer as secretary, Ray Falconer, Paul MacCready and Don Fuquay as professional staff.

In addition to what was to become Project Skyfire, we developed a mobile laboratory, and within a year had underway a major program of time-lapse cloud movies, taken in many parts of the Untied States and other parts of the world (such places as Germany, Hong Kong, Japan, Hawaii, Baffin Land, Antarctica, etc.). I termed this a dynamic cloud atlas.

In 1958, after the death of Albert Johnston, founder of the Munitalp Foundation, I was asked by his daughter to consider shifting the center of operations of the Foundation to Kenya. I was dubious about the wisdom of such a move to Africa, but agreed to make a visit to that country. During this period, I visited South Africa, Tanganyika (now Tanzania), and Kenya to explore the possibilities of establishing a headquarters there.

1958-1962

After returning from Africa and having some meetings with the new principals of the Munitalp Foundation, I decided to remain in America and to leave the Foundation. I was retained for several years as an advisor.

With the occasion to grasp new opportunities, I became involved with the American Meteorological Society (AMS) in educational activities, including the preparation of motion pictures, and the development of a student program to encourage more young people to become interested in the atmospheric sciences. This effort developed into the highly successful Natural Sciences Institute.

I also became quite active in consulting activities with the Boeing Company, General Electric, Philip Morris, the United States Forest Service, the Bureau of Reclamation, EG&G, Arthur D. Little, Inc., and a number of universities, including Wyoming, Colorado State, Fresno State, Nevada, Rensselaer Polytechnic Institute, and South Dakota.

The success of the AMS Summer Institute for Secondary School Students, at the Loomis School in Connecticut, brought me into contact with Albany State Teacher's College, through Dean Oscar Lanford and President Evan Collins. This led to an offer of a position on the Albany faculty. This developed into a very satisfying relationship that extended for more than 16 years and which still (1979) continue. Within a year or so at the College, this relationship led to the formation of the Atmospheric Sciences Research Center (ASRC), the Whiteface Mountain Field Station and Observatory, the Yellowstone Field Research Expeditions, and the Natural Sciences Institute.

1962-1967

With the initiation of the ASRC and its survival after its first three years of existence, it became apparent that a real opportunity existed within the State University of New York (SUNY) to form an important research organization. This was a period of flux with the ASRC organization. Dr. Eugene McLaren replaced Dr. David Barry, who succeeded Dean Oscar Lanford as Director. During this period, in fact from its inception, I had served as

Director of Research and Chairman of the Scientific Advisory Committee. In 1965, I was offered the position of Director of ASRC by President Collins. Although I was reluctant to become involved in this type of responsibility, President Collins assured me that adequate support would be available to handle the routine administrative work required of this new position. Such support was in fact provided, although I found the freedom to conduct my own research was considerably curtailed. Fortunately, I have had my own private laboratory in my home for many years and thus carry out an active research program during evenings, weekends, in the winter at Yellowstone Park, and in the summers at our field station near Flagstaff, Arizona.

From my first contact with the University in 1959, I was convinced that the area of atmospheric research that needed attention and provided an excellent area for building up a strong professional group was in the field of atmospheric particles, i.e., fine (invisible) particles, as well as cloud condensation nuclei. The variable concentrations that occur in the global atmosphere were not then established, except in a rather vague way, and the complete spectrum and nature of anthropogenic particles in cities was not well established.

Thus, when I had this opportunity in the mid-sixties to build up a small but highly qualified research group, I was successful in assembling a remarkable staff. Through my work at General Electric and the Munitalp Foundation, I knew virtually everyone in this field, so I carefully prepared a list of outstanding scientists and proceeded to recruit them. I was quite successful in this effort and believe that we assembled one of the best groups in the world.

1967-1972

During the next five years, we sought federal support for obtaining equipment and advancing our research capability, knowledge and scientific reputation. We were quite successful in developing adequate support and assembling whatever equipment and facilities were needed to conduct the type of research of mutual interest to the ASRC staff. During this period, I phased out the Natural Sciences Institute and the Yellowstone Field Research Expeditions, since I felt we had achieved the objectives of these two programs. At this time I realized that if we were to build our Research Center into a stable and worthwhile organization with the SUNY system, it was highly desirable to seek my successor. Accordingly, I polled all of our professional staff to determine whether we had someone within the organization who would be acceptable. This led to the eventual appointment in 1976 of Volker Mohnen as Director. For five years prior to this official designation, I had gradually shifted administrative responsibilities to Volker, so that there was no apparent change in policy or program during this period.

1972-1977

As Volker assumed more and more responsibility for the continuing development of ASRC programs, I was able to shift a considerable amount of my effort toward the energy field with the main emphasis on solar energy. I was convinced that this was a logical sequence to our continuing studies of aerosols and gases. I was able to help some of our staff to become actively involved in the alternative energy field. In 1974, we designated a portion of our Five Rivers Field Station at the Schenectady County Airport as our Energy Sciences Laboratory. As part of this activity, we constructed a minimum energy building using funds supplied by the New York State Legislature for this specific purpose. With Volker's assumption of full responsibility for leading ASRC and my designation as Professor Emeritus and retired Director, I withdrew from all policy-making decisions, including participation in staff meetings.

1977-1982

Upon retiring as Director of ASRC, I concentrated on completing my three reports on global aerosols and trace gases, which had occupied my interest for many years. When these were completed in 1978, I moved toward a return to consulting work. My first priority, however, is to be available for any advice that may be sought relative to the University and ASRC.

In addition, I have spent the past several summers developing a large vegetable garden and an experimental solar greenhouse. My consulting work involves developing new approaches to solving the pollution emission problems associated with coal gasification and diesel engine operation, and in heat storage, insulation improvement, and the controlled movements of hot and cold air.

My newest research interest involves the nature and structure of rocks of the earth. By cutting quite large but very thin (8" x 12" x 0.060"; 20cm x 30cm x 1.5mm) slices of rock so that they are translucent, I can show the great beauty and intricate structure of many of the sedimentary and metamorphic and igneous rocks of a region. This is leading to a new "look" at geology, as well as producing objects of great beauty. Exhibits are being developed for educational purposes in museums. I am also producing artistic assemblies for aesthetic enjoyment.

One of my most ambitious projects was the design and fabrication of a six-foot diameter church window, which has been installed at the St. James Catholic Church in North Creek in the east-central Adirondacks. This was done as a memorial to my parents who, for many summers, attended this church that is located about eight miles from their summer camp. The window is quite beautiful and was made of Adirondack rocks, many of which were found in the surrounding mountains.

Early in January of 1979, I led a group of old time (and younger) hikers to Moccasin Falls in the Yantaputchaberg[1] west of Schenectady to celebrate the 50th anniversary of the founding of the Mohawk Valley Hiking Club.

In 1980, I was granted the Laureate Award by the University Foundation. That same year I was elected to become an Honorary Trustee of the Institute on Man and Science–now called The Rensselaerville Institute.

---

1 Yantaputchaberg is a combination of Dutch and Native words which means "John Ear of Corn Mountain" and is located in Rotterdam Junction, Schenectady County, New York

On March 19, 1981, I received one of the first copies of the *Field Guide to the Atmosphere* printed by Houghton Mifflin Company of Boston. This is Guide No. 26 of the Peterson Series and was written by me with the collaboration of John Day. I had signed a contract to write this book more than 25 years ago and was pleased that the company waited! It is now in its fourth printing and has been issued in a Spanish edition.

For many years I had a dream that in some manner Vroman's Nose in the Schoharie Valley would become public land. I tried various procedures and nearly was successful in having the Nature Conservancy acquire it. I had routed the Long Path of New York over its summit and during World War II, we at the Research Laboratory used it as a vantage point to observe and photograph our artificial fog smoke screens. In 1983, a combination of the descendants of Adam Vroman, a private philanthropist, and funds from the local service clubs and individuals, formed the Vroman's Nose Protective Corporation, in cooperation with the Schoharie County Historical Society.

In the spring of 1983, I was approached by the General Electric Research & Development Center to become a consultant on a special project. This agreement has continued.

During the spring of 1985, Lois and I purchased the Doug Morehouse property along the Edwards Hill Road west of Bakers Mills in the Adirondacks. This raises our total holdings up there to 150 acres.

In the spring of 1985, I was made Honorary Trustee of the E. N. Huyck Preserve and in June received the Citizen of the University Award of the SUNYA Alumni Association.

On July 27, 1985 Lois and I, with our youngsters and their children, gathered at our home, "Woestyne South," to celebrate our 50th wedding anniversary. Our brothers and sisters were also present, and a large group of friends and neighbors joined us for a gala celebration. Altogether we had four parties, two of them being up in the Adirondacks. What a good time we had.

On February 6, 1986, more than ninety of the staff and friends of ASRC gathered in the Patroon Room at the university to celebrate the 25th anniversary of the founding of the Center. It was a fascinating and joyful affair.

Over the past five years I have prepared museum exhibits of my large, thin rock sections for the Boston Museum of Science and Schenectady Museum. These have proven to be quite popular, to the extent that I was encouraged to prepare a permanent exhibit for the Schenectady Museum. About the first of March 1986 it was completed and is proving to be quite a unique exhibit. Meanwhile, I have just completed building a major exhibition piece, made up of hundreds of rock specimens that has as its primary theme the Divine Proportion.

At the end of May 1986 the 10th meeting on Weather Modification is being held in Washington, DC. This will be the 40th anniversary of my cloud seeding discoveries, and consists of a four-day conference at which dozens of interesting papers will be delivered. It begins to appear as though these activities are finally coming of "age!"

My Preparation for Joining Langmuir University

My life has profited from serendipitous events. One of my first memories is of Halley's Comet when, in 1910, our landlord closed the wooden shutters in the house we lived in, since he was taking no chances that the windows might be broken when the comet "came." Then, six or seven years later, while returning home from picking strawberries, I found a flint arrowhead. A year or so later I organized the Mohawk Tribe of Lone Scouts and, shortly afterward with three other of my Scout friends, helped in the publication of a tribe paper that we called *Archeological Research*. This paper was published over a period of four years, during which time we published 21 issues.

"AR," as we called it, opened channels to a number of illustrious persons, including the mayor of the city, the state archeologist, local historians, and Dr. Willis R. Whitney, Director of General Electric Research Laboratory.

"Doc" Willis R. Whitney examining a turtle in his home library on May 20, 1950. Courtesy of Schenectady Museum Archives.

I had hoped to become a forester; a profession I later discovered had been that of early ancestors of my mother's family, the Holstags, of Zutphen, Holland. Then, halfway through high school (Schenectady High) I had to leave school to help support our large family, due to the sickness of my parents.

At that time my two uncles– brothers of my mother– urged that I become an apprentice machinist at the local General Electric Company. This I did, beginning my apprenticeship in the fall of 1922.

During my second year of apprenticeship, Dr. Arthur C. Parker, State Archeologist, invited me to spend a month with him on an archeological expedition to central New York. I was granted a leave of absence by Superintendent Marquis and spent an idyllic month at Keuka Lake, helping to make an archeological survey of Yates County. Returning to my apprenticeship, I found my interest focused on scientific studies, due to my intimate contact with Dr. Parker, a highly educated man who was also a naturalist.

Toward the end of my fourth year as an apprentice, I requested a transfer to the Machine Shop of the General Electric Research Laboratory headed by my friend "Doc" Whitney. While I didn't seek his help, Superintendent Marquis finally granted my request, and I finished my final month of apprenticeship in the Research Laboratory.

After a year in the Laboratory Machine Shop I discovered that, being at the bottom of the payroll, I was the recipient of many dirty jobs, such as machining blocks of graphite, and shaping test bars of very hard steel. I became discouraged with the future as I saw it, and left the Laboratory to join the Davey Institute of Tree Surgery. This was an opportunity to travel, to work in the out-of-doors, to learn about trees and shrubs, and to continue to earn money while attending the Davey Institute during the winter of 1927.

I learned a great deal about trees at the school, and in the spring, started work in the big trees of Michigan.

By the end of the summer I had become disillusioned about a future in tree surgery but continued working while requesting a transfer to the East. This came through, and in the fall I returned home and began to work in the Capital District area, the first big job being in Troy. I worked throughout the fall, and then in the winter got a job in the orchid greenhouse of Danker's Florist Company in Albany. After spending the spring with a landscape gardening crew at Danker's, I decided to go "on my own" and secured a tree-pruning job with the Town of Niskayuna. During the summer I did some orchard work, and then in the fall secured employment as a landscape gardener at the Cushing Estate on the Troy Road. In the spring of 1929, after raising plants in a greenhouse on the Estate, I then planted the flower gardens and did some tree pruning. Early in the fall, I was approached by a horticulturist from Schoharie, who wanted me to go into partnership in developing a tree and shrub nursery near Schenectady. I sought advice on this proposed venture from Mr. Robert Palmer of the GE Research Lab, who was Superintendent of Services. He advised against the investment and offered me a much better job in the Machine Shop than the one I had left two years before. I hadn't expected the offer and didn't go to see him with any thought of job-seeking in mind. However, since the time was just before the great

1929 Stock Market Crash and the start of the Great Depression, Mr. Palmer's advice saved me a great deal of money and frustration and opened the channel to Langmuir University! It was one of the most important serendipitous events of my life.

A selection of photos from Vince's time at the Davey Institute of Tree Surgery in Kent, Ohio. Top left: Vince, Ray Proxmire, Bob Lewis, and Byron Bridges on a twig hunt near Willow Creek, four miles SE of Kent, February 5, 1928. Top center: The Davey "Skinners," Vince, Bill Mathias and By Bridges back home. Top right: Vince at Davey Arboretum on December 29, 1927. Below: Vince with Davey friends and an unidentified ranger. Courtesy of Jim Schaefer.

A month after rejoining the GE Research Laboratory "The Crash" occurred, and during the next three years, I was one of those who held his breath as Ginger Adams, head of the Machine Shop, made the rounds every Friday afternoon with our paychecks– that occasionally included a dismissal slip!

During all of this time I had accumulated a fairly large library of books on many subjects, including history, exploration, nature study, mathematics, physics, chemistry and geology. I read many of the books acquired and learned a great deal about the natural world and the world of science. As soon as I had enough money to begin a savings account, I made it a practice to buy any book that appeared to be interesting.

During this period I also became quite involved in civic activities, in organizing programs, and in self-education. I organized the Mohawk Valley Hiking club in 1929, the Van Epps-Hartley Chapter of the New York State Archeological Association in 1931, and the

Schenectady Wintersports Club in 1932, and helped in the formation of the Schenectady Museum. In addition I developed adult education programs on such things as birds, geology, trees, plants and mosses. In the process of doing these things, I did a great deal of writing and public speaking.

While carrying out such activities, I became acquainted with most of the leaders in the Schenectady area, and in this manner met John Apperson and Irving Langmuir.

While in the Machine Shop I requested permission from the boss of the Machine Shop to attend the Friday afternoon colloquia held by the Laboratory each week. I also became involved in cave exploration and met a young friend of "Doc" Whitney's, who invited me to participate in an intensive course in Cytology (the study of cells) that he had arranged, so that he could become trained in the process of preparing the sections of onion root tips using a microtome, after collecting, processing and embedding the samples. In the process I learned a great deal about these techniques, the mystery of cell division, the use of a compound microscope, a *camera obscura*, and eventually the preparation of photomicrographs.

I had become very active in studying the history and prehistory of the Mohawk and Hudson Valleys and the Adirondacks and in participating in photography, skiing, hiking, mountain climbing and exploring as well as in conservation, what is now called ecology, and in geomorphology. All of this knowledge eventually proved to be useful to me. Thus by the time I was invited by Dr. Langmuir to become his assistant I had acquired, mostly by self-education, a rather broad knowledge of the natural world, an appreciation of books, an ability to write, to organize and to impose self-discipline so that when I was given a Laboratory Notebook I knew how to use it, to organize a plan of action and to listen to people who know more about a subject than I did!

I also retained the freedom to operate any vacant machine in the Laboratory Machine Shop. In this manner I was able to quickly fabricate equipment which either Katy Blodgett, the "Boss" or I might wish to have. Best of all I had a host of friends in the service shops– machine, glass blowing, welding, photography, carpenter and the stock rooms, and knew who the best artisans were so that urgent needs could be met almost immediately. I knew my way around the General Engineering, the Works Lab, the News Bureau, the Appliance Department, the Varnish and Insulation shops and, of great value– the long, multi-tiered racks of old equipment that could sometimes be modified to meet an urgent need to explore a new "hunch."

I found that, by having a vague idea about something of interest, I could explore the racks of old equipment, wheels, gears, rods, sheets, pipes and countless other things, and save hours and even days by adapting scrounged things to try out the glimmer of an idea to find out if it had some merit, or deserved to be discarded without further ceremony.

My Parents

I was blessed with two wonderful parents. My dad was of German ancestry, his folks having come to America about 1850 by way of New Orleans. On their way north on the Mississippi River, my great-grandfather died of yellow fever and was buried along the river. In some manner his widow managed to get to Albany where relatives lived. One of her sons became a tailor and had five children, among them my father, Peter Aloysius. As a youngster he helped his father in making suits– having the job of sewing buttonholes. This was a chore that my father detested. He was the top student at the Christian Brothers Academy, and then went to Canisius College in Buffalo to study for the priesthood in the Catholic Church. From there, after graduation, he enrolled in the University of Innsbruck in Austria. While there, he did some mountain climbing. However, his health failed and he was forced to give up his plans to become a priest and returned to Albany.

My mother's family was of Dutch ancestry, having emigrated from Zutphen, Holland, where they owned considerable forest land. They also came to Albany about 1850, probably through New York City. One of their sons, Henry, became a gardener, and as a young man worked for the Sanders family in Scotia, New York. He married Wilhelmina Merkley and they had (11-13) children, among them my mother Rose Agnes. She was born at 13 Sanders Street in Albany, the family home, and attended a parochial school on Second Avenue.

My parents were married on September 30, 1903, and came to Schenectady, where they rented a flat in a two-story house on Elm Street, where I was born. My father became a clerk at the General Electric Company where he worked until retiring in 1930. During this period, and until 1924, he rented our homes. At that time, he decided to buy a home in the suburb of Schenectady called "The French Settlement." Prior to that move, I had entered my apprenticeship at General Electric. I lived in our new home on Augustine Avenue for eleven years.

In 1922, when it was apparent that it would be necessary for me to help with the finances, I am sure that both my parents regretted this necessity. In fact, a few years later when I was honored by the Schenectady Jaycees to receive their Man of the Year Award, the newspaper story highlighted the fact that I had not finished high school. My mother reacted by writing a letter to the editor that said:

> In the recent article about Vincent J. Schaefer, General Electric Co. scientist, "Snow Maker Quits School at age 14," I would like to make a correction. He did not leave school at age 14. He had passed his 16th birthday and was ready to enter third year high school. He did not wish to leave school, but he saw the need of his support for his home and family.
>
> He generously put aside his desire to continue school, entering the General Electric Co. as an apprentice metallurgist, finishing a four year course in three and one half years. When starting to earn a living, he did not put aside his school books, but applied himself far into the night with

more intensive studying than the high school curriculum and soon his superiors saw he was accomplishing more than had he stayed in school. In his late twenties he had an opportunity to go to college to get his degree, but felt he would be losing time for research.

For the good of the youth in our schools today, Vincent should not be eulogized for leaving school, but for applying himself more intensively when he felt the sacrifice of leaving school was necessary. Beside his intensive study of higher education, he took up many hobbies. Archeology began at the age of 13– nature study, birds, flowers, caves, rocks, etc., also sports.

So perhaps this true statement of how Vincent spent his time after he left school at the age of 16 years will disabuse the minds of any youths that school and book learning are not really necessary.

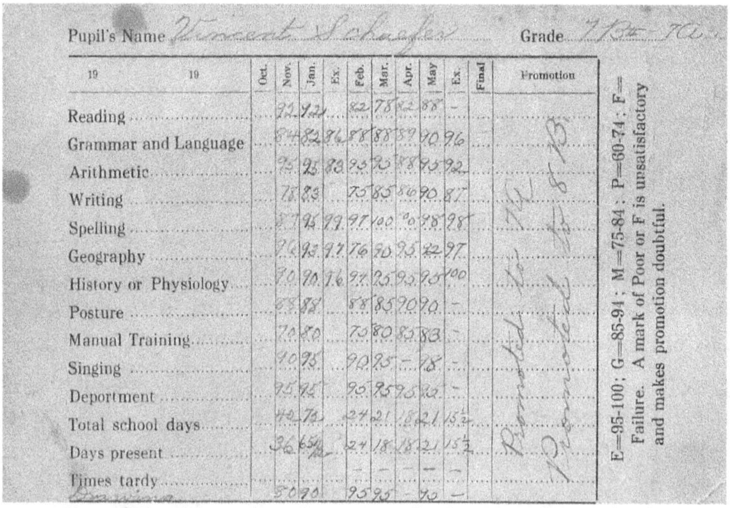

Pupil's Name Vincent Schaefer Grade 7B – 7A

19 19	Oct.	Nov.	Jan.	Ex.	Feb.	Mar.	Apr.	May	Ex.	Final	Promotion
Reading		92	92		82	78	82	88	-		Promoted to 7A
Grammar and Language		84	82	86	88	88	89	90	96		Promoted to 8B
Arithmetic		95	95	83	93	95	88	95	92		
Writing		78	83		75	85	86	90	87		
Spelling		87	95	99	97	100	90	98	98		
Geography		76	93	97	76	90	95	92	97		
History or Physiology		70	90	76	97	95	95	95	100		
Posture		88	88		88	85	90	90	-		
Manual Training		70	80		75	80	85	83	-		
Singing		90	95		90	95	-	78	-		
Deportment		95	95		95	95	95	92	-		
Total school days		40	75		24	21	18	21	15½		
Days present		36	65½		24	18	18	21	15½		
Times tardy					-	-	-	-	-		
Drawing		80	90		95	95	-	90	-		

E—95-100; G—85-94; M—75-84; P—60-74; F—Failure. A mark of Poor or F is unsatisfactory and makes promotion doubtful.

One of Vince's seventh grade report cards. Courtesy of Jim Schaefer.

My Mother, Rose A. Schaefer

As a youngster I was fortunate that we lived in the suburb of Bellevue. This was within walking distance of the General Electric works (under two miles) and was surrounded by fascinating streams, swamps, ponds, fields, woods, hills and farms. As youngsters, we quickly discovered the places to sled, ski, swim, fish, explore, play ball, and have a good time in general. My parents permitted me to do all of these things, provided I first took care of my chores and homework. Both of my parents loved Nature and encouraged my two sisters, two brothers and me to become involved with such interests.

The one and only time I can remember when I was considered to be "out of line" was after a fishing expedition to the Norman's Kill. We were absent a good part of the day, and as we headed home at dusk, our route happened to go past the Cameo Theater on Broadway. It was a very warm evening and the theater doors were wide open so that the screen was visible from the street. Without a thought, we stood on the steps and watched the show. When it was finished we headed down Arthur Street to home. There we were greeted by frantic parents who couldn't understand our tardiness. While I didn't receive any physical chastisement, my father broke my favorite fishing pole! While it wasn't a "boughten" pole, it was a "lucky" one made from a sapling. This loss was one that hurt, and I have not forgotten this action of my father that, I must say, was well deserved.

Vince's Arthur Street Home. Courtesy of Jim Schaefer.

Vince's childhood home at 2637 Augustine Ave. Courtesy of Jim Schaefer.

Although my father was not able to become a priest, he was a religious man and insisted that all of us receive a good religious education. This has been a sustaining force in my life and that of my brothers, Paul and Carl, and sisters, Gertrude and Margaret.

My mother had a beautiful soprano voice, and from an early date sang in the choir and played the church organ. She was an accomplished pianist and had a fine piano that she played with vigor and finesse. She was responsible for organizing and training the church choir while we lived on Cherry and Arthur Streets. Upon our moving to Augustine Avenue, she played the same role at St. Madeleine Sophie Church, located on the road to Guilderland, that she could reach by bus. This she continued to do year-round, on a daily basis, until a year or so before her death at the age of 87.

Since both of my parents were in fragile health much of their life, a considerable amount of my Dad's income was spent on doctor's bills, medicine and special hospital care. During her early life, my mother suffered from hay fever in the summertime that escalated into asthma and eventually into tuberculosis. After she recovered from the tuberculosis at the Ray Brook Sanitorium near Saranac Lake, my father, upon the advice of my mother's brother, Frank, decided to take the family into the Adirondack Mountains during the hay fever season, first to Tripp Lake near Chestertown, and then to a rented room in Georgie Morehouse's house at the Edwards Hill settlement above Bakers Mills. Since this area was free from ragweed that produced the pollen that gave my mother hay fever, she was quite free of this annual health problem. The following year Dad learned of a nearby house that was for sale, and in some manner raised the small amount of money needed to purchase it from Charlie Reese. Thus began the saga of Camp Cragorehol that became, and still is, one of our extended families' retreats from the hectic life of the city. Altogether, with my brothers and sisters and their families, we now own fifteen land parcels, including some ten camps where we all spend varying amounts of time as the spirit moves us.

Both of my parents encouraged me to become familiar with Nature. My mother frequently took us to the New York State Museum in Albany. She shared my enthusiasm for writing, reading, acquiring books and accumulating a considerable collection of archeological specimens. She greatly appreciated the pussy willows, arbutus, violets and other wildflowers that I presented to her, and the many wild berries I picked in the surrounding countryside. She encouraged me to take up the violin, something her brother Frank excelled in, but I never got beyond grammar school orchestra.

Despite their continuing health problems, both of my parents lived to a considerable age. My dad was 92 when he died, and two weeks prior to that he visited his Camp Cragorehol. My mother was 87 when she died. Until the last, her indomitable courage and spirit were evident. On her deathbed she announced that she would not die until May 1st, which was the Blessed Virgin's Day. That is what happened!

It was in their memory, that in 1982, I designed and fabricated the six-foot diameter window for St. James Church in North Creek where, for many years, they worshiped in the summertime. This window is made of large, thin slices of translucent Adirondack rocks, featuring the native garnet, anorthosite, sandstone and crystalline limestones that occur in the region. It is quite beautiful.

Vince and Lois on their 50th wedding anniversary, July 27, 1985, at "Woestyne South." Courtesy of Kathie Miller.

In retrospect, I guess by present economic standards, our early family life would be classified in the poor category. While my father was never out of work, his salary was not very great, and consequently, with continued living expenses and the almost constant series of health problems, my parents were never able to save any money, but lived from week to week on his paycheck. At one time my Dad actually was able to buy a parcel of land near the end of Campbell Avenue, probably on the installment plan. When he finally had a clear title to it and was thinking of building a house, he was forced by economic pressure to sell the property. Neither parent was willing to go into debt, a discipline that I inherited and that has stood me in good stead over the years.

We never had fancy food, but mother was an excellent cook and found ways to make our simple fare delicious. From the beginning of her married life she baked whole wheat bread. This often sustained our family when other food was in short supply. She taught me to shop, and I frequently went downtown to the Mohican Market, where the low prices not only paid for my carfare but also provided us with excellent butter, eggs and meat. I also learned to bake bread, a procedure which I can still accomplish when necessary.

All in all, the way of life I learned from my parents has remained with me. Fortunately, Lois had the same type of wonderful parents and the sensible disciplines we brought with us to our married life have been of great value to us in living a fairly tranquil life!

The example that all five of us children observed in the frustrating rental problems that confronted our parents every month were so vivid that none of us wanted to go through that regime if at all possible. Thus it was that Paul, Carl and I all built our own homes upon our marriages. My sisters, Gertrude and Margaret, married husbands, then started marriages in their own homes.

My lack of a college education was a pattern followed by my other sisters and brothers. Paul left school to learn the carpenter's trade, and subsequently became one of the region's best designers and builders. Carl followed my step and enrolled in the GE Apprentice School ending his career as a floor boss in the Large Turbine Shops of General Electric. Gertrude became a secretary at General Electric before she was married, while Margaret became a Registered Nurse, graduating from St. Peter's Hospital in Albany, before becoming an Army Nurse and ending her nursing career on the hospital staff at the infamous Nuremberg Trials in Germany. Gertrude married an engineer and had two sons and four daughters. Margaret married a medical doctor and had two sons and two daughters. Paul married a farm girl who was a naturalist and fine musician. They have three daughters and one son, while Carl married a graduate of Oswego Normal School. They have four sons and a daughter. Lois and I have two daughters and a son.

Many of our youngsters have graduated from college– a number of them with advanced degrees. They are without exception "fun" people, scattered across America, but a joy to be with whenever we see them.

My Wife's Parents

Lois' folks were on a par with my own. Highly principled, industrious, fun-loving people, their Perret/Crane ancestors were of Swiss-French-English origin. Lois' mother was a Crane, of an illustrious manufacturing family that was well known throughout New Jersey, New York and New England. She died when Lois was eight years old, after which Lois lived with two schoolteacher aunts for six years.

Her father married again to Estelle Tiernan, a wonderful, compassionate person who provided a warm, loving home to his son and two daughters. The Perret family lived in their early days in the Newark/East Orange area. Charles Edward Perret was an insurance man working for Prudential Insurance Company. He rose in the ranks to become an Inspector, with his final operational area in upstate New York. During this period they lived in Troy, where Lois attended and graduated Troy High School. She then went on to become a Registered Nurse at St. Luke's Hospital in New York City. After graduation she became an assistant to Dr. Gulick in Schenectady, who was for many years a highly regarded eye, ear, nose and throat specialist of the area.

Her older sister, Eleanor, graduated from Columbia University and became a high school teacher specializing in history. As part of her profession she became a world traveler and spent many summers in Europe and Asia, her favorite area being China. Her younger brother, William, graduated from Massachusetts Institute of Technology and became a physicist specializing in seismology. During his professional career he worked for the Mississippi Valley Corps of Engineers on flood control projects, and then transferred to the Atomic Energy Commission, where he was deeply involved in the underground nuclear explosion operations. He retired from the Sandia Laboratories at Albuquerque, New Mexico, in the sixties.

Christmas Day, 1937. William R. Perret, Charles Edward Perret and Vince on Story Avenue in Niskayuna. Courtesy of Jim Schaefer.

As youngsters, the two sisters accompanied their schoolteacher aunts for the summer recess to occupy a rented house on Bailey Island in Casco Bay, north of Portland, Maine. In a nostalgic expedition, I accompanied Lois to Maine in 1984 to see if the lapse of seventy years had produced major changes on the old ocean retreat. Not only did we find Tip Top House as she had remembered it, but it was in better condition than when they had lived there more than seventy years before.

The Perrets were watchmakers from Chaux de Fonds, continuing their trade in New York City when they emigrated about 1850. As mentioned earlier, the Crane family were industrialists who have a reputation for constructing, organizing and operating shops and mills.

Frank Perret, a cousin of Lois' father, became a world famous Vulcanologist and worked for the Carnegie Institute. His book on Vulcanology, replete with many photographs that he took, is a classic in the field. He was involved in a study of most of the great volcanic eruptions between 1904 and 1943 and became famous for his analysis of the eruption of Mount Pele in Martinique that killed 30,000 inhabitants of that Caribbean island. A museum bearing his name was built at the edge of the sea where the disaster occurred. My daughter Sue and I visited this interesting museum in the winter of 1973.

Charles Perret's second wife was of Irish ancestry. She moved with him to Troy, and then to Schenectady where I first met her. Before marriage she was an executive secretary, very knowledgeable about finances and business operations. She was a warm, religious person, full of fun, along with her mother who lived in the Perret home. "Gram" was also full of fun and a delight to know.

Lois' father, Charles Edward, was an excellent gardener and cabinetmaker. I learned many things from him. He was also an avid fisherman, and thus a man who understood my hobbies and aspirations. In 1934, when he first visited the prospective site of our home, he immediately laid plans to provide us with tree saplings, perennial plants and berry bushes. The 1-inch black walnut tree sapling he gave me grew and is now some 30 inches in diameter. The peonies and iris he gave us are still flowering, while the blackberries have migrated from the side of the garage to the slopes and finally are flourishing on the Great Flats. As Lois's father advanced in years he finally gave up his woodworking tools and machines. For a modest sum I purchased all of his equipment that I have now installed in a complete workshop in our North Addition that we built in 1953. Similarly, he gave me his gardening equipment, much of which I have used for the past twenty-five years!

He died in 1954 after an extremely active and mostly enjoyable lifetime. Upon his death, Lois' mother came to live with us and remained with us until her death in 1968. The Perrets were a delightful family whose friendship and love all of us thoroughly enjoyed. Our memories of them are still with us– vivid and soul satisfying.

My Hobbies

Throughout my attendance at Langmuir University I had many hobbies. They ranged through local history and archeology, to most of the natural sciences. Both my mother and father encouraged many of these, though the former was the person who most actively encouraged these activities. The most important organization in my early initiation to hobbies was the Lone Scouts of America. This national organization consisted of boys, ranging from eleven to fourteen years of age, who received the weekly Lone Scout Magazine, written mostly by boys. Their contributions ranged the field of the natural sciences, with hiking, camping, and exploring as a major aspect of their essays. We were also encouraged to correspond with others of the same age and interests. I developed friendships with about a dozen like-minded boys. I am sure this activity was a great help in introducing me to the use of words and the development of a degree of writing ability.

The most important hobby of these early days was Indian arrowhead collecting. This introduced me to the great fun of exploring, and searching for camps and villages of early prehistoric people who had inhabited the forests, flats, and terraces of the Mohawk and Hudson Valleys. As part of this activity I learned how to interpret the local topographic maps issued by the US Geological Survey.

This sense of exploring introduced me to local history, geology, Dutch barns, trees, flowers, ferns, birds and many other fascinating elements that existed on all sides in our region. As my knowledge increased, I became involved in imparting such information to others. With the additional time that became available on the part of most everyone, due to work hour cutbacks caused by the Great Depression of 1925-35, we found our offerings to be highly desired by a host of people and our public meetings and lectures were attended by hundreds of our townsfolk. The friendships that developed during this period were enduring ones and are still in existence.

The marvelous thing about hobbies is the complete absence of exterior pressures that are exerted on the participant. The many excellent books written on most of the subjects of the natural sciences provide an introduction into the fascinating world of Nature. At an early date I acquired the Putnam series of nature guides on the trees, flowers, insects, birds, ferns and other subjects, and used them a great deal. The fascinating thing about developing such interests is the widening vistas that appear, the friendships that develop, and the intellectual challenge continually presented. In later years I acquired a complete set of the Peterson Field Guides published by Houghton Mifflin of Boston, having more than 25 titles, and actually prepared one of them (#26– *The Field Guide to the Atmosphere*).

By having a wide variety of such interests, I had a wonderful time while out with the hiking club and the archeological association, both of which I had organized. The knowledge that I was continually accumulating complemented much of my professional development. When I was invited to help establish the Science Forum Radio Program on WGY, it was my assignment to handle as best I could the questions concerning the Natural Sciences that were posed by our listeners. In addition, I subsequently became involved with other subjects, such as meteorology, air quality, atmospheric optics, surface chemistry and such pseudo-science subjects as dowsing, orgone energy and pathological science.

Our region is rich in colonial history, the Dutch fur traders having been active in the area from the early 1600's. In 1662, Dutch farmers from the newly established village of Schenectady quickly settled the Great Flats, the islands, and other flood plains along the river. Their huge, beautifully built barns became a subject of great interest to me and I photographed and measured many of them before they disappeared by fire or roof neglect.[2]

In addition to my interests in local archeology, I had from an early period a fascination with rock forms and types, caves, fossils, waterfalls and everything pertaining to geomorphology. This was an interest encouraged by Mary Clark, an outstanding teacher in the Schenectady High School. The interest she engendered in me by her class in physiography has never lagged, and is still an area of great fascination to me. On my travels to various parts of the world over the years, I have found this to be a subject of never ending variety and attractiveness.

The ultimate in this subject has developed during the past decade when I obtained suitable equipment to cut large, but very thin, translucent slices of my vast collection of rocks, which disclose the crystalline patterns that exist in producing their intricate structure. Over the years I have located more than 250 locations where interesting rocks can be collected. From these I have cut thousands of slices that provide a nearly endless variety of

---

2 *Dutch Barns of New York: An Introduction,* by V. J. Schaefer, Purple Mountain Press, 1994.

color, structure and pattern, that can be used in fashioning beautiful art objects that are not only of great scientific value, but are also unique in their appearance.

By having a wide variety of hobbies one is never at a loss as to what to do with spare time. While some individuals permit their hobbies to take precedence over everything, I have found that a degree of "time discipline" is desirable so that a proper balance is achieved in the use of time. When properly achieved such a regime comes the closest to Doc Whitney's edict that in everyday life one should have fun!

Twenty Years at Langmuir University

One day in the early 1930's, as I was working at my bench in the Machine Shop of the General Electric Research Laboratory at the Schenectady Works, Dr. Irving Langmuir approached me. He asked me if I'd like to work for him in his laboratory! He had recently returned from Sweden, where he had received the Nobel Prize in Chemistry.[3] He was the first scientist of an industrial laboratory to be so honored.

Irving Langmuir conditions a slide for detection of tiny and invisible substances present in liquids, 1941. Courtesy of Schenectady Museum.

3 Langmuir was awarded the Nobel Prize in Chemistry in 1932 for his work on surface chemistry.

I was quite surprised and elated when Langmuir asked me to "come upstairs." I had been building equipment for him and his associate, Dr. Katherine Blodgett, for a year or so not knowing, of course, how the equipment was being used. With his invitation he told me that he would show me how to use the equipment to uncover scientific relationships in the field of surface chemistry. Thus began a happy and highly enjoyable relationship that lasted for more than twenty-five years, until Dr. Langmuir's death.

For a number of years I wondered "how come" I was picked to become Langmuir's research assistant when he could have had the "pick" of the young PhD's of the world. Academically, I had a very low score, since I had to leave high school at the end of my sophomore year at Schenectady High. Thirty years later, on February 11, 1963, at a retirement party held for Katherine Blodgett, Mary Christie, a mutual friend, was there. She had been Dr. Willis Whitney's secretary for many years. In the early days of the laboratory when I was there it was sometimes said that the laboratory would fall apart if it weren't for Mary Christie! As private secretary to "Doc" Whitney, she was the type of individual who seemed to know everything that was going on in the administration and operation of the Laboratory. A few days after the party I received the following letter from Mary:

> Dear Vincent,
>
> It was a pleasure to see you the other evening at the splendid ovation for our Katy.
>
> It took me back many years to a morning when Dr. Whitney, having completed his usual visits about our old lab (Bldg. 5), came downstairs and paused at the door of the office, looking very thoughtful. He said, "Langmuir wants someone to work with him– someone with good hands, who can make the kind of devices he's likely to need. But it has to be someone capable of taking personal interest in his ideas, helping to work them out."
>
> I may have shown the astonishment I felt, but all I said was "Why Dr. Whitney, Vincent Schaefer." He nodded, and said, "I'll see what Langmuir thinks" and returned to the fourth floor. The rest is history.
>
> You should know, if you don't, why you instantly occurred to me. It was your interest in artifacts (archeology, if you will). You were the only man among the excellent workers in the Machine Room who had any interest outside his work and his home life.
>
> Thank you for so amply justifying my faith in you and for being one of the bright spots in an ancient crone's life.
>
> Sincerely, Mary Christie

In 1922, after finishing my sophomore year at Schenectady High School, my mother informed me that it would be necessary that I leave school and get a job to add to the family's income. We had all planned that I would go to college, but continuing sickness, doctor, and hospital bills posed an emergency situation. Her two brothers, Joseph and

Frank, when told about the situation urged that I seek a trade at the General Electric Company in Schenectady. In 1922 I enrolled in the General Electric Apprentice Course, a four-year training program involving school courses and intensive instructions in the proper operation of all types of machines and tools. Graduation from this course resulted in the rating of journeyman machinist, toolmaker or instrument maker. This turned out to be an interesting and challenging experience. During my fourth year on the course, I made several attempts to get an assignment at the Machine and Instrument shop of the General Electric Research Laboratory. Although I had a personal friendship with Dr. Willis R. Whitney,[4] organizer and Director of the Laboratory, through a mutual interest in finding Indian arrowheads, I did not solicit help from him. However, my efforts with the Superintendent of Apprentices paid off, and a month before the end of my apprenticeship, I was transferred to the Laboratory.

After spending a year or so in the Laboratory after graduation from the Apprentice Course, I discovered that being low man on the payroll entailed getting an unending sequence of uninteresting, dirty, repetitive jobs that made me feel that I was wasting my time. Accordingly, I began to explore the possibility of abandoning my trade for a very different line of work. I learned that the Davey Tree Expert Company of Kent, Ohio, ran a training institute for tree surgeons, and that they not only had an interesting training school, but paid the select recruits while they were attending classes and learning all aspects of tree care. The job also involved traveling to Ohio and other places– something quite attractive to me at that time in my life. I signed up and journeyed to Kent, Ohio.

At the end of a winter of schooling, I began working in the big trees of large estates in Michigan. After a summer of hard outdoor labor in a number of Michigan towns, I became homesick and managed to get transferred to a Davey field crew near my home in eastern New York. After working for Davey for several more months, I was offered a job as a landscape gardener, including the operation of a greenhouse. The next Spring I was approached by a professional nurseryman who asked me to become a partner in developing a large tree and shrub nursery near Schenectady.

While this seemed to be an attractive opportunity, I decided to seek advice from the Superintendent of Technical Services at the General Electric Research Laboratory. After listening to my enthusiastic story, he cautioned me against starting a business, telling me that it looked like there were serious economic troubles ahead. This was in the summer of 1929! As an alternative, he offered me an attractive job at the Research Laboratory, much better than the one I left two years previously. Thus I returned to the Laboratory a few months before the Great Depression hit. During the next few months I survived the many layoffs that decimated the ranks of the laboratory's technical staff.

Although the depression kept worsening in the early thirties and austerity swept through the entire Laboratory, shortened days preserved a goodly portion of the jobs and I received the momentous visit from Dr. Langmuir.

---

4 Whitney is considered "The Father of Basic Research in Industry."

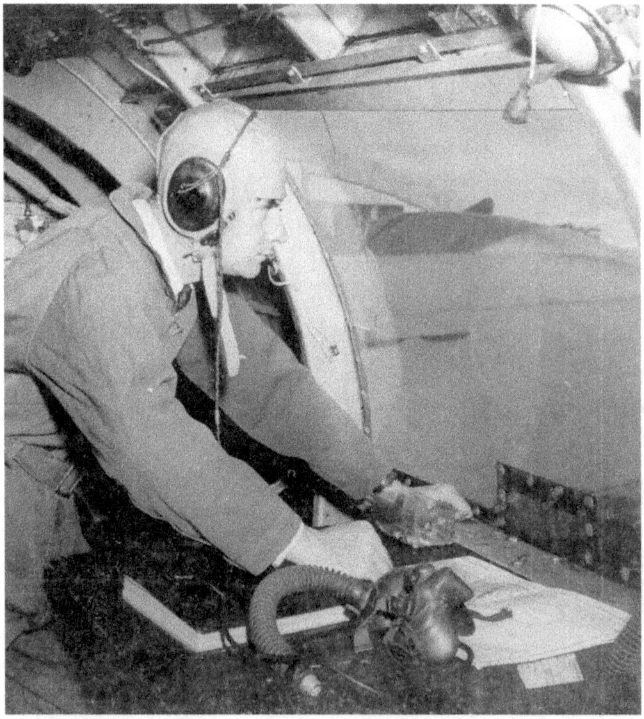

Vince Schaefer on board a B-17, ready for a Project Cirrus hurricane seeding mission. Little did he know to what heights his association would Irving Langmuir would take him. Courtesy of the CECOM LCMC Staff Historian's Office, Aberdeen Proving Ground, Maryland.

II

The Langmuir Years

Chapter 2
Langmuir the Boy, the Man, and the Scientist

Irving Langmuir, during his lifetime of 76 years was a remarkable person. As a boy he had traveled in Europe, had climbed a mountain alone when he was twelve years old, and had a home laboratory at the same age where he conducted experiments. He attended a variety of good and poor schools, did a lot of mountain climbing, returned to America to finish his pre-college schooling, and then entered Columbia University from which he graduated in three years as a metallurgical engineer. He then went to Germany, and at Gottinger in 1906 received his doctorate in physics under Professor Nernst.[5]

Returning to America, Langmuir obtained a teaching position at Stevens Institute in New Jersey– a job that was neither a challenge nor an attractive activity. After three years he decided to leave, and by good fortune in 1909 he obtained a summer job at the newly formed Research Laboratory at General Electric in Schenectady, New York. Dr. Whitney, who was in the process of staffing the Laboratory, recognized in Langmuir an outstanding young scientist and talked him into remaining at the Laboratory when the summer term had ended.

As a young man he had many interests, most of them related to the out-of-doors. In 1910 he formed one of the first troops in the Schenectady area of the Boy Scouts of America. About that time he met John Apperson, an electrical engineer at General Electric who was an avid outdoorsman. They developed a lifelong friendship. They climbed Mount Washington in the White Mountains of New Hampshire, Mount Marcy in the Adirondacks and other mountains in the wintertime. Langmuir may have introduced skiing to the Schenectady area.

Between the two of them they developed a group of dedicated skate sailors, who used the wide stretches of Lake George for winter skating.

Langmuir became a good photographer, having a darkroom and making many pictures, some of which he entered in competitions at the local photographic society.

On April 27, 1912 he married Marion Mersereau, who was a perfect companion for the next 45 years. She was a marvelous person. They both loved skiing, skating, climbing, sailing, and exploring, their Lake George camps, and trips by sailboat off the Maine coast. Together they took motor trips across the country, visited relatives and explored the Indian Country of the Southwest. Then, during the next forty years, he carried out a wide-ranging program of scientific exploration that ranged from basic studies of high-vacuum and

5 Walter Hermann Nernst (1864-1941) developed the first institute devoted purely to physical- and electro-chemistry.

gas-filled lamps, plasmas, atomic hydrogen, binaural sound, atomic structure, surface chemistry, artificial fog, aircraft icing, cloud physics and weather modification.

Throughout his lifetime he was a very kind man, with a love of children. He had a continuing high level of enthusiasm, an abiding sense of wonder, and a driving urge to solve scientific problems. At General Electric he enjoyed complete freedom to follow his inclinations with whatever support his research needed. This was never great since he devised elegant and simple solutions to the problems he tackled, and was never happier than when confronted with the unknown.

When he received a Nobel Prize in 1932, it was for his imaginative approaches to Surface Chemistry, even though a host of his other achievements might also have been cited. When invited to Russia in 1945, it was as the top scientist of the world.

During his lifetime he published some 230 scientific papers, received 15 honorary degrees, and was granted 22 prestigious medals. Unlike many other members of the world's scientific community, his scientific discoveries and achievements did not decline as he grew older but continued at a high level until his death.

It was my good fortune to be offered the opportunity to become his chief assistant a few years after I returned to the Laboratory for the second time. I became involved with him and Katy Blodgett, one of his colleagues, in 1931 and was invited by him to become his assistant shortly after he returned from receiving his Nobel Prize.

I did my best to justify his faith in my ability. From our first contact I could see that I was going to have a fascinating time with him. This close relationship continued until his death.

Mountain Climbing, Photography, and a Steam Whistle

During my early association with the Boss, I learned many interesting things about his boyhood. There was a period when his father's work took the family to Switzerland. Langmuir's father gave the boy a couple of strict conditions when he asked permission to climb the nearby mountains.

1) Always have a lunch along.
2) Get an early start.
3) Never climb longer than half the time between departure and darkness.
4) Wear warm clothing in a number of layers.

With these simple rules, the Boss told me he climbed all of the nearby mountains and developed a love of mountain climbing that never left him. While he never told me the names of the mountains he climbed, I suspect there were some fairly large ones. Whether the iron crampons which he gave me date to that period or later times I do not know, but I guess they were used during a later period, since they fit an adult boot.

In part of the collection of his effects that were given to me by Marion are three fine photographic enlargements of Swiss mountains that he may have climbed. I'm almost positive that he took the pictures, since the frames are handmade and the rear of the frames bear numbers indicating they were entered into a formal exhibit in Schenectady about 1920.

Langmuir on Mount Washington in 1944. Courtesy of Schenectady Museum.

Another marvelous story about his boyhood that he told me went as follows (as I remember it). He went with his family to a Canadian lake for the summer. In exploring the area he came upon an old sawmill that was powered by a large stationary steam engine. He looked it over and it appeared to be in perfect condition– the valves, though rusty, were still workable, the boiler appeared to be without cracks, the flywheel and governor intact, including even the steam whistle. He returned to the hotel, found some oil, some old rags, and a water bucket and worked hard cleaning up the engine and filling the boiler to the proper depth. Then he gathered a batch of firewood and started a fire in the firebox. After a while the pressure gauge showed a buildup of steam pressure. I don't remember whether he told me the flywheel turned, or if the reciprocating part of the engine's power train were activated, but he did tell me that the whistle worked fine!

I can just imagine seeing the old-timer desk clerk of the hotel dozing off, then suddenly hearing a whistle that had not blown for twenty years. He didn't immediately become alert but with the intermittent sound (Langmuir told me he was having a marvelous time blowing it and letting the steam pressure build up again) he suddenly hobbled out of the hotel as fast as he could to see what was going on. The local community was flabbergasted to find out that all the commotion had been caused by a young boy who liked to do things on his own, and especially liked to blow a steam whistle!

My Involvement with Langmuir's Friend, Apperson

One of Langmuir's closest friends was John Apperson, who we affectionately called "Appie." Appie met Langmuir when he first came to Schenectady, found him to be a kindred spirit and probably introduced him to Lake George, the place that was to become Langmuir's "second home."

From their first contact the two men found many of their long-held interests were identical. Thus within the year they were climbing mountains together, especially in the wintertime using skis. Appie had a substantial camp at a beautiful location on Huddle Bay just south of Bolton Landing. Langmuir acquired land adjacent to Appie on the south, and some years later bought a substantial part of Crown Island.

I first became involved with Appie early in 1929 before I was offered a new job at the Research Laboratory. Shortly after I organized the Mohawk Valley Hiking Club on January 6, 1929, I "discovered" Appie. He was a unique person and his character embodied all of the beliefs and attitudes about conservation that my brother Paul and I, and quite a few of our other members had. Within a few months we were involved in helping him with his projects– taking and editing films of lumbering devastation on Boreas Mountain, the erosion of the Lake George islands, the improper use of the Lake George Dam at Ticonderoga, the girdling of huge deciduous trees on private lands in the heart of the Adirondacks, and the terrible erosion still occurring on the glacial polished rocks of the steep mountain slopes in the Adirondacks following forest fires.

A number of us would spend our weekends at Huddle Bay with Appie, helping him with one or more of his projects. He had a routine that was quite predictable. On the way north in his car, he invariably would stop at a favorite butcher shop and stock up with lamb chops of first grade. He was a fine cook and we always had excellent meals between jobs. We all would get up about daybreak and head for a swim off his dock. We would then eat and find out what the project was for the day. Sometimes it would involve a rip-rapping job on a State island, it might be a quick trip to some mountain for photographs, or it might be an inside job of film editing, something that I frequently was asked to do. His fast Chris-Craft motorboat, which he had christened "Article VII Sec 7," referring then to the "Forever Wild" section of the New York State Constitution. It made our lake trips adventuresome and effective.

The last thing Appie did when we were about to leave on Sunday evening was to coat his huge trestle table of polished maple planks with a liberal covering of boiled linseed oil.

This treatment, carried out virtually every weekend of the year, had imparted an incredible polish to the table that seemed to be as hard as steel.

Appie was a bachelor (he often referred to Lake George as his bride!) and as fine an individual as I have ever known. His bachelor's quarters consisted of a substantial two-story house near the Ellis Hospital (just a few blocks from Langmuir's). One of its largest rooms was devoted to sleeping bag and skate sail fabrication, with a large rugged sewing machine in its center.

With the cooperation of Langmuir he developed the down filled sleeping bag built in such a way that no seams joined the inner and outer layers of the bag. Rather a simple and ingenious procedure for making muslin tubes joined the two layers together. After many field tests with thermometers placed in strategic locations, he and Langmuir had developed the near perfect bag. He then taught us how to make them. Before the year of 1931 had ended, our group had made more than twenty-five such bags using Egyptian cotton balloon cloth, three pounds of gray goose down for each bag, and a reversed vacuum cleaner to fill the muslin tubes. After 55 years my bag is still intact and serviceable!

He had also devised climbing ropes for skis, to provide reliable traction on bare icy slopes of the Adirondack peaks. These were made of 3/8-inch hemp rope, knotted in such a manner that they presented a criss-cross of rough hemp fibers that slid onto the ski, with a loop over the ski tip and firm fastening at the tail of the ski.

I had occasion to compare Appie's climbing ropes with sealskins when a Swiss friend and I climbed Mount Marcy in the dead of winter in 1933. As we started up the summit cone of Marcy, after climbing out of Panther Gorge to Four Corners Lean-to, my friends' sealskins came off just as we reached a critical slope on Marcy. Fortunately my ropes held and I was able to help get the sealskins back on, although they were nearly useless on the icy surface.

The Boss and Appie were a formidable team. Langmuir, especially after he received the prestigious Nobel Prize, had a name and reputation that was noticed by the newspapers. Appie would frequently work up news stories and turn them over to Langmuir for public presentation. Many successful projects developed with such cooperative effort.

Dr. Katherine Blodgett had a camp adjoining the north edge of Appie's Huddle Bay property, and we saw her frequently.

The conservation ideas espoused by Apperson were often very unpopular with land developers, some of them having power at high levels in industry and government. In a number of instances, such persons would try to have Appie fired from his engineering job at General Electric. Whenever this was tried Langmuir would defend his position and Appie would continue to operate as usual.

We all learned a great deal about politics, the environment, economics, publicity, and many other side issues in our association with Appie and Langmuir. These were exciting times, and we were quite successful in protecting the Adirondacks from the erosion of its protective clauses in the Constitution, and in the process had a fascinating time.

Sweetser: A Shadowy Figure

One of the interesting coincidences of my adventure at General Electric Research Laboratory was to discover that Langmuir's retiring assistant, who I was to replace when I went "upstairs," was a neighbor who lived on Arthur Street when we rented a flat in a two-story house on that street. To my knowledge he remains a sort of mystery man. When we lived on Arthur Street (that we left in 1924 to move to a house my father bought located in the French Settlement about a mile to the southwest) the Sweetsers, especially Mr. Sweetser, was an enigma. He was very small in stature, and thin, with a sandy mustache and a very mild, self-effacing manner. His wife was quite the opposite– tall, large boned, quite assertive and addicted to wearing large flowery hats and having a flock of noisy roosters that were all pets. The Sweetsers were English.

At the time mother was not well, and the noise of the roosters, especially very early in the morning, was a continuing annoyance. It got so bad that mother, who rarely complained about anything, reported the nuisance to the police. Action was taken and the roosters disappeared. That may have been the reason why, not long afterward, the Sweetsers reported my father to be a suspect as a German spy during WWI. My father had no love for the English (the nationality of the Sweetsers) and let it be known in no uncertain terms. Having studied at the University of Innsbruck in Austria as a young man, he entertained sympathies with that part of Europe prior to the declaration of war by the United States. After that he no longer voiced his sentiments and became a patriotic booster of the United States, though he never appeared to change his mind about England.

Fortunately my father was apparently highly respected at General Electric, where he worked as a clerk in the accounting department. His boss at the time– a Colonel Francis – defended him and nothing further transpired. It was thus a surprise to me to learn that Langmuir's previous assistant had been Mr. Sweetser. He apparently was involved with Dr. Langmuir in his high vacuum and gas filled lamp research. Sweetser instructed the glass blowers, who fabricated a veritable forest of glass tubing that occupied most of a room adjacent to the Boss's office. Sweetser was a shadowy figure in the laboratory. He apparently carried out the Boss's instructions to meticulous degree, but showed no assertiveness, had few if any friends, and after leaving, left no evidence of having been there except for the forest of glass! A year or so later the Boss asked me to dismantle the glass works.

Since I had a strong feeling for history, I felt that this glass assembly had a rightful place in the historical development of many of Langmuir's discoveries in the high-vacuum and gas-filled lamp field of research, so I arranged to turn over the glassware and its wooden framing to the Schenectady Museum.

Arthur Jones, its director, felt the same way that I did about the historic value of the material I was dismantling, and provided transportation of it to the museum. Whether it has survived the move from the old County Home where the Museum was then located to the Nott Terrace Heights, where it is now ensconced, I do not know, but will try to find out.[6]

When I arrived at Room 408, Sweetser had been gone for a year or so and I was located in a laboratory on the other side of the Boss's office. Thus I had little in common with the man I replaced and had quite a different personality. While I made sure that I carried out all the instructions I received from Langmuir, it soon became apparent that he hoped that I would go beyond the scope of his suggestions and I did so rather quickly and with a fair degree of success.

Langmuir's Fun with Airplanes

I have earlier described my flight with the Boss in 1931, when he took me up on a flight to search for ski slopes during the early days of the Schenectady Wintersports Club that I organized. I didn't mention in this earlier tale that shortly after we reached our cruising altitude of 5,000 feet, Langmuir had me take over the controls and soon had me in complete control of the plane. He had me use the stick and rudder controls to maintain a steady course, to bank, turn, climb and dive, and then to stall and recover. It all seemed so natural that I thoroughly enjoyed the experience. After that, he took over and for several hours we explored the area in the northwestern Catskills, which would have been available to a Schenectady Snow Train along the Delaware & Hudson freight line to Pennsylvania. He had bought a Waco F monoplane on September 27, 1930, shortly after completing his flight instructions, but before making his solo flight the following week.

At the end of October of 1930 Langmuir met Charles Lindbergh for the second time (the first time was in January of 1929, at that time they formed a great liking for each other). Lindbergh urged the Boss to become involved in some of the problems in navigation and told him there was a great need for better information on the optimum types of light systems, color of lights, sensitivity of the eye in picking up light signals such as flashing lights, etc., etc. Accordingly, it wasn't long before Langmuir and one of his laboratory colleagues had embarked on a series of tests and observations which resulted in a pioneering paper on light signals and navigation aids, published with W. F. "Westy" Westendorp in 1931. This was an elaborate paper on the fundamentals of light signals that pointed out that red was the light to be used for the greatest visibility and that flashing lights in fairly rapid succession provided the signals most easily detected.

During Lindbergh's second visit, the Boss told him he had purchased an airplane and was doing quite a bit of flying. It is said that when Lindbergh asked him which he now enjoyed most– flying or skiing– the Boss answered skiing! He was never one to worry about diplomacy! By 1934 Langmuir had sold his Waco F for a newer model, but before the end of that decade had sold it and abandoned private flying on his own. The reason for that decision was his complete disgust with the ever-encompassing and restrictive regulations that were reducing his fun of flying. When the regulations were received stipulating not only

6 The Schenectady Museum does have a number of items belonging to Vince and his colleagues.

the new log books were to be of a specific size, but also that the cover was to be of a certain color, he wrote a short article on this subject and sold his plane.

His love of flying continued unabated, however, and whenever he could arrange it, he would find ways to occupy the co-pilot's seat or plan his research activities to include some close contact with flight activities.

Nothing, however, replaced the fun he had cloud-dodging, making wheel tracks in the tops of clouds, or letting others feel the thrill of flying! After 54 years I can still feel the response of that little Waco to my stick and rudder controls!

Langmuir's Preoccupation and Other Characteristics

Some persons, not knowing Langmuir to any degree, thought him to have a cold and non-compassionate nature. This feeling was probably engendered by the fact that when passing in the hall or street he failed to recognize them even if he looked at them.

This, I am sure, was due to a degree of shyness that he possessed, along with a tendency toward high concentration of his mind when related to some current problem.

His concentration was so great that he often appeared to be oblivious to his immediate surroundings. At one time, for example, after he had returned from an extended trip on the previous day, the next morning at breakfast he left a generous tip to his waitress– Marion Langmuir!

Far from being cold and distant, he was the most delightful of traveling companions, or big boss of a small group (Katy and me). The twinkle in his eyes, the hearty laugh, and the deep compassion that were part of his complex character, were frequently manifested to those who knew him.

His association with youngsters was one of the nicest things about the Boss. Quite frequently he would stop at the Children's Home in Schenectady and take one or more out on an outing either swimming, skiing, boating or some other activity. He was much involved with the Boy Scouts, and I believe was associated in a major way with the first one formed in the city, of which he was Scoutmaster.

With our youngsters he acted like a favorite uncle. At Lake George he would bundle them up and take them for a fast spin in his Chris-Craft. They have never forgotten the fun they had at Crown Island.

The Langmuirs had several beautiful dogs and at Crown Island they had a wonderful time. The Boss had a large bird dog that seemed to be his favorite. One of his experiments with his dog was to smear peanut butter (which the dog was very fond of) between the dog's gums and the inside of the dog's mouth. When this was done the dog went into the strangest performances. I don't remember the details. Its actions seemed to have nothing to do with trying to eat the peanut butter but the dog would lie down, roll over and went through a whole series of actions that the Boss said was part of a routine that never varied. How he discovered this strange reaction or what he did in a follow-up eludes my memory, if in fact I ever knew.

Work Habits of Langmuir

The boss had an interesting work routine. Much of his productive work was done at home. There he had an office of sorts and in his large attic a laboratory workshop. There he made things of wood and carried out other very simple experiments. I'd say in general that he spent 80% of his time at home.

This practice left him free of distractions, visitors and other factors that would disturb his train of thought. From time to time he traveled to the west coast, Europe, and Japan either by plane, ship or train. These invariably involved giving lectures, at which he was extremely good. His infectious enthusiasm whenever he lectured was very popular. However, in his lectures he often abandoned a sentence in "mid stream" if he sensed that his audience was with him. As a result, when a transcript was prepared of a talk he gave it was a major job to retrieve well-rounded sentences in a rewrite. Having heard him lecture many times, I was never aware of this situation and didn't notice it even after my editing efforts had been sorely tried!

Most of his many scientific papers were also written at home. He wrote them in longhand, and when delivered to his secretary, rarely needed any corrections or additions.

His abilities to tackle new fields of science were awesome. While he appeared to have a good grasp of mathematics, some of the newer techniques were foreign to him. When confronted with the need to utilize some of the newer developments, he would approach one of the professional mathematicians in the Laboratory or in general engineering for guidance. I remember one time when he was concerned with probabilities and statistics. He approached one of the newer members of the Laboratory, who was a wizard with numbers and mathematical procedures and techniques. He arrived at Langmuir's office with an armful of books and spent a full day in intensive discussions. Langmuir left at the end of the day carrying some of the books home. When he returned early the next week he had apparently mastered the statistical process and seemed quite happy with the results. When I next saw his mathematics expert, he told me that over the weekend the Boss had found that the published tables did not cover some of the areas of the Boss's interests, so he obtained the original papers of the individual who wrote the book, learned his techniques for constructing the tables, and within a relatively short time, had developed a set of new tables that greatly extended the published ones! That was Langmuir's way of doing things.

He was greatly appreciated by the top administration of the General Electric Company. A historian friend of mine recently told me that he had seen a record of the salary received by Langmuir. He said that in view of the times it was a princely sum and one only equaled by some of the top officers of the Company. To know Langmuir and visit his home one would not expect this to be the case. While he and his family lived in a fine house on Stratford Road within the so-called GE Realty Plot[7] that the company had developed for their chief executives, engineers and scientists, their lifestyle was homey and almost austere.

7 The GE Realty Plot was purchased by GE in 1899 for housing for its senior scientists and employees. Langmuir lived at 1176 Stratford Road.

While Langmuir owned several small airplanes in succession, he drove a small car and enjoyed simple pleasures. He had two camps on Lake George– a small one near Apperson's on Huddle Bay near Bolton Landing and a fairly large rambling old hotel on the crest of a hill on Crown Landing to the northeast of Bolton, astride the wide waters of Northwest Bay. This island received the brunt of the severe windstorms that swept down out of Northwest Bay, and often became the shelter of canoeists who found the waves more than they could handle safely.

The high regard that Langmuir enjoyed from the officers of General Electric was a tremendous asset to those of us associated with John Apperson. Appie was a fighter for his causes. He didn't mince words and was a bothersome foe of those who wanted to repeal Article VII Section 7 of the New York State Constitution that protected the integrity of the Adirondack State Lands.

More than once his opponents did their best to have Appie fired from his job as an electrical engineer in the Company. In each instance, Langmuir interceded and generally was able to provide further evidence that Appie was not only right in his actions, but deserved widespread support. There were many skirmishes during the early thirties when the developers and other opponents of the strict provisions of the Constitution were met and defeated. Much of the credit for these results should be given to Langmuir as well as Apperson, though he preferred to retain a low profile in such confrontations.

The ultimate action on Lake George, which constitutes a fitting memorial to the success of the many fights in the conservation field in which Appie and the Boss were involved, relates to Dome Island. It is the largest island in the lake and has often been referred to as the crown jewel of the Lake. Appie was always concerned about its integrity and when, toward the end of his life, it appeared that Dome would be bought and a large hotel erected at the expense of the forested cover of the island, Appie bought the island and gave it for safe-keeping to the Nature Conservancy. I never knew whether or not Langmuir was involved in this action but I am inclined to think he was.

Langmuir the Wanderer

One of the fascinating activities of the Boss's was his love of wandering. From time to time, just he and Marion would take off in a small car and head for the open road. A number of times he would go by a circuitous route to northern Arizona, where he and she had a favorite place called Gouldings Trading Post,[8] run by Hosea MacFarland. This was a place in the central part of Monument Valley, within the Navajo Indian Reservation. One could get meals and rooms and hike or ride into the wild country to the west and north. I know that he and Marion thoroughly enjoyed themselves in that wild area.

His enthusiastic stories about the beauty of the region excited my interest and led me to spend many summers at a later time in that region. While I never stopped at Gouldings, I passed it many times on my way to Mexican Hat, Bluff, the Goose Necks of the San Juan,

8 The Trading Post was listed on the National Register of Historic Places on October 20, 1980.

Natural Bridges and Halls Crossing. Much of my exploring took place on Painted Mesa, Navajo Mountain, the Kaiparowits Plateau, Smokey Mountain and other parts of the Lake Powell country, as well as farther south in the vicinity of the San Francisco Mountains. It is a truly marvelous and beautiful country, and I can picture Irving and Marion in their glory.

They also, from time to time, sailed along the Maine Coast. The Boss would rent a small sail boat, stock up with provisions and take off, just the two of them exploring the coast, and as I remember some of his tales, they would go up as far as the Maritimes. I never sailed with the Boss, but I'm quite sure if I'd had the chance in those days to do so that I would have become as enthusiastic about it as he was. I have had only a limited exposure to the rocky headlands and islands of that region and can easily sense the reason for his enthusiasm. I once visited Rachel Carson at her camp on the coast at Boothbay Harbor, along with my wife Lois and two friends from Montana. She was the type that would enthuse Langmuir with her broad interests and sensitivity to ecology and environmental welfare.

Another kindred spirit was Roman Vishniac.[9] I never learned how the Boss got to know him except that it may have been that Roman spent considerable time at the Biological Research Station in Woods Hole. There is a wonderful story about their last meeting. It seems that they had not seen each other for some time and in crossing the street near the institute on the Biological Research Station, the two of them coming from either side met in the middle of the street, exchanged enthusiastic greetings, and began to "catch up" on each other's activities and current interests. They were both quite oblivious to their surroundings, even though the traffic had to drive around them from both directions!

Some of Langmuir's Research at Lake George

In 1927, during a European trip by sea on the S. S. Rotterdam, Langmuir observed long lines of seaweed oriented in the direction of the wind. At the time he concluded that the lining up occurred due to wind-driven ocean currents that converged and were forced downward to arise again at a distance half-way toward the adjacent line.

When he reached England toward the end of his extended 1927 tour, he discussed his observations and conclusions with Sir Hugh Taylor and Horace Lamb but neither had any answer to his query.

Langmuir continued his interests and field experiments and, in fact, one of my first assignments from him after starting my work in his laboratory involved a device for measuring underwater currents at Lake George. At about the same time he received the loan of a new sounding device, the bathythermograph, from his friend, Athelstan Spilhaus, then at Woods Hole. This was the second one ever built and Langmuir used it exhaustively

9 Vishniac (1897-1990) was a Russian-American photographer who filmed the Jews in Central and Eastern Europe before the Holocaust and was a biologist who contributed to photomicroscopy and time-lapse photography.

over the years. Something like 2,000 measurements were made from his boat and in the winter through holes in the ice.

In a summation of some of his Lake George experiments I wrote in 1957:

> The helical vortices he studied and described were only a small portion of his interest in the interaction between air and water. His largest unpublished work is concerned with the heat and energy budget of Lake George. Based on his Crown Island camp and using motorboat, ice skates, or skis, depending on the nature of the lake surface, Langmuir would measure the temperature of air and water, the wind velocity and direction, and other pertinent atmospheric and water conditions, during all types of weather, and at many stations on the lake. Since many of his stations were at locations where the lake was more than 150 feet deep, he used triangulation with shoreline and mountain top reference points to fix his positions. After World War II, a war surplus lifeboat sextant was a treasured instrument.
>
> Shortly before his death he assembled all of his notebooks, maps, slides, voluminous graphs and reduced data, apparently with the plan to write a paper on the labor of his love. This was not completed. All of the data is now in the collection of scientific papers that I assembled and which were donated to the Library of Congress by Mrs. Marion Langmuir. It probably represents the finest collection of such field observations in existence.

Some years later, some of my colleagues in the Atmospheric Sciences Research Center (at my suggestion) obtained some of this data from the Library of Congress and published some very interesting data on Langmuir Vortices.

One of the Ways Langmuir Measured the Wind

One of Langmuir's favorite hobbies was the demonstration to youngsters of many simple but provocative scientific principles. One of his favorite examples was the fairly accurate measurement of the wind using a piece of string, a toothpick, a small chunk of solder, a paper clip, a bit of glue and 3" x 5" file card.

The file card is bent into a bivane by folding in half along its long dimension. The paper clip is partly straightened and the straightened part glued into the crease in the file card, the curved end projecting from the lower end of the card. A three-foot length of string is strung through a hole made in the center of the fold, and a knot secures the end. A one-inch length of solder is then crimped to the end of the paper clip to serve as a counter-balance. The toothpick serves as a fixed stretcher of the file card, having an included angle of about 100°. Suspending the file card with the string, the solder on the end of the protruding wire is used as a counter-balance and bent to balance the card so it hangs with the crease in the card in a vertical position.

When held in the wind the bivane, when properly balanced, flies without wobbling and its angular departure from vertical provides a measure of the wind velocity. The unit may be calibrated by holding it out of the side of a car, beyond the slipstream, and noting the angle of flight at different speeds. The easiest way to do the calibration is to use a 5" x 5" piece of cardboard marked in 5° divisions ranging from 0° to 90°. The level is fastened to the top of the card and the string to the upper corner. Once calibrated at 5, 15, 25, 35, and 45 miles per hour, a graph can be constructed on the rear of the 5" x 5" card that depicts the air velocity as a function of the departure of the string angle from the vertical.

The Boss had a number of such simple experiments that he frequently cited to intrigue his young friends and me!

A Movie of the Boss

Sometime in the late winter of 1939, the News Bureau of the General Electric Company was approached by a movie producer who was interested in making a 35mm movie of Dr. Langmuir in his laboratory. I think the man's name was Virgil Kaufmann. He wanted to prepare documentary movies of all the living scientists who had won the Nobel Prize. This was to be the first of his series and thus would be of great value to the producer and a good publicity gimmick for General Electric.

Kaufmann arrived with good credentials and talked the News Bureau into using their film equipment and cameramen to assist in the production. I was asked to set up all of my Langmuir trough equipment over at the movie studios of the Bureau. Within a day or so I had everything set up, properly lighted, and we were ready to go.

As is always the case, many things went wrong, and it was several days before things ran smoothly; the Boss knew what was expected of him and eventually the end was in sight. This period of movie production was just after Katy Blodgett had discovered non-reflecting glass. Thus it was decided to include sequences with Katy preparing the skeleton films and wearing eyeglasses that were half coated, *i.e.*, half of each lens had a non-reflecting film coating, the other being untreated. This worked beautifully. Katy arrived in a classy dress outfit (not the smocks that she used routinely). She looked quite beautiful. This also went well, although a number of sequences were required until she got over her nervousness before a camera (a new experience for her). After about a week of movie activity, the producer figured that he had all of the footage needed. We left with the feeling of accomplishment.

The next day we were informed that the pictures and sound were out of synchronization and that the entire effort was ruined! Would we do it again? The Boss was quite aggravated by this turn of events, but being the scientist and having had things go wrong many times in the past, he agreed to perform again.

The second time I had our equipment in better shape than initially and the shooting went extremely well. The first sequences thus in a sense paid off since the second effort required fewer repeats, and we ended with a smooth documentary.

Several months later we were shown the results. The movie had been printed in a sepia tone and was very well done.

What happened eventually I do not know, nor do I know whether the movie is in the archives of the laboratory. It is something I must check![10]

A Postcard from Langmuir Skiing in Switzerland

Late in the winter of 1939, I received a postcard from the Boss dated March 11, 1939. The postcard was a photograph of a black squirrel being fed by hand with the title "Aroser Eickhornchen." The text of his comments follow:

> Dear Vincent,
>
> There are a lot of these interesting black squirrels (white tummies) with tufted ears like horns. Skiing is perfect, weather marvelous. Lowest temp during the 10 days we've been here has been +12°F, but even in brilliant sun on cloudless days no wet snow! No crusted snow. On 2-3 snowy days have been writing up lectures. English scientists as you see from recent notes in Nature, are dead set against cyclols! From every point of view! But their reasons are not very impressive. But they are aroused enough so we should see a lot of expl. activity– so the mess will be cleared up very fast.
>
> Irving.

This was penned from Arosa in Switzerland, his favorite ski area.

His reference to cyclols is related to the concept of Dr. Dorothy Wrinch, who originated the idea of the structure of globular proteins such as pepsin. She came to America and enlisted his support, and as a result we entered the field of the surface chemistry of protein monolayers. This engaged our attention from early in 1936, until the spring of 1939.

As soon as the Boss returned from Europe, he was persuaded to perform in the movie concerning his research in surface chemistry, that part of his scientific career that was cited when he was chosen to receive the Nobel Prize in Chemistry.

After the film was completed (twice!), we became involved in other research subjects and never returned to further studies of cyclols. It is my understanding that Dorothy Wrinch left England and settled in America to teach. She never visited the Laboratory again.

---

10 There was a documentary made called "Langmuir's World" in 1998 that featured Langmuir in archive footage and with an appearance of Kurt Vonnegut, Jr. The director was Roger R. Summerhayes, Langmuir's grandson. The Schenectady Museum does have the 1939 film that shows him demonstrating his surface film studies.

Hydrogen Tanks that Go Boom

Things were quiet that afternoon as Ed Hennelly was in the Boss's office, deep in a discussion. All of a sudden there was a tremendous explosion that rattled the windows and caused considerable consternation on the part of all.

Ed said that the only reaction from Langmuir was that he raised his head from his concentrated attention to the papers he was reading and said the single word– "Hydrogen"– and then went back to the subject at hand.

It was in fact a tremendous explosion that destroyed the large hydrogen storage tank located at the western edge of the GE Works, at the rear of the four-stack power plant. Although I rather doubt if the Boss had ever heard a similar explosion, his acute mind and analytical powers were so great that it solved that particular phenomenon in the wink of an eye!

Langmuir's Trip to Russia

Sometime around 1945 I learned that the Boss had been invited to visit Russia. I knew very little about the visit and still am quite vague about the details. I do know that shortly after receiving the invitation a letter came from someone in Washington that he should ignore the invitation. Since this was not to be an official visit, the Boss was quite angry about the Washington letter and decided to refuse to abide by that advice. He continued on to plan for the trip, obtaining his airplane ticket and left on schedule.

Arriving in Russia, the group was given the red carpet treatment, managed to see most of the important scientists, and had a number of private talks with them, who also refused to recognize any of the restrictions that fit the normal pattern for visitors.

After having a wonderful time visiting research laboratories and again having private talks with most of their top physicists and chemists, the group continued their journey east ending up in Vladivostok. Somehow they got to American soil and finally returned home.

I am sure from what I heard unofficially that the State Department was furious about all aspects of the trip. I heard repercussions from several of my friends; one in particular was Dr. J. E. Church of the University of Nevada, the inventor of the Mount Rose snow tube for measuring the snow pack and forecasting the hydrologic run off from melting snow. While he didn't provide me with particulars, his eyes would light up when he mentioned some of Langmuir's activities and observations.

While Langmuir was not a favorite with the State Department, he cared very little for pomposity or protocol. I believe the Boss prepared a lengthy report for his trip, experiences and observations. I never saw it. However I do know that he was highly impressed with the knowledge and ability of some of his Russian scientist friends. He reported that in some areas, particularly as they related to the efficiency of photosensitive surfaces, they were considerably ahead of our technology. He even brought back samples of some of their latest achievements in this field that they had given him. I'm sure if the KGB knew about it at the time, some of their top scientists would have been sent to Siberia!

The Boss told me about one fascinating sight he had seen near the North Atlantic Coast. A large cumulus was dissipating with the rain area showing a brilliant rainbow. Although the apparent precipitation zone was quite high in the cloud, the upper part must have consisted of snow, and at the melting point the rainbow appeared, extending through the falling rain to the ground.

For some time after the Boss had returned from Russia he tended to approve of the system he saw. He made an effort to promote further cooperation in friendship and scientific exchange. However, after a year or so the Boss became disillusioned about the Russians and their system and abandoned his efforts to promote cooperative scientific exchanges. While he may have had further connection with the Russians I was never aware of it.

An Adventure in New Mexico and Arizona

It was sometime during the early summer of 1949 when Jack Workman and the Boss decided to take a trip to Albuquerque to the Grand Canyon and invited Bill Camp, Steve Reynolds and me to join them. This was to be a fun trip. Workman had just received delivery of a brand new van made by General Motors. Heading west on Highway 66 a fairly heavy rain was falling, but our new vehicle was performing beautifully, and we went along without any worries.

About twenty miles west of Grants, New Mexico, we suddenly headed south for our first night's bivouac that was to be at an old Civilian Conservation Corps camp near Bluewater Lake, which the College used as a geology camp. We had hardly started on the road when I sensed trouble. The car began to slither and slide in a most astonishing manner. I had experienced similar sensations back in the Northeast during an ice storm or when we had an inch or so of snow on a cold road but– we were on a dirt road.

Workman was a good driver, so although we slid off the crown of the road toward the ditch, he made a last minute recovery and we headed back toward the center of the road. However, he overshot and we headed for the opposite ditch. Here again he managed to keep a forward motion only to repeat the maneuver. And thus we traveled six or so miles to Bluewater Lake.

After we had repeated these maneuvers a half dozen times, I noticed out of the corner of my eyes that Langmuir had pulled out his notebook and pencil and was making some calculations. After swinging back and forth a dozen or so times he announced that we were carrying on a cyclic maneuver having a frequency of 176 per mile or some such value! I'm not sure whether or not Workman appreciated this profundity, but it gave the rest of us nervous hysterics!

It was some fifteen years later that I held the wheel on a similar road west of the San Francisco Peaks in Arizona. I then realized the strain and good driving abilities of our chauffeur! At that time, in a sense I welcomed the experience, since it culminated in a study I had been making of hydrophobic soil.

This is a situation in which a very fine particle volcanic soil becomes coated with the wax from myriads of pine pollen grains until the soil becomes hydrophobic and refuses to soak up water or to let it permeate into the soil below. If the road had recently been graded so that there was a crown on the road, it is virtually impossible to keep a car in the center of the road. It will either slide to the right or the left and, if forward motion is retained, a cyclic or saw tooth pattern develops dependent on the speed of the forward motion!

The next morning, after a hearty breakfast, Jack Workman showed us how to gather and eat pine nuts. Bluewater was surrounded by an almost pure stand of piñon pine, and in a short time we all had a handful of nuts and were learning how shuck them in our mouth to extract the delicious meat.

The road had dried up overnight so we had no difficulty in the trip out. Our tracks of the afternoon before attested to the driving skill of Workman and the mathematical accuracy of Langmuir.

Our next stop was Meteor Crater west of Winslow. At the time we were the only visitors. Dr. Niniger, who had a small museum at the overlook and at the time was in charge of this remarkable natural phenomenon, joined us, and we had a delightful and spirited discussion.

Langmuir was much impressed with the crater and soon had his notebook out and was busily making calculations. When he completed his estimates, he announced in no uncertain terms that any effort to find a massive meteorite in the bottom of the crater would be a failure since it was quite likely that it was completely vaporized upon impact. Thus it is likely that the tiny nickel pellets that are found in considerable abundance in a southeasterly direction from the crater, is of a metal "rain" which followed this vaporization.

Leaving the crater we headed for Flagstaff, and then for the South Rim of the Grand Canyon. Arriving there, I was awestruck with the beauty and immensity of the canyon. Little did I realize that some thirty years later I would be a member of an Ecology Survey that rafted down the length of the Canyon making scientific observations of many kinds– measuring the air quality from Lees Ferry to Diamond Creek, a distance of 255 miles and through fifty-five of its rapids.

After watching the beauty of sunset in the canyon, we spent the night and arose in time to watch the sunrise– both of them magnificent sights. In the morning we visited a number of the viewpoints. At one place the Boss suggested that we prepare a series of stereo photos of the canyon. Both of us were using similar 35mm color cameras. Langmuir told me to select a prominent structure in the canyon, while he did the same while we were 100 feet apart. At a given signal we both clocked our cameras. Since then I have taken thousands of stereo pairs, mostly from airplane windows using 5 to 10 second intervals to get the extended base line necessary for a good stereo effect. I have also found it possible to prepare the ground stereos with a single camera so long as there is not a rapidly changing cloud in the canyon. The main precaution to follow is to be sure that one of the pictures is free of any local vegetation or rock in the foreground. Such pictures when viewed in a stereoscope are spectacular. By nine o'clock in the morning we had viewed the canyon from most of its vistas and reluctantly headed home.

On the way home, we amused ourselves by constructing a long letter to the president of General Motors, complaining about a series of defects our 3-day journey had uncovered in the vehicle. The most serious of these, as I remember it, was a very disconcerting tendency for a gear in the transmission to pop out while moving along at cruising speed. The letter started out, "Dear Charlie," and listed a half dozen or more complaints that had developed during our travel. I'm not sure that the letter was ever sent, but something happened, since we had many other adventures with it while working at the School of Mining in Socorro.

The Meeting of the National Academy at The Knolls

It was in 1950 that the new buildings of the GE Research Laboratory at The Knolls had been occupied by most of the personnel from buildings 5 and 37, that a dedication celebration of this new facility was planned. As part of the program, the National Academy of Sciences planned to conduct its fall meeting at the new laboratory and to have a number of outstanding scientists participate in a symposium. Dr. Langmuir was asked to speak and decided to talk about rainfall cyclic precipitation. He believed that periodic seeding at Socorro, New Mexico was causing this.

To emphasize his claim he had the US Weather Bureau precipitation charts for stations in the Midwest greatly enlarged so that even at the back of the large auditorium the seven-day rainfalls were visible. These records as printed had the days of the month for each station on the horizontal with the stations listed in the vertical. Consequently when no rainfall occurred, the column for a specific day had no entry, while if there was rain, the amount that fell was printed.

It was a spectacular demonstration. The seven-day cycle was so striking that no one could deny its existence. Ray Falconer was assigned by the Boss to gather the data and did an outstanding job.

To this day there has been no satisfactory explanation of this remarkable perturbation in the periodicity of the rainfall throughout the Midwest.

The Disposition of Langmuir's Scientific Papers

During the fall of 1957, following the death of the Boss, I was asked by Mrs. Langmuir and their family lawyer, Mr. Harold Blodgett, to assemble his scientific papers that had been requested by the Library of Congress. With the help of my secretary, Miss Alice Klopfer, we spent quite a few days assembling and boxing his voluminous correspondence, reports, notebooks and papers. It was a big job. Part of it involved searching his attic laboratory, where he had a bunch of simple tools and other paraphernalia that, from time to time, he utilized when doing things at home. Among these effects were some of his skis, his skate sails, skates, and other sports equipment.

There were great quantities of notebooks of the type he habitually had tucked in his jacket or shirt pockets wherever he traveled.

Some of the material was in a high degree of order, especially his Lake George data. It was apparent that one of his priorities after retirement from active work at the Laboratory was the preparation of a major paper on the energy budget of Lake George that would summarize the voluminous quantities of data he had accumulated over the twenty years and more of field observations made at all times of the year at the lake.

That the Library of Congress did a good job of cataloging his papers was apparent when, in the mid-sixties, Ron Stewart and Jon Scott (of our Research Center) went to Washington and obtained copies of some of the Boss's Lake George data in an effort to bring together some of his findings on Langmuir Vortices. This effort was quite successful and several papers were published on their efforts.

This effort to assemble his papers was an interesting but sad undertaking on my part, especially since it ended with a degree of finality my post-graduate career at Langmuir University!

Chapter 3
Early Activities: Years in Living and Learning

My Friend, Katy Blodgett

One of my marvelous experiences while attending Langmuir University was my close friendship with Katherine Blodgett.[11] She was a remarkable woman. I first met her when she needed some equipment built while I was working in the Laboratory Machine Shop. The first thing I built for her was a large brass tray that I later learned was called a Langmuir Trough (I still have it!). Since I had no idea as to how it was to be used, it is much more massive than it needed to be and its dimensions are wrong. However, I built it along the lines she suggested and it has worked beautifully as a demonstration device. After I was invited by the Boss to "come upstairs" and saw the need she had for carrying out her experiments I built another one that she used until her retirement some thirty years later. I made a similar one for myself that is still usable and seems to embody all of the requirements for an efficient and effective experimental trough. When I came upstairs I discovered that I was to share her laboratory and for sixteen years we essentially worked together. We shared our discoveries and our failures and had a wonderful time.

Katy was an ideal co-worker. She had received her Doctorate from the Cavendish Laboratory at Cambridge, England, the first woman to receive this advanced degree from that distinguished university. She never complained of any problems encountered there, though I suspect it was a lonely undertaking. The only complaint about the Cavendish Laboratory that I heard her make was that in the English winter it was often so cold in the lecture halls that the ink froze.

She was a tireless worker. I have seen her sit on a little stool and deposit built-up films until she amassed a total of 3,000 on a glass slide. Every layer had to be perfect. She did this for Dr. Clifford Holley of the University of Chicago, who wanted to see if such a film could be used in some x-ray diffraction studies that he had underway.

To make a perfect film was a fantastic and exasperating job. Time after time, as she approached the magic number, the film would suddenly crack to ruin the integrity of the film crystal. After many trials she discovered that it was necessary to add the tiniest amount of copper to the water bath in order to succeed.

---

11 Katy Blodgett never knew her father George. He was murdered by a burglar in their house on Front Street in Schenectady just before she was born in 1897. It is still considered unsolved as the man who was arrested and convicted may not have committed the murder. He hung himself while in jail in Salem, New York. She was the first female scientist at GE with a Doctorate degree

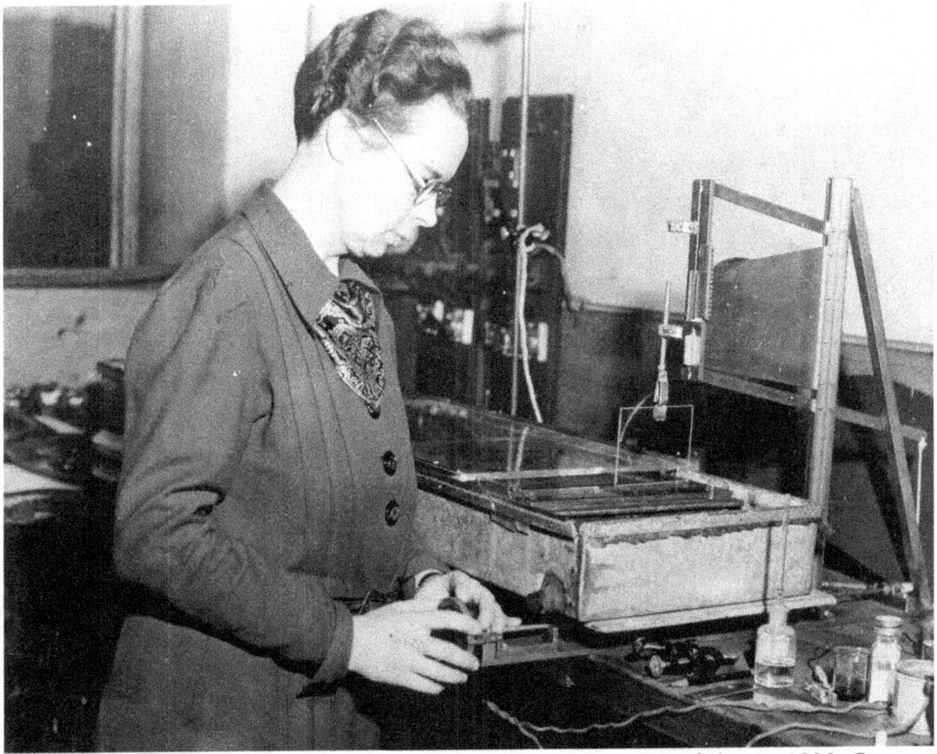

Dr. Katherine Blodgett demonstrates placing an anti-glare fluid on plate of glass in 1938. Courtesy of Schenectady Museum.

When I arrived at Room 408 that was to be our working laboratory for sixteen years she arranged to share a laboratory bench with me. I was given a small oak flat top desk that I fitted up near one of the two windows in the room.

Once Katy knew what the Boss wanted me to do she showed me the techniques to follow and guided me in many ways. We soon were sharing a waterstill and acid cleaning solution and other components of the laboratory. Throughout the sixteen years at Building 5 and the six following years from 1948 to 1954 at the new laboratory at the Knolls in Niskayuna, I never heard a cross word from her. Her cheerful, upbeat attitude was a marvelous attribute. We shared our problems, advances and occasional successes, as well as our personal life. She never married but had many friends. She took a great interest in my marriage, my wife Lois and the youngsters that followed. When our second-hand boiler cracked she was in the process of shifting her coal furnace to an oil burner and arranged for us to get her old furnace. After fifty years, having converted it to oil, it is still providing us with heat!

Katy lived in the "old town" of Schenectady that was originally within The Stockade. She was quite a successful gardener and had a deal with the GE Woman's Club to use a portion of their backyard that sloped down to the Mohawk River for her garden. She raised both vegetables and flowers, and I still have the blue phlox she gave me more than forty years ago. It is now flourishing in five or six places in my yard and in the flower gardens of our children in Sudbury, Massachusetts; Plymouth, Minnesota; and Richland, Washington. These are "Katy's flowers."

Katy was quite a cook. For a while however the trick of making good popovers eluded her. This finally challenged her scientific mind and she embarked upon a series of quantitative experiments. Before long her popovers were something to behold! She gave me a running account of her progress and finally brought in some beautiful products of her achievement.

She was a good friend of John Apperson and had a tiny camp next to his on Lake George. She spent many weekends with her close friend Edith Clark, a General Electric mathematician, and with the Christies, who lived nearby. She and her friends were strong supporters of Appie's conservation and environmental activities. In Schenectady she was very active in the Zonta Club, serving as an officer involved in many civic activities for the betterment of the cultural wealth of Schenectady.

I was in the laboratory when Katy discovered skeleton films wherein building films at the correct pH of the water, the monolayers consisted of half stearic acid and half barium stearate. She found that by soaking such films for a few seconds in benzene– a solvent for stearic acid, all of the stearic acid molecules would dissolve out of the built up film leaving behind a skeleton structure of barium stearate. She quickly discovered that the refractive index of the resultant skeleton film was greatly reduced. From this discovery came non-reflecting glass. I'll never forget the excitement that this discovery generated in the laboratory! Within a few days she had prepared her eyeglasses with such films and the effect was spectacular. The glassy reflection was gone and was replaced with a deep purple almost invisible sheen.

Unfortunately the barium stearate skeleton film was not of commercial value since it was a soft metallic soap. However, her discovery pointed the way to the large scale treatment of camera lenses, telescope optics and a host of applications using the basic principles of her discovery. None of them have been as perfect in the reduction of reflection as hers, but the millions of lens surfaces that are seen today on all camera lenses attest to her discovery.

As I was given freedom to do my research, my studies covered an ever-broadening array of activities. We remained in close association however, and I relied on Katy to witness in my notebook the discoveries and advances I made. Her first witnessing record on was October 29, 1934, while the last signature was made on January 11, 1954.

During the war years she went with us to Vroman's Nose to monitor our smoke screens and operated the radio at the summit of the nose. Shortly after our highly successful demonstration she constructed the color filters used to evaluate the performance of the smoke generators, using the sun's color to provide us with that information.

When I left the Laboratory to become Director of Research of the Munitalp Foundation, Langmuir had left and she joined forces with Tom Van der Slice. She was much impressed with Tom's ability and knowledge and told me that he might take Langmuir's place in the Laboratory. It was one of the few times I disagreed with Katy. In breadth of interest, enthusiasm, intensity, friendliness, contacts and influence, I could see no one who could take Langmuir's place at General Electric. Tom moved up the corporate ladder rapidly to become a Vice President, but then left to take on the Presidency of General Telephone. Since then I have lost track of him.

Katy retired from General Electric in 1963. We had a gala party for her attended by a large number of her friends. Subsequently she was injured in an automobile accident that made her a semi-invalid. Her mind remained clear and she retained her high spirits until her death in 1979.

At the present time there seems to be a strong revival of interest in Katy's research achievements. Her built-up films techniques are being followed in a number of laboratories. There is much of basic importance that remains to be discovered both in the physical, chemical and biological sciences, concerned with single layers of molecules.

Plants and Shrubs for Woestyne

As soon as I acquired a deed for our property on Schermerhorn Road and in the midst of our house construction, I took time out occasionally to plant perennials, shrubs and trees as they became available. Dad Perret made a rich and diverse supply available. He was an avid gardener and had quite a few plants and shrubs. Roots of peonies, lily of the valley, iris of several types, blackberries, rhubarb, phlox, and lilacs were among the plants he gave us, and after more than fifty years they are still thriving. The blackberry bushes I planted northwest of the garage site prospered to the point where I invited Katy Blodgett and others to harvest the berries for their own use. These plants slowly migrated downhill and onto the Flats below until now the best picking is to be found along the edge of our wetland. The same thing occurred with the phlox. It has become established in great concentration on the hillsides below the house and the garage.

I obtained a few clusters of blue flowering myrtle (or periwinkle). We now have great areas of our hillsides covered with a beautiful mass of this attractive plant, as well as in many areas around our evergreen trees.

I found a serviceberry (or shadbush) in the Pine Plains that has now become so large that its trunk has a diameter of nearly a foot. It is so high that I frequently am not aware of its blossoms until the petals begin to fall.

One day, while driving along Crawford Road, I found a place where a county road crew, in widening the road, had ripped out a hemlock tree. Taking a chance that it might grow, I rescued it and planted it near the garage. It is now nearly two feet in diameter!

One Christmas time I bought a small Norway spruce for our tree. Early in January, during a thaw, I was able to plant it. This tree is now of massive proportions with a diameter of two feet.

Dad Perret gave me a black walnut sapling of less than an inch in diameter. It is now the largest tree on our land, having a diameter of more than 24". Descendants of this tree (from sprouted nuts) now are scattered in various places and are the fastest growing of all our trees.

Also near the garage is a very tall white fir. It started out as a seedling less than four inches high. We also have seven large red pines that also grew from seedlings less than six inches high.

A mountain ash that I collected during a hike with Irving and Marion Langmuir is now more than 30" in diameter. Nearby are a black cherry, a white pine and a Norway maple, all of large size.

In 1936, while exploring the vicinity of the Little Nose, I found a spectacular tall plant growing in the waters of the old Erie Canal. I carefully obtained several specimens and planted them on the edge of a small pond I had formed by accumulating water from a small spring below the house. I discovered it to be purple loosestrife. It prospered, spread and has now established itself in most of the wetlands adjacent to the Great Flats Aquifer east and southeast of our home. My neighbor, who was a commercial beekeeper (having 100 hives), was greatly pleased with the loosestrife since it bridged the gap that existed between spring and early fall.

Aerial view of Woestyne South. Schaefer house in middle left. Vince built his home here where Native Americans lived for thousands of years before. Courtesy of Jim Schaefer.

I transplanted a variety of wild ferns– the Ostrich, Cinnamon, Royal, Evergreen, Christmas, Lady, Maidenhair– and have been pleased to see how well they have prospered. The same thing occurred with the wild plants– the Bloodroot, Joe Pye Weed, jack in the Pulpit, Canada, white and blue violets, Red Baneberry, the Trout Lily, Pink Dogbane, Blue and Yellow Flag, White, Red and Nodding Trillium and Giant Solomon's Seal. One of my great pleasures is to watch the wildflowers appear on the edge of my big mowed area that I let flourish as meadowland.

Fortunately for my plant varieties, the soil of Woestyne South varies from a sandy loam at the upper (west) end of our property, to a sandy clay loam at the lower portion, to almost pure clay at the edge of the wetlands. In fact, there are the remains of an ancient brickyard at one location in the latter area that is so saturated with broken brick, that an earlier attempt to make a garden there had to be abandoned. This is an area that should be explored using archeological methods.

The Local Geology Related to Woestyne South

The grassy knoll and ice cold spring that first attracted me to the place where we built our home was formed during the last 10,000 years, despite the fact that the underlying rock, which is quite close to the surface, is of the Ordovician Period, formed more than 300 million years ago.

After the great Continental Glacier melted that had covered our region with a vast mass of moving ice having a thickness of a mile or more, a huge lake formed that is now thought to have extended the entire length of the Hudson Valley, with an ice and gravel dam located in the vicinity of Staten Island. At one stage this lake had an elevation approaching 400 feet above sea level. It was fed by the meltwaters from the vicinity of the Great Lakes, since the St. Lawrence River was still blocked by the melting glacier. Approaching from the west, it was slowed down as it ran into the lake waters, depositing its load of boulders, cobblestones, gravel, sand and rock flour (clay) to form a great delta that extended from below Little Falls to the valley of the Hudson. In this manner it deposited the extensive mass of sand and gravel that at present constitutes the 80-foot depth of the Great Flats Aquifer that lies adjacent to, and directly east of, our property.

Over the top of this gravel deposit was a nearly pure deposit of fairly coarse sand, having an elevation of 350 feet above sea level, and extending some twenty miles to the southeast, and covering all of the present valley of the Mohawk and that of the Normans Kill.

When the natural dam confining the waters of Lake Albany broke, the rushing waters removed a considerable portion of the sandy delta and leaving the steep sides of the eroded deposit that now constitutes the Schonowe and Bellevue Bluffs. Sheet erosion modified the steep slopes to form terraces and it is on such a terrace that we built our home. Ten feet or so in elevation below the surface of the terrace innumerable springs occur– some tiny, but some of considerable size. Thus it is that two of them emerge on our property, but a big one comes out of the hillside about one hundred feet south of our southerly property line to form the Schermerhorn Spring. This has flow of more than ten gallons per minute and a temperature of 44°F.

This line of springs comes from a "perched" aquifer that extends all the way to the Hudson River in Albany. Emergence of these springs on the Hudson River side is blocked by an extensive band of clay that marks the edge of the Lake Albany sand delta. However, west of this clay toward Schenectady is the Pine Bush– an extensive region once covered with sand dunes and now marked by a growth of Pitch Pine, Scrub Oak, Sweet Fern and Huckleberries. The cold, spring fed trout streams that flow from the sand– the Hunger, Kaikout, Rainbow, Krum and Becker– originate at the same elevations as our local springs along Schermerhorn Road. Other large springs flow out of the sandy aquifer along the Poentic Kill in the vicinity of Tippecanoe and into the Davitie Gat Kill, its major tributary. The Ordovician shales and sandstone that underlie the region come to the surface along the Poentic Kill and the much smaller streams just northwest of Woestyne South, called Partridge and Hermit Thrush Brooks, that drain the Old Maids Woods area.

Incidentally, the rocky channel of the ancient Mohawk River has an elevation of sea level and is buried under the Great Flats Aquifer adjacent to Woestyne. Instead of following the present course of the Mohawk, it heads southerly and originally entered the Hudson River through the valley of the Normans Kill just south of the city of Albany. It is now filled with glacial till to a depth of more than one hundred feet on top of which is the sand and gravel of the large aquifer.

The Building of Woestyne South

When I was fourteen years old, I began to explore the Mohawk River Valley for Indian relics. My father, who loved the out of doors and had hiked in the Austrian Alps while a graduate student at the University of Innsbruck, used to take me on Sunday walks across the Great Flats and along the Erie Canal northwest of the City of Schenectady. In this way I became familiar with that region and frequently on days when he couldn't go with me, I would go alone or with one of my chums.

In this way I discovered a number of fine Indian sites, and was soon amassing quite a collection of Indian artifacts.

One day in the spring, I crossed the flats and hiked to the lower end of Schermerhorn Road where it joined the Kings Highway (also then called the River Road) at Lock 23 of the Old Erie Canal. At the southwestern corner of the road junction was a large field that had been recently plowed and rained upon so that conditions were ideal for finding artifacts.

Within a short time I found a dozen or more beautiful arrowheads as well as a magnificent round stone pestle or corn grinder. It had been broken by the plow, but I found that when glued together it constituted three quarters of its length. Several years later I found the missing piece, along with a beautiful small ground and polished stone axe.

With the discovery of the Bradt site (named after the nearby ancient brick house that was built in 1735 by Arent Bradt and that is still in use as a home) I discovered six other camps and village sites, most of which we later found were occupied several thousand years ago.

The best of these was the Schermerhorn Site that was almost contiguous with the Bradt Site and was adjacent to a beautiful large spring of very cold water.

This area was my favorite place for eating my lunch. A grassy knoll not far from the spring had several fine apple trees and the old Schermerhorn Family Burying Ground. It was a delightful place. Just below the knoll was a huge Dutch barn, built apparently by William Teller, a direct descendent of one of the original settlers of Schenectady. An ancient wood road ran from the old barn to the woodlands on top of the steep hill that sheltered the region from the cold winter winds from the northwest. This road cut across the grassy knoll at a moderate grade and swung past the burying ground. (Incidentally, about 1918 the gravestones and remains were removed and placed near the Cobblestone Reformed Church at the top of the hill where the Princeton and Putman Roads join.)

I was so taken with this grassy knoll, I resolved that some day I would acquire the land and build my house on it: fourteen years later that is just what I did!

After Dr. Langmuir asked me to work with him and it looked like it was time for me to leave my parents' home and fend for myself, I began to look around for someone to share it with. I had never had much interaction with girls. My first experience was one that made me leery of them! While in grade school a girl had apparently "set her eyes on me" and began to let her feelings become apparent. While she was a pretty girl and very friendly, I wasn't interested, and I guess I must have let her know of it by lack of response.

After this brief episode I didn't pay much attention to the opposite sex until after I rejoined the General Electric Company in 1929. The year before, I began to sense that there was more to life than finding arrowheads, running machines, "skinning" trees and raising flowers. I decided to assemble a group of outdoor enthusiasts to foster social contacts and experiences.

I soon found that there were indeed some very attractive girls joining the club having interests similar to my own. I made several attempts to find one that was compatible with my interests without much success.

Vince sliding down the Olympic ski jump in Lake Placid, NY on February 18, 1932. Courtesy of Jim Schaefer.

Then, on April 24, 1930– it happened. My sister Gertrude, in some manner, arranged for me to accompany her and a girlfriend she had met on one of our hikes to go to Proctors Theatre. Her friend was a vivacious redhead, thin and wiry and full of fun.

Three days later, being a registered nurse, she assisted me in putting on a first aid demonstration as part of an adult education program the hiking club was sponsoring during the Depression.

Later in the spring, the club conducted a weekend party at Lake George; swimming, mountain climbing and just having a good time. It was then that I realized that I had discovered an individual that might be the one I had been searching for.

Vince taking a pause along the Range Trail from Basin Looking at Saddleback Gothics, Armstrong, Wolf Jaws in the Adirondacks on September 2, 1930.

At the time I was still working in the Laboratory Machine Shop, hardly earning enough to support myself and someone else! Our friendship continued however, and it became increasingly evident that Lois Perret was someone quite special.

We saw each other on all of the hikes and occasionally took special hikes into the Helderbergs, to Indian village sites along the Mohawk, and to other areas of mutual interest. During the summer of 1931 I got more than twenty members of the club to start the fabrication of sleeping bags, with the advice and guidance of John Apperson. One reason for this large-scale effort was a plan that I had concocted to take in the 1932 Winter Olympics at Lake Placid. Since none of us had any money to speak of, I proposed that we get permission from the Lake Placid Club to use the Adirondack-type open lean-tos that were clustered near Adirondack Loj at Heart Lake at the trailhead for the hiking trails into the High Peaks of Mount Marcy, Colden and McIntyre. When winter arrived all of our bags were completed and, in the mid-winter of 1932, we headed for Heart Lake.

During the Olympic Games there were more than twenty of our club camped out in the open lean-tos. Since Lois had to work during the start of the games she didn't accompany us when we set up camp, but arrived with another group later in the week. After arriving and spending her first night, she was among the early risers the next morning as we struggled to get a fire going to boil the coffee. Despite strenuous efforts we never did get a suitable supply of good firewood, thus our morning fire was a rather feeble one. As the water began to simmer and everyone was getting impatient, Lois, trying to be helpful, managed to dump the nearly boiling water onto the fire, effectively quenching it! Needless to say she was not very popular for a few minutes! To this day– after more than 55 years– the episode is still recounted with glee!

The Olympic adventure was a highlight in the club annals and is a story worth telling in a succeeding chapter.

Left: Bart, Herm, Vince and Frank on Mount McIntyre in the Adirondacks in February 1932. Right: Vince on Mount McIntyre the same day. Courtesy of Jim Schaefer.

Sufficient to say, on July 4, 1934, my birthday, at the falls of the West Branch of the Sacandaga River, in one of the wild and beautiful places in the Adirondacks, I asked Lois to share my life and she accepted– probably saying to herself, "it's about time." We received our parents' blessings and proceeded to plan ahead.

In thinking back to those times, I am intrigued with the fact that it never occurred to me that we should rent a house after getting married. All three of us brothers just assumed that we would build our own homes.

Thus, I wasted no time in attempting to acquire title to the grassy knoll that I had selected for my home site when I was fourteen years old. I learned that the land was owned by Mr. Clarence Schermerhorn, who lived in the brick house built in the late 1700s and located close to the big Schermerhorn spring that had been so attractive to the early Indians.

After considerable persuasion on my part, Mr. Schermerhorn agreed to sell me a parcel of land adjacent to his property and extending along the Schermerhorn Road to the old Carriage House of the Schermerhorn Mansion, then occupied by two maiden sisters related to Clarence.

I encountered great difficulty getting a deed for the land. After agreeing to sell it, Clarence had second thoughts and waffled back and forth, finally agreeing to have a deed prepared after removing a considerable piece of land at the lower (eastern) part of the parcel. Since I was desperate to get the deed, I agreed to this deletion, finally getting it on October 11, 1934. (Incidentally, some years later, after Mrs. Schermerhorn died, her estate was purchased by a young man who was a hard worker and did a fine job of restoring the old house. He approached me about purchasing the land sloping down from my home that included the site of the old Dutch barn and extending down to the lands owned by the City of Schenectady. I acquired it and now have a large garden, a patch of white pines, a wild region of wetland and a large field of more than an acre that grows grass and many wildflowers. It is beautiful!) On October 13, 1934, I entered into a purchase agreement with Mr. William Keyser, my next-door neighbor, to buy from him a large barn located in back of the Carriage House that he had converted to living quarters. I later discovered that a provision of his deed received from the Schermerhorn sisters stipulated that he had to remove the barn by the end of 1933. Thus he was nearly a year late in carrying out this provision of his agreement. It explained his readiness to sell me the barn for $200.

On the 27th of October we laid out the corners of the house. It was to be on the brim of the grassy knoll that had so attracted me some fourteen years previously. We positioned it so that the two apple trees were preserved, the one, a Rhode Island Greening, would be to the north of the house, the other, a Northern Spy, would be west of it.

Meanwhile Lois and I had spent many enjoyable evenings designing our home. We decided to copy the lines of the ancient Dutch farmhouses, combining features and lines of the Bradt Homestead within sight of our location, and the Abram Yates House in the Stockade area of Old Schenectady. Our design included a very steep slate roof, the "mouse teeth" in the brickwork and the loopholes in the attic area.

We developed our initial floor plan to have an area of 900 square feet, with a full basement and two bedrooms on the second floor. We drew many plans, discarding those that for one reason or another had shortcomings. In profile the house, with its steep roofline, covered an area of 600 square feet, with a lean-to extending out over the remaining 300 square feet on the eastern side. Since the basement was open at ground level on the eastern (downhill) side, this exposure appeared as a two-story elevation.

Our final plans were reviewed by Mr. Robert Bowen, my brother Paul's architect. Except for a few minor changes, he approved our plan.

The foundation was dug partly by hand, and then by horse and scoop shovel. As we dug into the sandy loam that made up the terrace below the steep hill on the west, I kept my eye open for prehistoric and historic artifacts. Except for a few flint chips, nothing of Indian occupation was uncovered. However, I did find a perfect European type clay pipe of the type made for Indian trade. It had the initials on the bowl of "R. T." (R. Tippet). Although I had found pipe fragments similar to this at other historic sites, this was the best I have ever seen.

Another interesting feature uncovered during our excavation was areas of hardpan. These consisted of irregular chunks of extremely hard clay containing glacial gravel. They were so hard and difficult to handle that even a heavy pickaxe had little effect on them. We had to dig around them and handle them like a rock. When the excavation was completed a wide footing of concrete was poured, blocks were laid, and we soon were on our way.

Meanwhile we had purchased the large barn and sought help in razing it from our friends in the hiking club. A large group showed up early one Saturday morning and by evening when we had a hearty supper for everyone most of the structure was dismantled, carried to our building site and put into many piles.

The barn had been fashioned with a heavy frame of 8" x 10" timbers fastened together with mortise and tenons with hickory pegs. These pegs were carefully removed so that the structure could be reassembled in framing our house.

I discovered that my brother Paul was a master at assembling such a structure, and before long the house began to take form. The barn had a tremendous amount of excellent wood in it. In fact, the only wood we had to purchase in constructing our entire house was the oak flooring that was used in our big living and dining room and our two upstairs bedrooms. The barn siding consisted of long 1" x 12" white pine boards that we used in sheathing the sides and roof of the house and for all the underflooring.

We both love fireplaces, so one of the first things we did once the house was framed was to get my brother Carl to put in two of them– one in the living room and one in the basement. The latter one was made of limestone rocks of the Helderbergs that over the years I had collected because of their unique fossils or other properties. The one in the living room was made of old bricks from the same source as those used in the outside–second-hand bricks retrieved from old houses by Ernie Michels. While they still had the mortar remaining on them following demolition, Lois and I got quite adept at stripping the bricks of old mortar thus taking them back to their original form.

One day after Carl had finished building both fireplaces, my brother Paul told me that he had found a fireplace mantel in an old tavern near Pattersonville. It had been ripped off the fireplace and nailed up over an opening to keep the snow out. He had found the owner and was told that he could buy the mantel for five dollars. We got the mantel and brought it to our new house, measured the space and found that all we had to do was to trim one inch from its top. Even the arch of the bricks that Carl had made fit the curve of the opening in perfect proportion! At about this time I decided to approach our bank to obtain a mortgage, since some major expenses loomed ahead as we began to plan for central heating, plumbing supplies and such items.

Despite the fact that neither Lois or I had any funds in the savings bank, we were both working and thus were able to buy the cement and nails and other basic items needed to keep Paul in working supplies. The purchase of the barn for $200 had exhausted my savings!

Everything was OK with the bank officials until I mentioned that we were planning to build our own home out of timber from a barn. That was a fatal statement. I was immediately turned down!

Without funds from the bank, the progress of house construction slowed down but didn't stop. Paul had other jobs that kept him in working funds, while Lois and I put every penny we could spare into paying for the second-hand brick, slate and other items, including an old coal furnace, second-hand sink, and the cheapest bathroom supplies we could find! On July 1, 1935, Paul and a helper finished building the stairs that went into the basement and up to the two bedrooms on the second floor. My Uncle Joe Holtslag, a master technician who worked at General Electric, did the electrical wiring and plumbing for us. He spent many evenings on the job assisted by his oldest son, Joe Jr., who still talks with pride about the job they did for us. Before putting on plasterboard, we insulated the space between the studs using 3" batts of rock wool, the only insulation material available in 1935.

By the spring of 1936 the house was approaching completion. My farmer friend, Rob Hartley of Minaville,[12] gave us a heavy colonial door. It had been saved when the family homestead was taken down and replaced by a more modern house. The door fitted perfectly into the space we had allowed for that purpose. The two doors opening out onto the east terrace from the basement were barn doors that I hung using ancient iron strap hinges salvaged from the barn, and found in a junk pile back of the Teller-Schermerhorn Dutch barn that was located below the eastern boundary of our property.

As the main part of the Schermerhorn barn was taken down and moved to our building site, quite a number of the larger timbers not needed for framing the house still remained. Bill Gluesing, one of our hiking club friends, had given us an old Model A Ford touring car. While the body was in sad condition, the engine still worked well. I found it to be possible to place a rope around one of the timbers, and on a rainy day, discovered that the mud and wet road provided suitable lubrication so that by getting a good start, I could snake the timber away from the barn site, up to the road, and from there up to our land. Eventually, I had a large pile of these huge timbers. We also had a pile of thick hemlock planks from the barn floor, as well as enough framing timbers to erect a second building that was to become our two-stall garage.

We dug another foundation space in the hillside and, using the same procedure as with the house, prepared footings, but then constructed forms and poured a concrete foundation for a two-stall garage. I then placed the planking for the floor on the concrete walls and piled the big timbers on top of one another so that eastern face of the garage looked like a log cabin. Instead of a steep roof, like the house, we designed a gambrel roof for the garage. When this was sheathed, I put a slate roof on it using again the old, wide slates that we got from Ernie Michels. I had watched the slater who covered the house, obtained a slater's hammer and anvil, and had no trouble at all in laying in the roof.

Clarence Schermerhorn, from whom we purchased the land, died before we finished building our home. Shortly afterward, I purchased from his widow the huge Dutch barn that was downhill from our land. To my dismay, the roof had been leaking for a number of years so that the irreplaceable roof rafters had rotted out. Thus there was nothing to do but to salvage the wood in it.

Except for the massive anchor beams, columns and plates, the remaining timbers of the barn were mostly the extremely long roof rafters, all of which were in poor condition.

---

12 Minaville is a small hamlet in Montgomery County and was the birthplace of Sheldon Jackson, author of *Alaska and Missions on the North Pacific Coast* (1880).

As a challenging project, I decided to see if I could dismantle this huge barn single-handedly. I first removed the deteriorated tarpaper covering, and then the wide original sheathing boards. These were identical to those used in the Mabee House at Rotterdam Junction that has an erection date of 1670.[13] This was a precarious operation, but one that I managed without any problem. I then cut the connections of the roof rafters that, according to the ancient architecture, had no connecting roof ridge board. While the rafters were more than sixty feet long, I found that they were balanced perfectly on the purlin plates that, in turn, were supported by the columns holding the huge mortised anchor beams. I found it possible to cause them to teeter-totter, and then by suddenly giving them a twist, they would roll down the remaining rafters and fall to the ground. Due to their rotten condition they would often break into two or more pieces as they hit the ground.

From time to time I found myself stymied as to the next step to follow in the dismantling process. When this occurred I would stop my operation while I thought over the problem. Invariably the solution would come to me, and I was soon confronted with nothing but the massive thirty-foot long anchor beams and the columns holding them. At this point Paul told me that he would like to have them for framing a large library addition that he was planning to add to his limestone house that he had built on St. David's Lane. This solved my problem completely, and within a few weeks all of the Teller-Schermerhorn Barn was being used effectively.

I found the pine siding of the barn to be in excellent condition and that it was a beautiful wood for making cabinets in the kitchen, the bedrooms, and my basement laboratory. While the outside of the boards was weathered, the inside had been smoothed and, as a result, only a light sanding followed by a coat of linseed oil produced a golden brown finish for these cabinets.

After our home was finished with our beautiful random width flooring, we realized the need for a good-sized corner cupboard. I thought of making one, but we decided to check our favorite antique shop in Fonda, New York, run by "Pete" Morford. He was an old friend of ours and we knew if there was any chance of getting such a piece of furniture at a price we could afford, it would be at his shop.

We went there, told him what we were after, and he immediately took us to the back of the shop and showed us a cupboard he had just obtained from the Southern Tier of the state. It looked to be perfect, and in asking about the price, was quoted one that was within our budget.

I can't remember how we got it home, though I think we might have put it in the back of the Model A Ford that was still our only vehicle. After unloading it and moving it into the dining room, I was dismayed to discover that the windowsills would prevent it from sliding into place. Measuring things carefully, I found that by cutting horizontal slots

---

13 Subsequent dating using tree ring dating techniques at Cornell University place the date of 1705 for construction of the Mabee Farm.

into the angular sides of the cabinet, it would fit. I cut these slots with great care and was delighted to find that the cupboard slid into place perfectly!

The capacity of this ancient, solid, white pine cupboard is large enough to accommodate all of our good china, platters, silver, vases and all other items that have accumulated over the years. A recent appraisal of this object indicates that is now has a value of more that thirty times what we paid for it!

Shortly after we were married on July 27, 1936, we had a visit from the Boss. I'm sure that he had been keeping an eye on our house-building activities from the beginning. After enthusiastically viewing the house, he told me that he and Marion would like to take a hike with me to look at the site of the ski our Wintersports Club was developing on the slopes of Yantaputchaberg, the highest of the Rotterdam Hills. We did so. While exploring the region I found a small mountain ash sapling that was in the way of one of our trails. Since I knew that it would be cut down, I managed to uproot it, get it home, and plant it all within an hour or so. It took root, grew, and after fifty years, is now more than two feet in diameter!

At about that time we needed to get a better means for approaching the threshold of the house. A stepping-stone seemed to be the ideal choice. I had found a large, cut, limestone slab along the Old Erie Canal then owned by the General Electric Company. I approached the head of their Buildings and Grounds Department and asked permission to take it. They were quite willing for me to do so. I managed to borrow a truck, and in some manner we slid it into the bed of the truck, and then slid it into place in front of the door. It is a beautiful relic of the Old Erie and is especially interesting since it has a hexagonal array of vertical cracks that must have been formed in a shallow sea many millions of years ago.

Shortly after our house was finished, I decided to build a stone wall along Schermerhorn Road. I knew that the Town Highway Department had been reconstructing Gordon Road, on top of the hill heading toward Mariaville, using the old stone walls that had been built to separate the wood lots in that region. My old friend "Sime" Schermerhorn once told me that all of the stone walls of our region were built by professional wall builders who quarried the Schenectady Sandstone outcrops in the area in order to have stone for the construction.

I found that in tearing down the walls for ballast on the road, the highway crew had left several layers of the walls behind. I salvaged these rocks, piling them into the rear of our Model A Ford. I did this every morning before breakfast, unloading the car, having breakfast and then walking into work (about two miles away). In the evening I'd fit the rocks together in dry construction as the original walls had been. In this manner, in thirty days I had built a wall that is 3 feet high, 2 feet wide and 150 feet long.

Portions of the wall collapse from time to time due to pressure from the snowplows but it takes only a few minutes to get it back in shape. After fifty years it remains as the only stone wall along Schermerhorn Road and the only one I know of which contains the rocks quarried by those old fence builders!

During the early spring of 1936, a mutual friend from the Schenectady Wintersports Club, Dr. Klaus Sextus, approached me with the proposition that he'd appreciate it if while he was in Germany on a protracted visit we would take care of his car. One stipulation was that we actively use it.

Since our only transportation at the time was the old Model A Ford that Bill Gluesing gave to us, this offer from Klaus was very welcome! His car was a fairly new coupe with a rumble seat, and for the next year or so our transportation problem was solved.

When Klaus returned from Germany and repossessed his car, we wondered what we would do until we could afford to buy another. Looking at the Model A, it became obvious that it was our only possibility. Accordingly I got my shovel and hoe and managed to remove the soil and weeds that had accumulated. We then hosed it down, fixed the seats, checked the engine, tires, oil, radiator coolant and battery, and found we had transportation! While it wasn't elegant, it provided us with a practical solution to our transportation problems for several years!

Early in 1953, while in the throes of deciding on my future, we made plans to put an addition on the north side of the house. The greening apple tree had died and the wooden porch that extended out from the kitchen had deteriorated to the point where it was no longer useful.

We made plans for a basement with a large room on the main floor and a bedroom on the second. The rooflines were to be a gambrel, which was a roofline that succeeded the steep-roofed early Dutch design. A dominant feature of this addition was a large picture window facing the view across the Great Flats toward the southeast. Below it was to be a large Dutch door I had "rescued" from the Tymerson House along the Mohawk in Niskayuna. I had found it dumped on the slope behind the demolished house. It was in excellent condition. When the elevation of the house was worked out, it appeared that because of the size of the door the floor would need to be higher than that of the kitchen in the main house. Since I could not consider trimming the Dutch door we decided to construct a ramp between the house and the new addition. This was done, and although the ramp is sometimes a nuisance, I'm glad we decided to preserver the integrity of the door.

The basement room was divided into a nice workshop, an insulated vegetable "cellar," and a place to keep our frozen food cabinets.

About fifteen years later another addition was designed and built, this one on the south side. It consisted of a screened sun porch and an outdoor patio on the main floor level. In the basement I have space for my "new" laboratory. There, in addition to shelves for chemicals, I have my Langmuir troughs, cold chambers, air quality instruments, high voltage generators, and other equipment for experimentation.

The main part of the basement has, in addition to my big limestone fireplace: a dark room, a film editing area, a microscope room, a 24-inch diamond saw, and adjacent to this area, a furnace room and laundry. It was a great day when I decided that some day I would build a home on that grassy knoll along Schermerhorn Road above the Great Flats of the Mohawk!

The adventures we have had over the years in fulfilling my original dream have been fun. Starting with nothing more than a desire to have our own home, and desirous, if possible, to have it without a mortgage, we achieved this goal, and on July 27, 1986 Lois and I, with all of our three youngsters, Susan, Katherine and James, and the seven grandchildren, celebrated our 50th wedding anniversary with a series of four parties!

The Historical Environment of Woestyne South

The little settlement of Schonowe occupies historic ground. It was the outpost of the newly-formed village of Schenectady, when it was formed in 1661 by Arent Van Curler. The ten adult freeholders at that time were apportioned farmland on the Great Flats, divided into twenty parcels, with the Kings Highway (old River Road) serving as the dividing line. The person receiving a parcel close to the village also received one farther away. Thus each "hofstede" consisted of a parcel fairly close to The Stockade, with another one with less availability to protection. The rear or hindmost farms abutted on the bluff along the present Schermerhorn Road and a number of them, such as Arent Bradt, built their large Dutch barns in the western reaches of their land. Although there were five or six of these magnificent barns below Schermerhorn Road, all have disappeared either by burning or being demolished. The last of these was the Teller-Schermerhorn barn that I was forced to take down in the mid-thirties because of the rotted roof rafters that were so long they couldn't be replaced under the circumstances.

Originally, all of the early farmhouses that were built below Schermerhorn Road faced the river and had access to the Kings Highway following the farm boundaries across the Flats. These access roads still existed until Interstate 890 was built about ten years ago. However, these roads were pretty well abandoned when Schermerhorn Road was built and stabilized.

Having lived at Woestyne South and known a number of the old timers of Schonowe, such as Simon Schermerhorn, Fred Abel, John Marlette, Aaron Becker and John Meyers, I picked up many stories and personal reminiscences about the local region. In fact, a few years ago I prepared a map for the Schenectady Planning Commission that located about a hundred historical landmarks of the area bounded by the West Shore Railroad, the old River Road, the General Electric Works, and the Delaware & Hudson Railroad that constitutes the boundary of Schonowe.

The land I bought from Clarence Schermerhorn is the upper portion of Hindmost Farm No. 10 that was assigned to William Teller in 1662. The southern bounds (our expanded boundaries) includes the ancient road that crossed the Great Flats to the Kings Highway, the foundation of the old Dutch barn, the runoff of the Schermerhorn Spring, two campsites of prehistoric Indians, the western portion of the Schermerhorn Site, the site of an ancient brickyard, and the depression that marks the old wood road that connected the Dutch barn with the wood lot on top of the bluff that was adjacent to Old Maids

Woods. This latter is still owned by the City of Schenectady but is protected by an agreement with the Nature Conservancy.

The Name of Our Home

When our home on Schermerhorn Road was finished in 1936 it was such a fine structure that I felt it deserved a name. Since the land on which it was built was part of Hindmost Farm No. 10 assigned to William Teller in 1662, I felt that a name for the location should reflect some its early Colonial history. Our part of Schonowe was referred to as the "Woestyne."[14] I discovered that this meant "wilderness." Since our location is still on the edge of the extensive woods that stretch northwesterly across the Rotterdam Hills to the valley of the Sandsea Creek that runs into the Mohawk at Pattersonville, it seemed to me that the name Woestyne would be quite appropriate.

Some years later, when we acquired substantial acreage in the Adirondacks, we called our camp Woestyne North. Thus our home has now been called Woestyne South.

In Which I Established a Workshop and Laboratory

I have always had a work area in my home, starting from my early teens. When we built our home on Schermerhorn Road, I had the first opportunity to develop a substantial laboratory and workshop. This was augmented when I brought all of the tools accumulated over a lifetime by Dad Perret. He was an outstanding self-trained cabinetmaker and had quite an assembly of machine tools, primarily for woodworking. However, I found I could adapt most of them for metal working when needed.

Initially I converted most of the basement area to become a combination workshop and laboratory. Gradually I added to its versatility, first with a dark room, then a microscope room, and eventually a separate area for editing and storing my 16mm movies that began to accumulate during Project Cirrus.

I did much of my writing in front of an open fire in my limestone fireplace.

When our first furnace ended its useful life and needed replacement, Katy Blodgett heard of it and offered to give me hers, which she was replacing with a new General Electric oil burner. This was a coal furnace that we found worked better than our older one. Eventually we converted from coal to oil since it was possible to remove the grates and use the boiler without any problem. This move was of great advantage since, no matter how carefully I operated the coal furnace, a large amount of coal dust and ashes permeated all of the basement area. It was this conversion that permitted me to use some of the space for my microscope and film editing.

With the completion of our north addition, I now had adequate space for my machine tools, and was able to install my circular saw, turning lathe, drill press and sander, along with cabinets, shelves, and a workbench.

---

14 A 1735 Dutch dictionary confirms this. Also spelled *wildernis*.

When I left the Research Laboratory, I offered a token payment for all my laboratory equipment. This was accepted. Thus I was able to acquire my cold chambers, troughs, chemicals and all other equipment that had accumulated over a twenty-year period.

This equipment was invaluable to me. I knew that practically all of it would have been consigned to the scrap heap if I had not taken it. It has served me well over the more than twenty years since I left the Laboratory, and I am still using much of it!

My Library at Woestyne South

I have had a fascination over books all of my life.

I started collecting books in a serious way as soon as I began to travel alone to Albany. I would frequently take the interurban trolley to Albany, stop at the State Museum for half a day and then, before heading home, would go to Lockrow's Book Shop below City Hall to spend several hours browsing through the vast shelves of books in the shop. I was greatly intrigued with adventure books, especially expeditions to the Polar Regions. I read them avidly and could imagine myself in some phase of exploration.

I then heard of John Scopes, who had many more books than my friend Lockrow. He seemed to like me and gave me extremely good prices. He practically gave me a complete set of a leather-bound 1919 edition of the *Encyclopedia Americana* and, at about the same time a set of the Natural History Survey of New York State published in 1836. It is my recollection that I paid fifty cents a volume! These are now collector's items.

As I moved into science, I soon realized that books on science by reputable scientists represented a synthesis of much of their life work. Thus, by amassing a collection of books on subjects currently of interest, I found them to be of tremendous value in achieving an understanding of the scientific relationships within a body of knowledge. I eventually discovered however that, just because it might be described in a book, there was no guarantee that it represented Ultimate Truth. On the contrary, as "Doc" Whitney often said, "If you read it in a book, it isn't so!" What he was saying, of course, was that by the time something had been discovered, talked about, accepted as fact and established as truth and thus safe for publication in a bound book– that was about the time when new "facts" were discovered that raised a question about the validity or even the reality of an item that had been enshrined.

With a realization of the value of books I never hesitated to order a copy if I felt there was likelihood that it would help me in acquiring a better education.

I was particularly intrigued with the series of Field Guides on Natural History printed by such publishers as G. P. Putnam, the Audubon Series, and the Peterson Series, the latter published by Houghton Mifflin Co. Other nature books never failed to intrigue me, and I couldn't resist the urge to acquire a copy, no matter what the subject might be. With such a rich reference library, I made it a point to become reasonably well informed about most branches of the natural sciences, and found rich rewards no matter where I happened to be.

At one point Lois decided to prepare a catalog of all of our books. It was a tremendous job, since there are close to a thousand volumes in my scientific library, with two or three times as many on subjects ranging from exploration to history, and geography to poetry. She completed the job, but now it is sadly out of date, since it seems that our book-buying is on an ever-accelerating curve!

Although I have agreed to turn over my scientific library to the University of Albany and have already given them a goodly part of my collection, it seems that all of the shelves are full again and the stacks on top are approaching the ceiling of my office!

Vince (front left), Bill Gluesing (kneeling, center), Larry Shaw (far right) and other ski buddies near the Normanskill. Courtesy of Jim Schaefer.

Chapter 4
My Introduction to Surface Chemistry

In Which I Inherit a Slide Rule

After receiving a Research Notebook for keeping a record of my research activities, Langmuir called me into his office and presented me with a 12-inch slide rule. He also instructed me in its use.

At the time I was greatly pleased to receive this gift, but didn't think too much about the fact that this was his original slide rule and it had obviously been used a great deal. At later dates I noted (in his family photographs that showed his interest in encouraging his nephews to become scientists) that it was this same slide rule that appeared in all the pictures. Thus Alex, Robert and David Langmuir all were introduced to the realm of mathematics with this device. It is old and hand-polished with use, and has the Boss's name scratched into one edge of the rule in capital letters. It is quite possible that he used this in his undergraduate schooling at the Brooklyn Polytechnic.

To replace the slide rule he gave me, Langmuir acquired an elaborate 15-inch slide rule fitted with an enlargement lens on the slide. It was a beautiful tool.

I used the slide rule I inherited quite a lot and then, as I gradually became aware of its historical value, I acquired a 15-inch unit for my purposes, carefully preserving the smaller slide along with a small collection of memorabilia that has now been put on display in the library of the ASRC.

Throughout all of my tenure at the Research Laboratory, calculations of various types needed in our research experiments were carried out either by slide rule or mechanical adding machines of the Burroughs type.

Shortly after the Boss received the Nobel Prize, the Burroughs Company asked Langmuir if they could have his old machine in return for a brand new unit. He jumped at the chance, since the one he had used was a rather primitive device. They had him inscribe his name on the one they acquired and within a short time the latest version was delivered. This was used primarily by Katy Blodgett, who carried out most of the Boss's routine calculations when they were needed.

The only time when a more high-powered unit was needed was when the trajectory of particles was established under varying velocities and sizes.

This was done again by Katy using the extremely bulky, noisy, but highly accurate analog computer that was located on the top floor of the GE Athletic Club building that, when properly programmed, clanked out answers that still remain as some of the fundamental data in cloud physics.

Some of My Early Research Studies

One of my first assignments to carry on a research project involved a study of the lubricating properties of a single layer of molecules deposited on a very clean 1" x 3" glass slide. I learned from Katy Blodgett, Dr. Langmuir's research associate, how to clean the slides, spread a single layer of stearic acid molecules on a clean water surface, pressurize it, and then deposit a layer of molecules onto the glass slide. This was essentially a series of very simple, straightforward operations which, if done right, produced a compact monolayer of these straight-chain fatty acid molecules on the glass surface.

When this was done, a 1/8" diameter glass rod, 2" long, was bent to form a stirrup with an acute angle of about 20°. Its midpoint was bent so that when placed on the glass-coated slide, it rested on three spots. This was carefully cleaned and then placed on top of the single layer of molecules.

Previously I had constructed a tipping device, hinged at one end and fitted with an angle measuring scale. A horizontal slide holder could then be lifted so that the treated slide bearing the glass stirrup rider could be tipped gradually until the rider started sliding. When this happened the angle at which the movement started was recorded. The operation was repeated several times, the rider being moved to a new path before each test.

Similar measurements were made using cleaned glass slides and cleaned riders with both the slides and the rider coated first with single layers of molecules, and then with several layers. Within a short time I began using other types of molecules, and before long, became quite adept with the entire operation. I also built a rider that could be weighted. We soon established that a single layer of molecules was as good a lubricant as multiple layers, so long as the film was not mechanically disrupted. If this happened then seizure occurred, and I got values similar to that which developed when both surfaces were very clean.

These studies led to much more sophisticated equipment that ended with a high-speed rotating device in which a polished steel rod rotated within a split bearing whose upper half could be loaded with weights until metallic seizure occurred. One of the most interesting results of this high-pressure lubrication study was a run that used a colloidal mixture of stearic acid suspended in water used as the lubricant. With maximum loading the rod and bearing did not seize and at the end of the run the 1/4" rod had been squeezed so much that it was reduced to a 1/8" diameter rod whose surfaces were highly polished!

These interesting findings were suddenly interrupted by a more urgent project that turned out to be so fascinating and extensive that I never got back to high-pressure lubrication studies. Such is research.

The Role of Serendipity in Research

In all of my relationships with the Boss he treated me as an equal, and although I was well aware of his superiority in knowledge, expertise and professional background, I repaid his kindness in every way I could. It was a marvelous situation as far as I was concerned! Within a few years he gave me complete freedom to operate as an independent researcher. I made sure that my own ideas were subordinated whenever he wanted me to do something for him. His ideas were legion and could keep half a dozen assistants busy most of the time. However, it often developed that the results of an experiment would frequently nullify a series of suggestions so I tried to sort out some of the suggestions I received from him so that they led to a progressive movement on the subject under consideration. I also learned quite early that one of the maneuvers in tackling a research program was to do something no matter how crude it might be. I also was told that it was important to establish the "limits" of a phenomenon by casting far and wide of the imagined goal so that we might know how "good" or "bad" various procedures might be. This process often eliminated many possibilities, and often gave us leads and confidence to go in the right direction.

Finally, in conducting truly basic or pioneering research, it was not at all helpful to review the literature! If the phenomenon was referenced in any way then it was closer to applied research than achieving a new discovery.

An essential feature of such exploration is to nurture the occurrence of serendipity. This can best be achieved by having a broad understanding of the essential features of the subject under consideration, including a wide variety of what might be considered extraneous knowledge. One never knows when a bit of information will provide the bridge necessary to join diverse items into a useful finding. But above all, the key to serendipity is that the pursuer be blessed with the sagacity and the ability to sense a serendipitous event when it occurs!

Many, if not most, of the milestones in the long road to major scientific advances are the result of serendipity.

I once asked the Boss how many of his major discoveries were the results of serendipitous happenings. He grinned and said, "Every one of them."

Dr. Whitney, organizer and head of the GE Research Laboratory for nearly forty years, was the person who, more than anyone, popularized the concept of serendipity and serendipitous happenings in the industrial world. When I first met him, it would take very little encouragement to get him talking about the subject. He, more than anyone, injected the idea into the research method and I am quite sure he is the one who introduced Langmuir to an understanding of the process. The Boss thought about it a great deal and actually wrote a short paper on the subject entitled "Planning for Unplanned Research." This treatise, along with his talk on "Pathological Science" and his paper on "Convergent and Divergent Phenomena," have been dominant factors in my career and that of many of my students over the years.

My Trial with Melted Sodium Liquid

One day (June 2, 1935) Langmuir, having returned from a trip, suggested that it would be useful to know more about the conductivity and other characteristics of liquid metallic sodium. This was a silvery-appearing substance that was stored in kerosene. I had heard from time to time about the behavior of sodium, especially that if it was contacted with water. When this occurred it would oxidize with explosive violence. Joe Keenan of Ruggles' group showed me jars filled with round strips of the metal, all stored in kerosene. He recounted lurid tales of trouble with the material and told of tricks played by acquaintances, who would throw pieces of the metal into a puddle of water, while spitting to the consternation and great puzzlement of passersby. As a result I had an ingrown fear of the material that I had been asked to study.

Langmuir and I devised a test furnace and conductivity measuring device made mostly of metal. To heighten my consternation, I learned that I'd have to heat the sodium as high as 1,000°C, and that I also needed a water line for cooling the condenser. While I couldn't see any way in which the water could contact the molten sodium, its very presence in the experimental set up gave me some worries.

Once the experiments got underway, I was plagued with shorting problems in my measuring circuitry. The sodium vapor had a tendency to condense on the internal electrodes in such a manner that my measurements rarely measured the conductivity of the sodium vapor that was the objective of our experiments, but by condensing in the wrong places provided many run failures. These problems, plus delays in getting special parts constructed in fabrications over which I had no control, caused great delays. Fortunately there were many other experiments in surface chemistry that could be done during the waiting periods. One of these– the measurement of contact angles of oil droplets on an oleophobic surface– showed a strange phenomenon wherein the contact angle became depressed as a function of time. In studying its behavior I began to wonder whether it was modifying the oleophobic film. When the drop was removed it showed a definite change in the apparent thickness of the film on which it had rested. I showed this to Katy and between the two of us we discovered that some of the molecules from the built up film had dissolved into the oil! When another oil (having higher volatility but the same refractive index as the built up film) was placed on the mark that had been made, the apparent optical thickness of the multilayer was restored, but it then reappeared as the high volatility oil evaporated. Katy followed up on this finding and soon discovered the skeletonization phenomenon, wherein the films that we were normally building, which were half stearic acid and half barium stearate, could be greatly changed by soaking them for a few minutes in pure benzene. When this was done the refractive index of the built up film could be reduced from 1.5 to 125 or 1.3 by such treatment, as all of the stearic acid molecules were removed and thus replaced by air. To our collective astonishment, such skeleton films remained intact without shrinking. The refractive index could be shifted back to its original value by filling the skeleton "holes" with an oil having the right refractive index. When the oil evaporated, the optical properties of the skeleton reappeared.

After many delays, I finally had a new tube for the alkali metal studies constructed and its circuitry checked out. To my disappointment, the same shorting problems emerged as soon as the temperature began to reach a value of 500-600°C.

When I reported this to the Boss, I think he sensed my distaste for the entire experimental set up. At a certain point in the late Fall of 1935, he told me to abandon the experiment and to concentrate on the exciting developments that Katy and I were finding.

Following his suggestion I went into the lab where I had been conducting the liquid sodium studies. As I approached the equipment, something (I have forgotten the details) fell off the bench and produced a loud crashing sound.

A friend, who happened to be talking with the Boss when this occurred, later told me that the Boss grinned and said, "I guess Schaefer isn't losing any time in dismantling the experiment!"

This was the only activity I was ever involved in over my twenty-five year association with the Boss that I didn't accept with high enthusiasm and a sense of adventure. I had a hunch that he wasn't much interested in it either!

Splash Patterns of Water Drops on Soot

Shortly after Langmuir received the bathythermograph from Athelstan Spilhaus, he asked me to prepare some sooted slides such as were needed for recording the temperature profile of the lake.

When I followed the instructions for preparing a slide, I had a feeling that the method could be improved. I played around a bit and soon discovered that by preparing a hot rubbed ferric stearate monolayer followed by the passage of the glass slide over the small yellow flame of a Bunsen burner, it was possible to produce a sooted slide that had no visible graininess and a very uniform coating.

Another technique that was recommended was the coating of the recording with skunk oil. This was a rather foul smelling material that was supposed to toughen up the sooted slide so that it could stand abrasion. I quickly found that by dipping the slide in a 2% solution of polyvinyl formal (Formvar) in ethylene dichloride, a highly protective plastic coating was deposited and protected the slide completely from any other kind of damage.

About ten years later, while working at the summit of Mount Washington, he had need for a coating that could be exposed to a passing cloud and would be marked by the impact of cloud droplets. I prepared some tiny square pieces cut from a microscope slide, constructed an exposure device and sooted the tiny slides. When exposed in the cloud on the mountain, they revealed extremely useful and measurable traces of the cloud droplets. At about the same time I tried exposing larger sooted slides to rain drops and discovered that the larger droplets splashed on the surface of the slides and produced fascinating symmetrical patterns that resembled a spoked wheel. I eventually found that the grooves in the soot were produced by the air entrapped under the falling drops and that for some reason became divided into fifty or more tiny air bubbles. I established that by dropping a glycerin drop rather than a water drop. It splashed, but did not recoil from the hydrophobic surface, as did the water drop. The air bubbles, which produced the grooves in the soot, were still trapped under the disc of glycerin that remained and could easily be seen and photographed!

The Oil Guard Project

During October 1934 several persons approached us from one of the operating groups of the Company for help in solving a particularly troublesome problem. It appeared that the lubrication of motors, ranging from the tiny ones that operated clocks to much larger ones, would not stay in the bearings of the rotating shafts but would creep out of the bearings. Eventually the reservoirs would be dry and trouble would develop.

Since one of the features of our built-up films was that the surface was not only hydrophobic, but oleophobic as well, it was hoped that in some manner we could advise them in a way to stop the oil creepage.

Katy Blodgett started work on the problem, but it soon became apparent that more urgent research was on her mind, so I took over. I tried polishing the surfaces with buffing wheels loaded with ferric stearate in the hope of producing an oriented film that would resist creepage. Finally, I prepared resin solutions containing stearates that, when applied hot, would dry with an oriented oleophobic layer of molecules on the surface. This worked beautifully for a considerable period of time, but eventually broke down as the resin film deteriorated. I devoted a goodly portion of three months to this investigation and, although the final results were of dubious commercial value, the friendships that developed during the course of the investigation were of lasting value, as was the knowledge gained on the interaction of molecules, resins and surfaces.

The problem still exists, and I notice that some of the most recently manufactured clock motors still fail when they run out of oil lost by the creepage phenomenon. As often happens the clock motor, whose accuracy depended on the 60-cycle signal carried by alternating current in the electric network of the civilized world, is now being replaced by the even more accurate quartz clock that has none of the lubrication needs of the older system. So it often happens in science that when a new set of principles are discovered they eventually lead to practical applications of the new basic ideas.

The Measurement of Molecular Thickness

In collaboration with Katy Blodgett, I helped in the development of methods for measuring the thickness of molecular films. We would build a "step gauge" consisting of deposited multilayers. Approaching the thickness of a quarter wavelength of sodium light, we would then build steps of two molecular layers that would span the quarter wavelength thickness. These consisted of 41, 43, 45, 47 and 49 molecular layers deposited on polished chromium-plated steel slides, the step series beginning at 39 layers of acid barium stearate films transferred from a water surface. I should perhaps point out at this time the certain procedures we followed.

We would clean the chromium-plated slide, heat it so that a few flakes of ferric stearate, a metal soap, would melt on the surface. This was then smeared across the heated surface, rubbing it in with a clean linen towel. Our previous studies had shown that when this material was rubbed while hot until no visible residue remained, we ended up with a single layer of ferric stearate molecules coating the slide. This prepared surface was an ideal base on which we could then build a uniform multilayer. When pure stearic acid dissolved in benzene was placed on a previously cleaned surface of water, the molecules would spread with their hydrophilic hydroxy heads on the water surface, about half of them reacting with the barium ions previously dissolved in the water to form the barium stearate, the other half remaining as stearic acid. After such a monolayer had been placed on the surface it would be compressed by placing a small drop of oleic acid at the end of the water surface opposite the dipping well of the Langmuir trough. This liquid film, also a monolayer, served to provide a moveable piston that would continuously supply a pressure of 30.5 dynes per centimeter on the upper edge of the film.

Under these conditions, when the metal slide was lowered into the water, a single layer of compressed molecules of acid-barium stearate would become fastened to the rubbed monolayer of ferric stearate. The slide would be lowered into the water for a distance of 2" (5cm), and then moved upward out of the water. The strange and wonderful thing that had been discovered in earlier Langmuir-Blodgett experiments was that upon moving the slide out of the water a second layer would be deposited and it would be completely dry! The strongly hydrophobic surface of the acid barium stearate layer shed every trace of water during this maneuver. When the process is repeated during each dip and retrieval of the slide, the same effect occurred. In this manner, in 19 round trips 38 monolayers are deposited on the slide. These, plus the rubbed-on film, produces an "optical" thickness of 29 layers or approximately 980Å.

In making a step gauge the 20th round trip was not lowered as far as the 19th, and with each successive round trip, the slide was lowered to a lesser depth in intervals. In this manner a step gauge was constructed with successive step intervals of 2 x 25.2 = 50.4Å (1Å = 1 x 10^{-8}cm).

With such a step gauge it is possible to place the slide on a tilting platform, with it illuminated with a sodium light. At a particular angle, two adjacent steps will suddenly appear black. That is the angle at which the light rays passing through the adjacent steps produce what is called destructive interference. It is possible to trace the light rays and show that they have the same length because of the near and far proximity of the steps from the light source.

With such a mechanical set up it is possible to coat half of the step gauge at right angles to the directions the steps were originally constructed with another film. When this is done, the angle of destructive interference from adjacent steps changes. When this angle is noted and used in a simple formula, or applied to a graph constructed by Katy Blodgett, the thickness of the newly deposited film can be determined quite exactly.

To substantiate these relationships it is possible to make these measurements in a much simpler manner. While not as exact it provides a very effective rough approximation of the thickness of an unknown.

If instead of using a monochromatic light source, such as produced by a sodium lamp, white light is used and the quarter wave length step gauge is examined with polarized light, the successive steps are seen to go through a series of interference colors starting with yellow and grading through red to blue. With the polarizer at right angles, the colored steps disappear. Thus, if a molecular film of unknown thickness is deposited over half of the step gauge in the direction of the increased thickness and examined with the polarizing film, it is often possible to match the increased thickness with the uncoated step gauge, comparing the colors. In this manner it is often possible, with a little practice, to measure thickness to an accuracy of better than 5Å.

Once a person gets used to estimating thickness with such a color gauge, it is possible to make similar measurements by a far simpler technique that I discovered during this period of joint research.

I found that by dipping a rubbed chromium plated slide into a 1% Formvar solution of the type I used for making pellicles and snow crystal replicas, and withdrawing the slide from the solution with a smooth motion and then letting the solvent evaporate while holding the slide in vertical position, a beautiful colored film would form with the identical color range found with a molecular step gauge in the quarter wave length thickness range.

A molecular film of unknown thickness could be deposited on such a film again at right angles to its increasing thickness. When examined with a light polarizer, the displacement of apparent optical thickness could again be extrapolated and an estimate made of the thickness of that deposited film. Both of these techniques could also be used to evaluate the uniformity of a coating film. At the critical thickness range between violet and blue, the tiniest of irregularities could be seen.

Ed Land of Polaroid

Among the early participants in the Science Forum, in which I was a participant from 1935 until it ended in 1955, was Dr. Edmund Land, who made his first major invention in the form of Polaroid film. This was a polarizing material that was a most intriguing material. It was a light-brownish film having a thickness like a sheet of paper, and except for its brownish cast was quite transparent. When two sheets of this polarizing film were placed together, either nothing happened to light transmission, or it blocked virtually all of the light transmitted by the double film.

Polaroid film was a tremendous boon to the field of optics where polarized light might be important. Heretofore, it was necessary to use a Nicol Prism– a very expensive device made of a perfect tourmaline crystal that was cut in a very special way so that light transmitted was polarized.

While Dr. Land was in Schenectady, he also presented a colloquium lecture before the staff of the Research Laboratory– an event that was traditional and was inaugurated in the earliest days of the Lab. While at the Laboratory, Dr. Land visited a number of friends to see what was going on. When he entered our laboratory he was carrying a roll of polarized film. When he asked Katy Blodgett how much could she use, she was just about speechless, since it was sold at a substantial sum per square inch. He peeled off several feet of the film, cut it, and presented it to her.

When he left, I immediately set about to prepare pairs of film holders for mounting the film so that we both could have film analyzers. I used thin sheets of black Bakelite and within an hour or so had the films mounted and in use. They were fantastic! Where heretofore we had been limited to using tiny Nicol prisms for examining our films, we now could immediately look at them and perform light transmission experiments that had been virtually impossible before.

We saw Ed Land several other times, notably when he came to our lab to demonstrate his brand new invention– the "instant" camera– and I visited him at his Cambridge laboratory several times. I will never forget the sensation he privileged us to experience with his Polaroid film. The Boss was as excited as we were and agreed with us that it was a great day in surface chemistry!

Bubbles in Ice and Other Things

One day the "Boss" drew my attention to the bubbles in ice. Previously I knew that ice contained bubbles, but I let it go at that. He told me that one day at Lake George the bay had frozen over with a clear sheet of ice so clear that it was like glass and appeared black in reflection, taking the color of the bottom of the bay. The ice was four to five inches thick and thus safe for walking or skating. As he was skating he noticed that while clear, the ice also had myriads of bubbles, most of them in vertical arrays. Upon closer examination he saw that in addition to the vertical distribution, there was a wavering continuous threadlike array of small uniform sized bubbles that looked unusual and intriguing. He followed the line, and in one direction it descended in a very gradual slant until at the "end of the line" there was a tiny water bug. The bug from all appearances was emitting the bubbles.

However– I have since observed similar threads of bubbles, many of them being formed by organic substances such as tiny filaments of algae or other floating objects. They apparently serve as nuclei for the gathering of gas molecules in the freezing water that at the bottom of the ice is warmed slightly by the heat of crystallization. This slight amount of warming releases the CO_2 or O_2 or some other gases dissolved in the water at a saturated concentration.

It would be a fun and a challenging experiment to find such a line of bubbles and with a sensitive chromatograph identify the gas or gases present in such lines of bubbles.

Langmuir was an avid ice skater and was especially proficient with a skate sail. In the collection of sports equipment I found in his attic laboratory, when assembling his scientific papers for the Library of Congress after his death, a number of fine sails with their framing poles. I don't know what eventually happened to these along with his skis.

He was part of a group of sailors, including John (Appie) Apperson and Harry Summerhayes who, whenever possible, used the wide waters of Lake George for this exhilarating sport. Although I had a sail, I can't remember sewing it, so it might have been given to me by either the Boss or Apperson. Unfortunately, I never had the sail with me when the ice was clear. Instead of sailing with skates I occasionally sailed with skis from Appie's Camp on Huddle Bay. This, too, was sport, but physically tiring since my skis were made of wood and without metal edges. Thus to steer a course it was often necessary to edge the skis. In contact with the abrasive granular snow that dominated the wind blown snow cover the wooden edges quickly rounded and thus were not easy to maneuver while tracking the wind.

Another cherished possession in my collection of mementos are Langmuir's ice climbing crampons. These consisted of a group of ten wrought iron spikes fashioned so as to be fastened to the bottom of climbing shoes or boots that make it possible to traverse or climb slick layers of ice. The spikes on these particular crampons are shorter than most, being only about an inch long. In addition they consist of three separate assemblies of spikes arranged four, two and four. I believe they are superior to the more modern versions that have only eight spikes and a single hinge.

These crampons were used by Langmuir in his winter climbs in Switzerland and were probably part of his climbing gear when he ascended the High Peaks of the Adirondacks in the wintertime.

Another part of his outdoor gear was a down-filled sleeping bag. He and Appie (John Apperson) developed the chambered tube bag in which the bag was designed so that no seam went from the outside of the bag to the inside but rather had a series of muslin tubes sewed alternately on the outside and then the inside of the long fiber cotton balloon cloth of which the bag was made. In its early development Langmuir devised a series of temperature measurements to establish the best configuration using goose down and eider down.[15] They quickly found that nothing was better than eider down, but even in those early days (the '10s and '20s of the 20th century) that down was prohibitively expensive. The next best insulation was first-quality goose down, either white or gray. The latter was less expensive so it is the material we concentrated on using.

In the early thirties, after I had formed the Mohawk Valley Hiking Club and became acquainted with Appie, he taught us how to make sleeping bags. He had a room in his bachelor's quarters having a husky sewing machine that was a Mecca for outdoors people. They were welcome day or night, and there were many projects for using the sewing machine and other equipment, including tarp tents, ground cloths, parkas, sleeping bags, ski-climbing ropes, and skate sails. The design of the sleeping bags made during World War II was identical to the bags designed by Apperson and Langmuir thirty years earlier. More

---

15 Eider down is down feathers from the eider duck (*Somateria mollissima*), also called the Common Eider. It is a sea duck that lives in the northern coasts of Europe, Siberia, and North America.

than thirty bags were made following this design by members of our hiking club, a number of which are still around.

The only weakness in the construction of the bag was the under part, since the down was compressed and lost most of its insulating capabilities. I solved this problem by substituting a layer of tougher sailcloth for the inside bottom layer of the bag. On either side at the outside edge of the bag I sewed tubes out of the projecting cloth into which I would put my skis while at the top and bottom I made tubes into which I put my ski poles. By using these poles as stretchers I fashioned a taut stretcher bed that permitted the down to hang uncompressed under the stretched inside cloth. Using logs or blocks of compressed snow I thus had a highly comfortable warm bed that could be used in winter to forty degrees below zero using a bag containing three pounds of down.

Monolayer Skins as a Microanalytical Tool

During my intensive studies of distilled water quality and differences, which I carried out in collaboration with Katy Blodgett, I noticed frequently that after spreading a monolayer, in the act of removing it by use of the cleaning barriers, the surface appearance of the monolayer showed the distinctive patterns as the monolayer was crumpled by the sweeping action of the barriers. Distinctive crumple patterns began to be noticed that seemed to have some correlation to the purity and pH of the water.

It then occurred to me that his crumpled film might possess distinctive features. On March 20, 1935, in my Laboratory Notebook, I outlined a procedure for preparing a monolayer skin. Thus I began to skin the crumpled film after it had been compressed from 50cm to 1cm. To do this I fashioned a tiny scoop of a thin layer of platinum sheeting having the dimensions of 3/8" x 1" x 0.030". This could be quickly cleaned in a flame and then when passed along the length of the barriers, which trapped the crumpled monolayer, the scoop would gather the folded film and easily remove it from the surface of the water. I would then squeeze out the entrapped water and consolidate it into a tiny pellet and put it on a microscope slide so that it could be melted and placed on the stage of a polarizing microscope.

The results of these simple operations were spectacular!

When the monolayer consisted of pure stearic acid molecules spread and then retrieved from a substrate of very pure doubly-distilled water, the crumple pattern appeared as many tiny star-like patches distributed uniformly over the film area. When crumpled, skinned, melted and then observed to crystallize under the polarizing microscope, the crystals appeared as whorls of brilliant blues, green and red birefringent colors. The effects were identical to those observed when stearic acid crystals were treated in the same manner from the bulk substance.

If, however, there were a very small amount of metallic ions dissolved in the water, the crystallized skin would show an entirely different crystalline habit. Thus with a trace of copper in the water the compressed skin would appear green; with a trace of aluminum in the water no birefringent color whatever would appear in the solidified melted pellet, but rather the crystalline pattern was dominated by a series of Liesegang Rings exhibiting black and white contrast patterns.

A survey made of the effects of traces of metallic ions of a wide variety of metals showed that each element produced its own distinctive combination of effects when observing the crumple pattern, the color and/or bulk of the skin, the melting point of the pellet and the appearance of the solidified residue. This was a fascinating area that would probably have led to important techniques for micro-analysis if followed up in detail. However, as with so many of the fascinating discoveries we made, other subjects were of equal or greater importance so we had to content ourselves with publishing an account of our observations with the hope that someone would find them of enough importance to conduct a further exploration.

Some years later an enterprising mining engineer approached me with the question as to whether or not the copper skin method could be used for prospecting. He postulated that the undisturbed soil above a rich deposit of copper ore should contain traces of the mineral whose concentration would be controlled by the proximity of the soil sample to the Mother Lode. I encouraged him to prepare a grid work of sampling spots in the general vicinity of the suspected ore body. Our procedure could then be tried to see whether or not isolines could be drawn that might provide prospecting information. I told him to try this out with a known ore body to see if it might be useful. I think there might be considerable merit in his speculation, but I never learned whether he made an effort to get such a study underway.

Although I had every intention of doing further work in this fascinating area, and in fact at several periods I managed to spend further research on the subject during the next two years, the fascination we found in our pioneering work with protein monolayers diverted our attention to such a degree that we never were able to continue the study. Then, as our protein studies were opening vast new areas of opportunity, World War II intruded and before we realized it, our Langmuir Troughs were pushed aside as urgent defense problems demanded attention. Then, as that war ended, we became deeply involved in cloud physics and weather modification activities so that except for brief utilization of surface chemistry techniques in some of our studies, we never got back to it. There are still many areas of research we encountered that have never been followed. It is a rich area of research and I trust it will again be explored!

The Science Forum

One of the unusual aspects of attending Langmuir University was the many opportunities that from time to time occurred because of the unusual sets of circumstances that were continuously evolving because of my proximity to the Boss and the "visibility" which that entailed.

So it was when the Science Forum evolved by Emerson Markham of WGY. He had developed the Farm Forum, a highly successful program that became a favorite of the farm families of the listening area that covered most of eastern New York. He proposed developing the Science Forum pretty much along the lines of his earlier success, except that instead of using one person to extemporaneously answer questions sent in by the listening audience, Emerson proposed having four scientists on the answering panel, and instead of having a chance to look up the answers he suggested that they be sprung on us without any previous knowledge. He figured that this would add "life" to the program, which it certainly did! In seeking to select four participants he decided to identify a chemist, a physicist, a physical chemist and a natural scientist, the latter to answer a wide variety of questions about the natural world around us. He sought all of his panelists from the Research Laboratory and selected Dr. Winton Patnode, Dr. Lewi Tonks and Dr. Frances Norton, and for the naturalist he selected me. I felt quite honored to be selected but felt that I could hold my own with the others and so it turned out. We met once a week and had a wonderful time. From time to time, when any of the "regulars" were sick or traveling, Emerson had a group of alternates who performed. We enjoyed the program so much, however, that we often scheduled our trips so as to not interfere with the scheduled performance!

Ice Flowers, February 25, 1943, 5:45 PM, Baan Farm near Plotterkill, NY. Vince inspects a field of ice crystals left upon grass and weeds after the sudden subsidence of flood waters from the Mohawk River. Courtesy of Jim Schaefer.

Previous to our part of the program a top scientist, generally one who had just been in the national news because of some new discovery, presented a 13-minute talk about his recent achievement. This would be the same individual who had been selected by the Laboratory to visit it and present its weekly colloquium. We would then be on the air for another period of the same length of time, after which we would go together to a fine restaurant to have an unhurried evening meal. What fun we had!

An interesting sidelight to this fascinating evening was the fact that in a number of instances, after returning home and going to sleep, I'd wake up suddenly at two or three o'clock in the morning with the solution to a problem that had been bothering me at the Laboratory, or with an alert mind full of all sorts of new ideas.

I believe that these occurrences were not random accidents, but the result of the intellectual stimulation received the evening before. I have experienced a similar reaction many times since leaving the Laboratory. This is an important aspect and mechanism of the Bosses thesis of "Planning for Unplanned Research" and an extremely valuable technique.

The final program of the Science Forum occurred in 1955, at which time I had left the Laboratory but was still a member of the original panel.

A few years ago the old crowd reassembled at the invitation of WGY radio. All participants in the program over the years were in attendance. From the time the first two returnees appeared until we left the Van Dyck Restaurant about eight hours later, there was a continuous stream of animated conversation. What a good time we had!

The Early Days of Calgon

Some time in early 1936 (50 years ago) we were visited by an inventor by the name, I believe, of Partridge[16], who brought with him a clear glass-like water-soluble substance that he claimed would remove Calcium ions from water. He had coined the name "Calgon" and told us it was a form of sodium hexameta phosphate. He thought we would be interested in its properties, as indeed we were. From that time forward we became extremely interested in its ability to sequester calcium ions. Over the years we found Calgon to be extremely valuable in helping us to solve effects that we suspected were caused by traces of calcium in our distilled water, and for other uses in surface and protein chemistry.

Katy Blodgett had discovered a very strange type of built-up film that we called the "X" film. It deposited on a slide only when the slide was lowered through the compressed monolayer. Unlike what we called the "Y" film in which, when a second layer was deposited as the slide was raised to emerge dry, the "X" film did not deposit on the raising cycle, but the film also emerged dry. Thus in a "round trip" two monolayers were added with the "Y" film but only one was added with the "X" film. I discovered that when a particular solution was producing "X" type films the addition of a very small amount of Calgon converted the water substrate from one that produced "X" films to one that produced "Y" films (the more normal type).

I discovered another fascinating phenomenon. I found that when a small amount of Calgon was added to milk the whiteness of the milk disappeared like magic leaving the milk clear but with a yellowish cast. I also found that when a small amount of pepsin (a

16 This was probably Everett P. Partridge who joined Calgon in 1935 and was Director of Research. He was Vice President of the company when he retired in 1966.

protein) was added to the clarified milk it curdled the same way as when ordinary milk was similarly treated to form what we called pot cheese. With the cleared milk a curd would form that looked very different. Years later I was told that someone had discovered that this transparent curd was an ideal food for babies who had problems in digesting ordinary milk products. At the time I just considered the treatment as a rather spectacular demonstration of the calcium ion sequestering power of Calgon.

As everyone knows, Calgon is now a highly successful water conditioner that readily converts "hard" water to "soft" water and thus permits easier clothes washing, the reduction of scale formation in water boilers, tea pots and similar situations where high calcium concentrations in water causes problems. Another discovery that I made in the process of working with Calgon occurred one day when the sun was streaming into the laboratory window, and I had retrieved a spatula from a test tube containing unmodified milk to which I had just added some pepsin. As I glanced at the layer of milk, which coated the shiny spatula, I noticed a vigorous, scintillating optical effect going on within the milky layer. I suddenly realized that the scintillation must be due to Brownian motion. I was familiar with this phenomenon as seen under the microscope, but until then had no idea that it could be seen with the unaided eye. As I was watching it the scintillation suddenly stopped! I then realized that this was due to the pepsin that I had placed in the milk a few minutes before and which had curdled the milk, thus producing a gel-like mass that stopped all the liquidity of the milk.

I have often wondered if our early friend, Mr. Partridge, persisted and made a fortune from his discovery of Calgon.[17] I certainly hope so since he was a very fine individual!

The Adsorption of Water Soluble Molecules

There are a number of organic molecules that have important specific chemical reactions that cannot be spread as monolayers but are of great interest. Such things as toxins and anti-toxins, catalytic reactions and the behavior of antigens and sterols can be studied quantitatively by depositing them by adsorption on an acid barium stearate thickness gauge.

It was found, upon Langmuir's suggestion, that if such a multilayer were put into an aqueous solution of thorium nitrate, that instead of emerging dry with a strongly hydrophobic surface, it would be wetted with water. If washed in running distilled water, dried and measured, the step gauge showed the presence of another layer of molecules probably, thorium oxide.

Before drying however, if the wet slide was covered with a solution of say diphtheria toxin molecules, and then after washing with another solution of diphtheria anti-toxin, then washed and dried, the new thickness would represent a new increment that probably represented the specific combination of these two reactants. If, on the other hand, another type of anti-toxin was used, no combined reaction could be found. The details of the

17 Calgon is a portmanteau of CALcium GONe.

preparation of such a conditioned surface are given in a joint paper by me and Langmuir, "Improved Methods of Conditioning Surfaces for Adsorption."[18]

Our Contact with Dr. Harry Sobotka and Cholesterol

Starting the first week of May 1937 we had a visit from Dr. Harry Sobotka[19], a medical doctor from the Mount Sinai Hospital in New York City. Dr. Sobotka was a research physician much interested in and an expert on bile acids, including cholesterol and its family of chemicals. For the next week or so I instructed him on our surface chemistry techniques and found that it was possible to build films of cholesterol and conduct reactions between it, its related modifications, and digitonin. He returned for another session in June.

At the end of this visit we had amassed enough interesting findings to write a paper that was published later in the year. It was entitled "Multilayers of Sterols and the Adsorption of Digitonin by Deposited Monolayers."

After Langmuir's death Dr. Sobotka wrote "Langmuir during his almost meteoric appearance in the field of organic monolayers created a whole body of novel techniques and enunciated ideas that had been foreshadowed to some extent by himself and by others but which he spelled out with courage and originality."

That our instructions, demonstrations, findings and encouragement paid off in the case of Dr. Sobotka is shown by the fact that he embraced the field of surface chemistry and for more than twenty years carried out intensive studies within his field of interest. In his memorial tribute to the Boss he cited eleven of his published papers in illustrating the progress that had been made, and more than fifty scientific articles that related in one way or another to our publications.

Among these were three others from two of our former "pupils," Drs. A. Rothen and E. Porter. Our contact with Dr. Sobotka and the impact it had on his subsequent career, and that of many others, illustrates the effect that even a slight contact with "Langmuir University" had on those who had that good fortune!

My Contact with Eliot Porter the Photographer

As I became proficient in surface chemistry techniques, the Boss invited researchers to Schenectady and had me spend a week or so with them teaching them the basic procedures for producing, depositing and measuring single layers of molecules, and their physical and chemical nature.

It was fun to do this since it broadened my horizons and always led to glimpses of their activities in Science. One of these was Dr. Eliot Porter[20] of Harvard University, who was

---

18 *J.A.C.S.*: 59, 1762 (1937).

19 Sobotka served as chemist for Mount Sinai for 37 years. He died in 1965.

20 Porter went on to become an acclaimed naturalist photographer. He was director of the Sierra Club from 1965-1971.

interested in diphtheria toxin anti-toxin reactions at the molecular level. On May 27, 1937 we started to give him a thorough indoctrination in surface chemistry techniques lasting about a week. However, I wasn't sure it would do much good, since on leaving he told me he was struggling with the problem of leaving the health field at Harvard in favor of following a burning desire to become a professional photographer.

I lost track of Eliot for many years. A few years ago I purchased a large book of beautiful color photographs of the Glen Canyon that has been submerged in the formation of Lake Powell in northern Arizona and southern Utah. I was among those who were unsuccessful in preventing the drowning of this magnificent canyon. After I had enjoyed looking at the pictures and realizing that they were the work of a master photographer in league with the best, I noticed that it was the work of my former "pupil." During the past summer, after rereading *Summer Island*, it occurred to me that I should write the photographer to find out if in fact he was my long ago friend. After a considerable lag in time, since my letter had to be forwarded from the book publisher, I received a letter from Eliot affirming that he in fact remembered quite clearly his sojourn in Schenectady. He told me that two years after his visit he made the break and has never been sorry. At the time in 1939 his friends told him that he made a crazy decision but he said that he has no regrets. In fact he told me that a book of his black and white pictures taken in the southwest shortly after his departure from Harvard is about to be published (Fall 1985).

With the death of Ansel Adams recently I feel that Eliot must be reckoned to be among the top living photographers of our present era.

Some Physical Characteristics of Monolayers of Plastics

During the last week of 1937 and the first week of the new year of 1938, I made an intensive study of the surface properties of polyvinyl acetate (Gelva) and polyvinyl formal (Formvar). In general I found them to have a number of features in common in both the surface areas occupied under similar pressures and thickness of the built up film.

In some respects they were similar to protein films, having an average thickness around 10Å. The surface area curves had some strange features, and I suspected these properties were related to the physical conditions when they were liquid and semi-solid.

With polyvinyl acetate, when the final transition state occurred at a surface pressure of 25 dynes per centimeter, I discovered that it was very easy to pull fibers from the compressed monolayer. By inserting a clean glass rod below the water surface and moving it parallel to the compression barrier for a short distance, and then lifting it away from the water surface, an extremely fine (nearly invisible) fiber could be pulled from the compressed film. This was quite uniform in diameter and looked very much like a strand from a spider's web. Nothing further was done relative to these observations, although it provided me with still another aspect of knowledge regarding the fascinating field of surface chemistry.

Attempt to Obtain Data on Tobacco Mosaic Virus

In 1937 Dr. Langmuir received a sample of Tobacco Mosaic Virus from Dr. Wendell M. Stanley, who had isolated this protein-like substance. It had a very high molecular weight, so we were interested to see if it could be spread as a monolayer.

All efforts to make a monolayer failed. I tried every one of the methods for spreading, which were quite successful with the smaller globular proteins, without success. I did find however that by placing it in a test tube as a 2% suspension and viewing it between crossed Polaroids, the colloidal dispersion exhibited a beautiful example of streaming birefringence. This demonstrated quite nicely that the colloidal suspension was uniform in size and that the very large molecules were asymmetric and mono-dispersed.

The development of this observational technique was an excellent experience and paved the way for my later research activities with mono-dispersed bentonite.

I did find it possible to spread a thick film of Tobacco Mosaic Virus on top of a salt solution, and in the process of doing so developed a two solution Langmuir Trough for such studies. By building a relatively long trough and putting a physical barrier at about a third of the long distance I found it possible to have a strong salt solution in the smaller segment with a standard solution for building monolayers or only distilled water in the other segment.

If after forming a salted-out film on top of the strong salt solution I wanted to transfer it to the adjacent liquid surface, I took a U-shaped piece of platinum or nickel sheeting about ahalf of the length of the barrier, flame cleaned it out, and then lowered it onto the barrier. It immediately became wet and served as a bridge between the two adjacent baths. The salted out film could thus be easily moved across the water bridge to the adjacent water surface by the movement of a surface barrier.

While nothing of particular significance was added to knowledge about Tobacco Mosaic Virus, we at least knew that it didn't possess the surface behavior of the smaller globular proteins, that it showed streaming birefringents, and that it could be spread as a thick uniform salted-out film and could then be "washed" by moving it to a distilled water substrate.

Thus I learned that the lack of understanding success led to the establishment of useful knowledge, and also to techniques that might not have occurred if we hadn't tried!

Our Studies of Artificial Mica

In 1937 we had a visit from Professor Ernst Hauser of MIT, who brought with him a strange material called bentonite. It was a substance that swelled greatly in contact with water. Professor Hauser had the idea that since this substance consisted of platelets that, with the proper treatment, could be oriented so as to overlap each other and thus, when dried, might serve as an excellent substitute for mica that was in great demand for electrical insulation. He had found a way to treat the bentonite mineral so as to retain only those particles of a particular size. He gave us some of his mono-dispersed material. I placed a small amount of the powder in distilled water and was amazed to see how rapidly it soaked up the water. It took at least ten times the water that would ordinarily be required to form a slurry. When I finally got it into a colloidal solution it still displayed a high viscosity. Placing some of it in a clean test tube closed with a rubber stopper, as he instructed us to do, I tipped it back and forth between two crossed Polaroid sheets and was thrilled to see the magnificent colors produced; what is called streaming double refraction. The bentonite, consisting of uniform platelets, twists light in such a manner as to cause the optical effect.

I then proceeded to prepare a series of concentrations of the material centered around one part bentonite to on hundred parts water. After preparing the mixes, I placed them in glass vials and had a glass blower seal the ends. I still have them, and between crossed polarizers they are still beautiful. An interesting feature of a number of the samples was the manner in which their sedimentation produced a semi-solid liquid crystal. All flow disappeared until they were shaken vigorously, after which they flowed as a colloidal liquid.

Unfortunately for Professor Hauser, the affinity of the bentonite for water foiled his hope of creating a substitute for sheet mica. Moisture is the enemy of good insulation of electricity and to my knowledge he was never able to eliminate this problem.

Some years later I was given a rock from one of the river valleys of Alaska. It appeared to be hard and brittle although it had a "soapy" feel. This was from a deposit of bentonite, and when a small chip of it was immersed in water it began to swell and slump to form a gel until finally, with sufficient water, it became a viscous liquid similar in most respects to the samples we received from Professor Hauser.

The swelling characteristics of bentonite suggested to hydraulic engineers that this substance would be an ideal material for sealing irrigation ditches and other structures to prevent the leakage of water. It works fine for such an application but unfortunately must at all times be underwater. Once permitted to dry, tremendous cracks occur which, when wet again, do not seal effectively unless mechanically worked.

The great value that this episode had for me was to introduce me to polarization, streaming double refraction, and similar optical phenomena that became an extremely valuable tool for me in later years.

The Affinity of Ice for Metal Surfaces

In 1937 we were approached by engineers in the refrigerator department. They wanted to see whether any of our knowledge and techniques concerning the surface properties of molecules could be utilized to make it easier to remove frozen ice cubes from freezer trays. Ordinarily at that time ice adhered so tenaciously to the aluminum surfaces of the trays that the ice splintered when an attempt was made to produce nice ice cubes from the cold trays where they had frozen. All sorts of mechanical methods had been tried without success.

Accordingly, I obtained some strips of aluminum of the type used to fabricate the ice cube trays and froze a large drop of water on its surface. I found it took about eight minutes for the drop to freeze. I let the sample with its drop remain in the frozen condition for an additional two minutes and then removed it and quickly oriented it to a vertical position, at the same time placing a weighted sharp pointed probe in contact with the edge of the large frozen droplet of ice. It was left there until the frozen drop was released from the surface on which it had frozen. The slide was placed on a metal knife-edge at right angles to minimize warming of the slide. The room air environment was identical for all tests.

A number of surface treatments were applied to the aluminum-surfaced test slide and quite a spread in time of release was noted. None of the treatments showed an instantaneous release of the ice as was hoped for. A built up film of 50 layers of acid barium stearate was one of the poorest treatments for release.

It was found that there was a definite relationship between the steepness of the contact angle of the water drop with the aluminum slide. The best release occurred from a hot rubbed monolayer of ferric stearate.

At the time there were none of the plastic materials available of which the current trays are made. I am sure if they were that the results of the adhesion tests would have been more encouraging.

When one releases a whole tray of ice cubes with a slight twist of the wrist it is hard to recall that fifty years ago such a successful achievement was unheard of and only an optimistic dream.

Record Keeping in Little Notebooks

One of the characteristics of the Boss was his meticulous and consistent use of Field Notebooks. He used the 2" x 3" ring bound notebooks of the type commonly found in five-and-dime stores. He had devised symbols as a sort of shorthand. Few aspects of a given situation missed his observant eyes and inquisitive mind. A 35mm Leica camera and a pencil completed his field equipment.

I copied his techniques and soon found that the value of using his methods was inestimable. I also followed his example of using the General Electric Research Laboratory Notebooks. These were of two types; a large one best suited for recording regular daily laboratory activities, and a smaller one that I found most convenient for special projects. My first Notebook was issued to me on October 14, 1934. It was a large one, 9" x 12" and consisted of 150 pages bound in leather and numbered consecutively. It was completely filled in eighteen months; the final large one extended from July 19, 1948 to my last day at the Research Lab that ended at 3:30 pm, February 26, 1954. During this approximately twenty-year period I filled ten of these large notebooks with observations of experiments and data recorded. Many of my observations were illustrated with photographs. The notebooks spanned the years without a break. Several years after I left the Research Laboratory, I learned that all notebooks had been microfilmed. I requested the originals and was pleased to receive them.

The Laboratory also furnished a smaller-sized leather bound Notebook approximately 5 x 8 inches and consisting of 333 numbered plates. I used five of these smaller notebooks on special subjects. None were entirely filled with data; a few were hardly used at all.

It was the practice of the laboratory scientists, initiated probably by Dr. Whitney, that whenever a new phenomenon was encountered some other laboratory coworker would be asked to sign the lab entry with a statement at the end of the lab description of "Read and Understood" with signature and date. This was of course for possible patent protection at some future date. I was fortunate to share my room with Katherine Blodgett. Many of my discoveries– good, bad and indifferent– bear her signature.

Chapter 5
Intensive Studies of Surface Chemistry and the Beginning of the War Years

Our Contact with the Electron Microscope

Early in 1938 an electron microscope was delivered to Dr. David Harker of our Research Laboratory. It was the second commercial unit delivered in the United States and I was fortunate to become involved in its operation. The role I played was that of specimen preparation. While the operation of the electronic and vacuum systems that produced the image on the internal fluorescent screen, and then on a photographic plate, was tricky and in the early days of the art, the success of the unit depended on specimen preparation.

Since I had just recently perfected a method of preparing very thin films of Formvar on frames, I was eminently prepared to produce replicas of metallographic specimens as needed. After giving a specimen an extremely light acid etch, I'd coat the clean surface with a dilute solution of the plastic. Thus, when the solvent evaporated it left a thin layer of Formvar on the etched metal surface. I then would strip this film from the metal surface onto a cleaned water surface. One or more small metal screens were placed on top of the floating film and then removed, dried and placed in the electron microscope for examination and possible photography. Unlike other preparators, I did not find it necessary to coat the replica film with evaporated carbon, gold or some other electron absorbing substance. The different thicknesses of the replica were sufficient to provide the contrast needed to get good electron micrographs. Dave Harker and I made a variety of experiments.

In one instance we had a visitor from the Mayo Clinic who claimed that with the advent of antibiotics he was finding that there was a substantial increase in new mutations–that those bacteria that survived the new chemicals were especially virulent and included some giant forms. Since he had difficulty in convincing some of his colleagues of these observations, he sought help from us since our microscope could bridge the limitations of the light microscope. Dr. Rosenow[21] spent a week or so with us and we were able to provide him with convincing electron micrographs that there were in fact giant forms in his collections.

---

21 This was Dr. Edward C. Rosenow, head of Experimental Bacteriology at the Mayo Clinic (1915-44), who went on to become a leading researcher in bacteriology, especially with influenza and encephalitis. He died in 1966.

As with so many other new discoveries, the discoverer always has to overcome the disbelief and often ridicule which surfaces among more conservative scientists when new ideas are forwarded that challenges the complete truth of the accepted science.

Dave and I published a rather nice paper on our findings, after which others took over the operation, following training on our techniques needed to get successful results.

A Brief Encounter with a Carcinogenic Chemical

On August 20, 1938 we had a visit from a Dr. and Mrs. Dodge of Washington University of St. Louis, Missouri. I spent half a day with them showing them our surface techniques. They were interested in determining whether or not a carcinogenic substance that they brought with them named chrysene could be handled with our regular techniques. This chemical was a four-membered benzene like substance having the four rings fastened together and having a melting point of 255°F.

After showing them our techniques I took the chrysene dissolved in benzene, and then spread it on a cleaned water surface. It behaved very much like cholesterol and I found it could be easily deposited as a typical Y film.

I mention this to illustrate the manner in which we cooperated with other scientists and our manner of treating them. Once they saw the simplicity of the equipment and the ease with which molecular films could be produced and used, they were enthused and grateful for our cooperation.

From my personal standpoint I welcomed such brief diversions, since my vista was broadened in yet another area of organic chemistry. We never heard from them again but we were finding that such follow through was not unusual.

Protein Monolayer Studies

By 1938 I had become quite proficient in studying single and multilayered films on water as well as on glass and metal surfaces. I had developed special trays, force and viscosity measuring instruments for the quantitative measurement of single layer molecules.

At that time Dr. Dorothy Wrinch[22] of England came to our laboratory seeking help to evaluate her ideas on the structure of globular proteins. I had recently constructed a surface tray that had an elegant method of illuminating the water surface. It consisted of a sheet of milk glass illuminated with a cluster of fluorescent tubes and inclined outward so that the highly illuminated light-diffusing glass was reflected on the horizontal water surface of the Langmuir Trough when viewed at a convenient working attitude. The trough had a well in it near one end that permitted me to lower a 3-inch slide to any desirable depth below the water surface. The trough was fitted with leveling screws and in turn placed in a slightly larger stainless steel tank that collected any overflow from the Langmuir Trough that might

---

22 Wrinch was a famous mathematician contributing in the areas of math, philosophy, physics and biochemistry, and in 1931 switched her interested to molecular biology.

occur in filling it or in cleaning the water surface. The inside of the solid brass trough was coated with bitumen, a black stable film. Dr. Blodgett had discovered that by blending highly oxidized automobile oil (oxidized by boiling it) with Nujol or Petrolatum obtainable from a drugstore, oil films could be spread that were of a specific color and without variation. I soon found that to achieve the greatest contrast a silvery gray film seemed best. This had thickness of about 500Å.

A monomolecular film with a thickness of 20Å, when formed in the center of the silvery gray oil film, pushed it aside and appeared as a visible film that was delineated by the edges of the visible oil film even though it could not otherwise be seen.

These techniques set the stage as it were for the study of protein monolayers. I soon found out that most monolayers of a protein had a thickness of about 10Å, and that many of them could be spread by touching the end of a clean moist glass rod or flame cleaned platinum wire to a small quantity of crystalline protein such as pepsin or egg albumin, and then touched to the central region of an oil-coated water surface. When this was done, a monolayer of the protein immediately formed on the water surface pushing away the oil film as it spread across the water surface.

After establishing the thickness of such films and some of its other physical properties, I discovered that if a clean wire was dipped into the indicator oil and then touched to the surface of the spread protein film, a unique pattern would form. Some of these patterns were star-like, others round, and still others would produce irregular outlines. I also discovered that in some instances the edge of the colored oil film would show a gradation in color. This effect I found to be a rough measure of the partial solubility of a component in the protein material that was probably an impurity.

I also found that the expansion pattern was a highly sensitive means of following the progress of the denaturating of a protein. Thus I found if I dissolved crystalline pepsin or egg albumen in water, I could spread it on water and it had all the physical properties of a film spread from dry crystals. If, however, I warmed the crystalline solution and then spread a film from it, the viscosity of the monolayer was lowered until finally, after further heating, the viscosity became similar to a highly liquid film. Testing it with oil, the expansion pattern went from a crisp star-like pattern, to a hardly recognizable star, to a completely circular pattern.

A Method for Peeling Replicas for Microscopic Studies

During the period when I was working with Dr. David Harker[23] on techniques for using thin film replicas in the electron microscope, one problem often encountered was the difficulty of stripping thin Formvar replicas from the etched surface of a metallographic specimen.

---

23 Harker became the head of the Division of Crystallography at the GE Research Center in 1949. Harker came to GE in 1941.

I finally discovered an elegant way to do this. The screens used as sample holders in the microscope were tiny thin discs of copper 1/8" in diameter. After casting a thin film of Formvar on the lightly etched surface of the specimen to be replicated, I would place one or more screens on the areas of interest, cover them with a layer of Scotch tape, and then peel the Scotch tape from the surface. The entire film, including the screens plus the film replica, would generally adhere to the sticky side of the tape. The sample screens could then be peeled from the scotch tape and mounted in the electron microscope.

A modification of this technique was also used in working with organic surfaces if difficulty was experienced in removing the replicas. A long triangular hole was cut into the Scotch tape before placing it on the surface being replicated. Upon retrieval the replica was peeled from the surface starting at the apex of the triangular hole. The replica generally was lifted in this manner and then mounted on a glass slide if the light microscope was to be used, or a circular electron microscope screen was positioned on the area of interest and then removed with the film replica adhering to it.

The Effect of Ultraviolet Light on Acid Barium Stearate Multilayers

On December 19, 1938 I embarked on an intensive study of the effect of ultraviolet irradiation on built-up acid barium stearate multilayers.

During the following two months I carried out 60 different experiments that were extremely interesting. These were carried out completely on my own but with cooperation of Mr. Frank Benford, Dr. Ralph Johnson, and the very active help of the glass blowing shop under Bill Ruggles, who constructed glass tubes needed for the study of the effect of various gases of high vacuum, and of various gaseous discharge tubes.

The primary technique I followed was to place a built-up multilayer near a strong ultraviolet source, irradiate a portion of it for a specific length of time, and then measure the radiation effects (if any) by optical measurement.

The most significant experiment occurred when I placed a built-up film at the focal point of ultraviolet rays emitted by the Uviarc[24] and reflected from a diffraction grating. My friend Frank Benford had a light emission testing laboratory room close to mine.

A total of twenty-two lines in the ultraviolet light region were found to be cut into the film by UV radiation as observed by the light scattered from the affected film. The 2,537 Å radiation line caused by far the most striking effect.

The conclusion from my intensive experiments was that the UV radiation broke up the stearic acid monolayers, the fragments then evaporating from the surface of the multilayer which thus was skeletonized in a manner similar to that found when such a film is placed in a benzene solution that dissolves the stearic acid molecules from the multilayer.

This was a fascinating experience as I worked my way through a whole series of definitive experiments and learned a great deal in the process.

---

24 The Uviarc was a mercury vapor arc lamp used as a source of UV light, patented in 1926 by the Cooper Hewitt Electric Company.

One of my crucial experiments was to start the skeletonization process by an intensive radiation of an acid barium stearate mixed film, and to then measure the rate at which the molecular residue evaporated from the irradiated film. When the evaporation was complete, I then filled the resulting skeleton film with a straight chain hydrocarbon having ten carbon molecules (decane), and found that the evaporation rate of the newly introduced chemical was the same.

When my attention was diverted to other research problems, I wrote a paper on these ultraviolet effect observations that was published in *Science* in 1939 as my second publication without Langmuir's collaboration. During the next fifteen years I wrote 140 papers that were published, of which more than a third were in the reviewed scientific periodicals. This all was fun!

The Formation of Plastic Pellicles

On November 10, 1939 my GE Notebook #2991 shows that I had developed a very effective method for making thin pellicles (films) of Formvar by coating both surfaces of a cleaned glass slide, by dipping the slide into a 1% solution of the resin dissolved in ethylene dichloride, and then raising it from the solution with a smooth motion, after which it was swung through 90° and held in a vertical position as the solvent evaporated. The slide, after drying and both coated sides scored with a sharp scriber on the bottom and sides, was then lowered through a cleaned water surface, held underwater for ten seconds, withdrawn and then slowly lowered through the water surface a second time. When this was done the thin film coating both sides of the glass slide peeled away from the glass and remained floating after the glass slide was again removed from the water. A suitable frame was then coated with a thin layer of rubber cement and lowered into contact with the dry upper surface of one of the stripped films. This could then be raised from the film surface with a tipping motion and the film would shed water and come out dry. The film was always taut on the frame. The frame was then rotated 180^0 and lowered with a smooth motion into contact with the second film. Since in the method of drying in a vertical position the film had a wedge-shaped cross section when the second film was contacted by the first and oriented at 180^0 from the first wedge shape, the second wedge-shaped film made optical contact with the first and the resultant two-layered film then exhibited a uniform thickness! Not only did this provide a much stronger film than would be the case with a single film, but any imperfection that might be present were invariably backed with a perfect film. Such pellicles were extremely easy to make but they were much more rugged and tough than any film I could find of any other material. Films as thin as 50Å (0.0050 microns), and as thick as a half micron, could be made in this manner in a couple of minutes.

Many applications ranging from extremely sensitive radiation detectors, condenser microphones and pressure sensitive indicators were made in this manner in the few months that I devoted to this study and development.

The resulting films were so tough that I could put 10 cc of water on the surface of a doubled film mounted on a 4-inch diameter circular frame, hold it over a gas flame and boil the water on it!

Early in January 1940 we were approached by an electronics engineer with a problem that was closely related to my earlier development of thin Formvar plastic films for use in the electron microscope.

In developing a new type of television camera there was a desperate need for a better beam splitter wherein the image viewed by the camera was split into two images with one of them going to the view finder and the other to the photosensitive tube. The beam splitters available commercially were only about 60% efficient. One we obtained and measured was coated with aluminum and showed a transmission of 13% and a reflection of 46% and thus an efficiency of 59%, despite the manufacturer's claim of 14% and 60%.

After trying several aluminized coatings evaporated onto my Formvar pellicles, I found the transmission reflectance characteristics to be about the same as the commercial pellicles. I then remembered that I had some zinc sulfide semi-transparent films that I had prepared for Dr. Charles Bachman in some fluorescent studies I had done with him a few months earlier. I found the slides, measured them, and found I had a beam-splitting efficiency of more than 90%.

Accordingly, I quickly prepared some Formvar pellicles on a rectangular frame that had been supplied by television camera friends, and arranged for them to be coated with evaporated films of this material that was not as easily done as with aluminum.

There was a very strange behavior of the zinc sulfide molecules. They showed a tremendous amount of mobility and, instead of depositing on the thin pellicle, they seemed to go everywhere but onto it.

We finally discovered that the radiation from the crucible in which we had to melt the zinc sulfide crystals was the cause of our problem. When this radiation problem was understood and baffles were provided, the deposits were excellent and of optical quality. I soon had pellicles made for the camera that had a transmission reflectance efficiency of 72% and 18% or an overall efficiency of 90%. These were ideal and were soon incorporated into a successful camera.

I then became involved in a surface chemistry study of sodium dioctyl sulfo-succinate, the newly discovered wetting agent that caused a duck to sink in the water.

The Surface Force Exerted by Indicator Oils

One of the great advances made by Katy Blodgett was the development of the Indicator Oil. I could see immediately that it was a powerful new tool and spent quite a bit of my time developing and exploiting this material. I made up a series of combinations ranging from blends, which at zero pressure produced colors which ranged from the first order silver gray, to dark yellow, to blue and blue-green, and to second order yellow. These five blends spanned the useful range of interference colors that, when pressurized by compression, thickened and produced new colors that could then be used to tell the pressure they exerted on a surface film in contact with their expanded edges in terms of dynes per centimeter.

In time I found that the silver gray and dark yellow blends were the most useful and they quickly became one of my most valuable surface chemistry "tools." These along with a group of purified oils, which we called piston oils, were indispensable substances. The piston oil concept was developed by Katy Blodgett, who found that oleic acid was an ideal material for producing a constant pressure against the edge of a monolayer on water. Since the oleic acid molecule was essentially the same in molecular weight as stearic acid (except for a double band located in the midpoint of the hydrocarbon chain), it was similar and highly surface active and insoluble in water. Its virtue consisted of its ability to spread as a single layer of molecules on water exuding from a drop of oil placed on the water surface. As the other film was removed from the water by deposition on a glass or metal slide, other oleic acid molecules would stream from the oil droplet exerting a constant pressure of 29.5 dynes per centimeter against the other molecules. Katy used a waxed thread to serve as the visible contact between the two films, while I found it feasible to use either a circular sheet of paper or a narrow strip.

In my exploratory studies I encountered films that would collapse at the oleic acid pressure but were quite stable at lower forces. Accordingly I made a survey of other pure oils and found four others that behaved like oleic but exerted lower pressures. They are listed in the table that I prepared in a joint paper with the Boss, "Properties and Structure of Protein Monolayers."[25]

Experiments with Non-Glare Surfaces on Glass

In June of 1939 I took a sheet of glass, ground its surface with a fine abrasive, washed off the grinding powder, lowered the glass into concentrated hydrofluoric acid for ten seconds, removed it, washed off all the acid and dried the slide. I found that while the treated glass still transmitted light quite effectively, all traces of image reflectance from the glass surface had disappeared. Instead the surface had a satiny feel and under the microscope I found that it was covered with tiny lens-like depressions that had been etched by the acid. I soon found that the dimension of the concave depressions could be controlled by the size of the grinding powder used to grind the surface.

This discovery was the result of a request of Larry Hawkins, our Laboratory engineer, who had a request from the GE Meter Department to explore the possibility of equaling or surpassing a so-called non-glare treatment of glass meter covers made by a commercial firm. When my surfaces were compared to the commercial samples they were found to be quite superior.

The only drawback in appearance from this treatment was the limited depth to which an object such as a meter dial could be positioned before the optical properties of the tiny

25 *Chem. Rev.:* 24, 181 (1939).

concavities caused the object to appear fuzzy. However, for all meters that were being built at the time, the treatment was quite effective.

At the present time non-glare surfaces of the type we developed are used in large quantities for protecting watercolor paintings and other valuable works and documents. While expensive, such glass treatment is quite effective in eliminating image reflection. They are not as good as the non-reflecting glass that Katy Blodgett discovered. Using harder materials that produce the same reduction in refractive index achieved by the skeleton films, they are used in treating practically all of the camera lenses of the world.

The Study of Light Scattering from Thin Films

Early in 1939, while the Boss was away on a trip, I became greatly interested in light scattering from single layers of molecules and other surfaces. It seemed that a host of problems and possibilities converged to make such a study quite relevant to basic, as well as, applied research at that time.

I assembled an optical system that was extremely sensitive and quite simple. Using an intense source of light produced by a small coil of tungsten filament and a double convex lens of short focal length, I focused the light in a conic light trap of the type invented by R. W. Wood of Johns Hopkins. Such a trap was easily made by heating a 5/16" diameter glass tube until it was workable, pulling it to form a cone, and then suddenly swinging the closed end by 90°. The light trap was then painted black. At a distance midway between the convex lens and the light trap, a second similar lens was focused and at its other focal point a selenium photocell was located. It in turn was connected to a Galvanometer.

Thus anything that scattered light at the location in front of the light trap would be focused and would activate the photocell. This arrangement was so sensitive that it could measure the light scattered by built up monolayers and multilayers, smoke, fog or any other light scattering material located at the front focus of the second lens.

At the time there was great interest in improving the contrast of pictures displayed by television tubes. An intense spot of light activating a phosphor (such as an ion beam) would produce a circular zone of brightness due to internal reflections that were devastating and greatly degraded the television image. I discovered that if the activated image could be formed of a thin membrane suspended inside the tube within a half-inch of the outer surface, the problem would disappear and the image had beautiful contrast, except for several other factors related to other physical problems.

I rigged up a simulated example and demonstrated it to key electronics personnel, as well as to Dr. Whitney. They were all quite impressed and when Dr. Langmuir returned and saw it, he too became enthused, especially when he discovered that I had worked out all of the optics and demonstrations on my own.

This became the start of an intensive study that I did on my own, which eventually ended with my first major paper, "Studies of Surface Properties by the Light Scattering of Deposited Liquid Films."[26]

While my solution to the internal reflection did not lead to a practical commercial application, it served admirably to separate out three phenomena that plagued the television image. Once these were separated and could be evaluated objectively it turned out that the "sticking potential" and the loss of light by backward light scattering became the major problems needing resolution. These were then solved when I found it possible to coat the back side of the phosphor screen with a reflecting film of evaporated aluminum.

On this I was successful in obtaining one of my first patents.

While this particular problem of light scattering had little to do with surface chemistry, an exposure to its various features was invaluable. When the time arrived for more basic studies, my mind was well prepared to evaluate the different features that appeared. My intensive studies of the scattering from molecular films ended early in 1940, at which time I became diverted with another intriguing problem related to the splitting of an image so that it appeared at two locations.

Meanwhile, from the earliest observations of light scattering, I gradually became aware of the effect of condensed liquids on the surface of molecular films. I soon discovered that the intentional deposition of such films was a powerful method for evaluating the uniformity of surface films.

The so-called breath pattern is an excellent example. Thus if one exhales a breath onto such a film, a layer of condensed moisture droplets may momentarily appear on its surface. The length of time such condensation persists depends in a large measure on the temperature of the slide. In fact if the temperature is low enough, such condensate will naturally form. This deposition of moisture marks the "dew point" of the surrounding air. The temperature at which it forms is a measure of the humidity of the air. The droplets that form appear as spherical lenses of water if the surface is hydrophobic. If, on the other hand, the surface is hydrophilic, water will deposit but light scattering will not occur, since the water builds up as a uniform film rather than light-scattering droplets.

It occurred to me that it might be possible to deposit hydrocarbons rather than water on such films, since acid barium stearate had been found to be both hydrophobic and oleophobic (water and oil repelling), due presumably to the dense uniform packing of the CH_3 "heads" of the stearic acid molecule.

My first effort to do this involved heating up some oil and letting the vapor deposit on the surface. This worked after a fashion, but produced great variations in the deposited droplets due to the convection currents in the air carrying the vapor. It then occurred to me to dip a built-up film into a uniform layer of indicator oil. When I removed the slide after dipping it, a thinner layer of oil returned to the water surface but the slide showed a fascinating array of oil droplets. The distribution of oil droplets revealed under low power

---

26 *J. Phys. Chem.*: 45, 681 (1941).

magnification was extremely interesting. The pattern on a particular step was quite uniform but in the successive steps it was quite different. The most striking effect was next observed with a step film of successively deposited single layers of molecules of the X type. Every second step was identical to the other, while those in between were markedly different. Again, under magnification the alternate steps produced the largest amount of light scatter and showed relatively large oil droplets, with steep contact angles and diameters ranging from 10 to 20 microns in diameter, arranged in polygonal arrays having a cross section of 70 to 100 microns. The steps in between had droplets so small that even at a magnification of 500X the droplets could not be resolved. The most interesting feature of this built-up "X" film is that the alternating banding that went from 1 to 30 layers showed similar alternations of light scattering from the tenth to the thirtieth layer.

Eventually I discovered that what was being dramatically illustrated were degrees of oil repellent characteristics suggesting that the surface properties of the exposed film were varying over wide limits.

Further exploration showed that the large droplets in polygonal array indicated a densely packed molecular surface of CH_3-ended molecules, while the tiny droplets in uniform distribution showed a layer in which the CH_2 side chains were dominant.

These findings strongly indicated the presence of active mobility of the molecules in a multilayer. After further exploration with a large series of step multilayers, evidence accumulated that suggested that once transferred from a liquid water surface to a solid, the molecules moved around filling voids or compensating for the stretching that took place during the transfer, and that this continued until some sort of equilibrium was attained. With this in mind my observations indicated that there was a multiple of eightY layers involved between the maximum and minimum of light scattering. With this indication I prepared a multilayer in such multiples and to my delight the alternations were just as sharp and repetitive as expected.

It seemed that each day brought a new facet of the widespread implications of the different aspects of these exciting findings.

Unfortunately the shadow of World War II was beginning to fall on the Laboratory.

Long before the actual declaration of war, we found ourselves being diverted into new areas of research. Such problems as particle filtrations, infrared detectors, and television cameras were proposed. We hoped to get back into the promising field of light scattering, but this never happened.

Ed Hennelly: A Fun Person

One of my good friends in the laboratory was Edward Hennelly[27]. He was an engineer type who, from time to time, collaborated with the Boss on some of his projects. During

27 Hennelly worked for GE for 42 years, retiring in 1954. He helped pioneer the radio vacuum tube and created one of the first radios that is now preserved at the Smithsonian Institute. He died in 1981 at the age of 91.

World War I he worked with the Boss in the development of the hydrophone that was then used in locating submarines by using binaural direction finding of sound. Hydrophones were placed on a pivoted arm, with one mounted on each end of the beam. Each hydrophone was connected by wires to another set of phones that were then connected to an air channel whose length could be varied by twisting a wheel. In this manner the location of a moving object could be ascertained with considerable precision using a pair of earphones connected to the varying air channel. Ed told me many interesting stories about their experiences off the coast of Boston testing hydrophones.

With the onset of World War II, the large concentration of German submarines off the East Coast, and the general lack of preparedness of the United States, it was decided to use every workable device available to fill the many gaps that existed in our defense system. A much more sophisticated detection method was under development, called SONAR, but it was not ready for deployment.

Ed Hennelly in some manner located a World War I binaural device that was the major part of the underwater detection system. I set it up in an adjacent room in the laboratory, installed a loud speaker on a pulley system so that the sound source could be moved. Then I started testing the sensitivity of the device for locating a moving object. This was the start of some fascinating research studies. I found that the acuity of the binaural sense was remarkable. It was discovered that I could make it even more sensitive by adding some "white sound" to the signal from the moving target. I then decided to use some hydrophones in a water system and found a place in the Erie Canal near my home that still contained about 5 feet of water. I used a single hydrophone activated as a sound source and tested the accuracy of the binaural unit and my ears to locate the source. Everything worked well and our simple devices were soon in use in such places as Long Island Sound. Before the war ended it was rumored that at least one sub had been located in the Sound with our unit. Other subjects diverted us and we lost track of the project.

My friendship with Ed continued, and I frequently visited him in his laboratory. He was one of the first to grow beautiful germanium crystals in a glass induction frame. These were an essential component in the very early semi conductors that became so important in the post-war world.

Ed could always be depended on to look at the bright side of life. He was one of the old timers; an enthusiastic booster of the Boss. I visited him frequently and learned many things from him.

Studies of Sodium Dioctyl Sulfosuccinate

Shortly after reading about the new wetting agent and seeing a movie of a duck that sank into the water because of its potent ability to destroy the non-wetting nature of surfaces, we received samples of the material from the manufacturer that I believe was American Cyanamid. I received the material during the latter part of the February 1940.

I immediately tried to form surface films of the chemical and had some fascinating data on its behavior. I found that it readily formed films, but equilibrium occurred only after several minutes had elapsed. I also found that at low surface pressures the surface activity was much larger than with the much simpler straight chain fatty acids. As the pressure increased the molecules became more condensed into a different configuration. Impurities and dissolved ions exerted profound effects on the force-area curves. It was soon apparent that while it appeared that much data of interest could be obtained, the pressing need to obtain it was not sufficiently high for me to take the time to attempt a thorough study.

During these studies I wondered about the foam-forming capabilities of the material (which was called DT-100 for short). I dissolved a few milligrams in distilled water and put about 50 cc in a 100 cc clean glass bottle and shook it vigorously. A dense froth of tiny bubbles appeared that looked to be quite stable. I set the bottle down and was about to leave my desk when all of a sudden the froth "broke" and disappeared in a couple of seconds! I had never seen such a thing happen before so I shook the solution again, this time measuring the period before the "break." It occurred again after the passage of the same time interval! I then tried varying amounts of DT-100 and found that the more of it that was added to a particular source of water the longer it took for the "break" to occur. This suggested that the wetting agent was sequestering the surface-active molecules that, after diffusing to the surface and building up a critical concentration, caused the foam to disappear. I then put 10 mg per liter of DT-100 in Quartz distilled water (the most purified we had produced) and tried the stability of the foam. While the foam "broke" in 21 seconds, the second time it was made more than an hour elapsed before it broke.

I then intentionally contaminated the water by adding 1.5 milligrams per liter of lead chloride. This time the froth broke in 11 seconds the first time and 105 seconds the next. When 3 milligrams per liter was added, the first break occurred in 6 seconds, the second in 7, with each successive period requiring 5 to 10 seconds longer. With a concentration of 5 milligrams per liter, the first break occurred in 3 seconds and even after twenty cycles the period had only slowed to 4 seconds.

Thus it appeared that I had serendipitously discovered a remarkably sensitive method of detecting trace impurities in purified water. While the implications of developing this technique were great, other projects prevented me from doing anything further with such a development except that it again greatly expanded my horizons of knowledge.

The Replication of Snow Crystals

On the evening of February 1, 1940, while waiting for the Rotterdam bus at Erie and State Streets, a light snow was falling. As was my custom, I held out the sleeve of my coat to get a glimpse of the type of crystals falling. They were unusually large and beautiful–perfect stellar crystals of the type now often used on Christmas cards. Incidentally, these crystals are fairly rare, as only a few snowfalls in a winter are made up of them.

I thought as I saw this evanescent beauty that it would be wonderful if such crystals could be preserved. I knew vaguely that Wilson Bentley, a farmer living in Jericho, Vermont, had spent a lifetime photographing snow crystals, especially of the stellar form, although his collection included some of the other forms. I hadn't developed a technique for making photomicrographs, and in my thoughts that evening on my ride home I wondered about the possibility of coating the crystals with a plastic film.

After leaving the bus at the corner of River and Schermerhorn Roads, I had a five-minute walk down the road and up the hill to my home. In that period I devised a strategy for preserving the crystals that might work.

I would make a 1% solution of polyvinyl formal– a tough plastic that was used for coating all GE magnet wire. I had some in my home laboratory, along with its solvent ethylene-dichloride, since I was quite involved in making thin plastic films for use in light-splitting applications and preparing replicas of metal for the electron microscope. I had been playing around with these problems at home since I didn't have time to do so during the day at the Research Lab. I figured I'd find out if the 1% solution would remain liquid at temperatures below freezing, and if the solvent would evaporate within a reasonable time. I had found a short time before that benzene would crystallize before it reached 0° Celsius.

Vince sitting in his backyard capturing snow crystals and preserving them in plastic solution (Formvar). Courtesy of Jim Schaefer.

As soon as I reached home, I fixed up a solution of the plastic and then put a good-sized drop on a glass microscope slide and put it in the freezing compartment of our GE Monitor Top refrigerator.

After supper I opened the freezing compartment and was delighted to see that the solvent had evaporated, leaving a nice clear disc of dried plastic where the drop had been. Removing the slide and letting the dew evaporate from the residue I looked at it with a 10X magnifier and was thrilled to see that a frost crystal had fallen into the droplet while it was still liquid, and that I had a perfect ice crystal replica. This was a beautiful example of a serendipitous happening!

Although the snow had stopped falling, I spent several hours exploring different types of effects with varying concentrations. I found that the original solution of 1% Formvar (F 15-95E) was the best.

For the next several months I obtained many hundreds of replica samples, ranging from selecting single crystals and replicating them, to massive samplings where I caught the crystals as they fell on slides wet with the replicating solution. I wrote and published a number of scientific papers on the processes used and some of the results discovered.

The Boss was quite interested in my discovery and asked to have some replicating solution bottled so he could obtain samples at his Lake George camp. A few weeks later he presented me with some fine replicas that he had prepared from snow falling at Crown Island.

During this period I was having my lunch with several friends from the General Engineering Laboratory that shared Building 5 with us. One of my friends was concerned with improving the contrast of the picture that developed in a television tube. He mentioned that there were two basic troubles, the one being that the phosphor particles in the tube would become charged with static electricity so that when the beam of ions that was producing a picture left a particular spot, the picture persisted for a moment and thus smeared the picture. The other problem was that when the phosphor became brightened with the ion beam, a goodly portion of the brightness went back toward the beam source and thus was lost. What was needed was to have the phosphor film coated with a thin conducting mirror. This seemed to be a hopeless case since all efforts to deposit a reflecting layer of aluminum on the phosphor produced nothing but a gray powdery coating.

It suddenly occurred to me (on February 18, 1941) that my snow crystal development was the key to producing a conducting mirror film. I proposed that after the phosphor coating was laid down by settling from a water slurry and the excess water drained from the film, the wet layer would be frozen. Once this had been accomplished, a dilute solution of plastic would be flowed over the icy surface, and then the solvent evaporated leaving a coherent thin plastic sheet on top of the phosphor layer. The tube would then be evacuated of air and a thin aluminum mirror formed on it by evaporation. The tube would then be baked at which the plastic film would disappear leaving behind a conducting, reflecting film that would greatly improve the contrast as well as eliminate the image persistence. Trial runs were made and the results were even better than expected. A fairly basic patent was obtained for the process.

During the intervening years the use of Formvar for replication and for producing very thin tough films in air has become routine in many science laboratories. While I have tried from time to time to improve on the process, after 45 years my original findings are still the best.

The Replication of Living Surfaces

Shortly after discovering the technique of replicating snow crystals, I found many new ways to use Formvar for producing extremely detailed replicas of the surfaces of a wide variety of things. These ranged from metallographic specimens for viewing in the electron microscope, to 10,000 line diffraction gratings for calibrating electron microscope pictures. I found it possible to make much better records of finger prints, a method for using dyed replicas, to determine the profile of the etched surfaces. In the latter application, I collaborated with Dr. Herbert Uhlig of the Laboratory, who was interested in the Neumann Bands of metallic meteorites. I found that these bands were depressions, and by putting dye into the solution before making the replica, that the density of the dye could be used to determine the relative thickness of the grains in the band and its adjacent surfaces.

I then discovered that instead of using polyvinyl formal (Formvar) dissolved in ethylene dichloride, I could make excellent replicas of the water-soluble plastic polyvinyl alcohol (Elvanol). This material could be used to replicate organic substances such as leaves, insects, flowers and such delicate things. By preparing replicas of the stomata of leaves, I could determine whether or not they were closed or open. The structural beauty of the petals of flowers was breathtaking. I replicated the eyes of dragonflies. To demonstrate the sensitivity of the technique I replicated the surfaces of a leaf-mining worm and discovered that the cellular pattern of the tiny worm was so uniform it appeared as a diffraction grating!

Much of this activity was carried out as a hobby, with most of my photomicrographic work being done in the microscope room and darkroom that I had developed in the basement of my home on Schermerhorn road.

Some years later I used a Formvar solution to prepare replicas of the surface of large hailstones and found that the replica was much better than the original icy surface. Not only was the detail and contrast much better, but also the replicas could be used (as with the replicas of snow and frost crystals) to project the fascinating patterns of the growth structure when I presented a paper on the subject. The same success attended my studies of glacial ice that I carried out at the Snow and Avalanche Research Institute on top of the Weissfluhjoch west of Davos in Switzerland.

Dr. Marcel de Quervain, Director of the Institute, had invited me to spend a week or so at his laboratory. While there, I gave several informal talks about snow and ice during which I told the audience about the "ice borers" I had discovered on Mount Washington. The audience of snow and ice experts had never heard of such structures. After my talk I took some of them outdoors and showed them a number of such borings just outside the door of the laboratory!

Artificial Fog

The Boss returned from one of his fairly frequent trips to Washington where he was a member of an important Advisory Committee. He called me into his office and described an important development.

According to an intelligence report, he said his committee was told that one of the key warships of the Germans, the "Bismarck," was found to be using a Norwegian fjord as a base. It was frequently obscured with a smoke screen that appeared like a natural fog. This was so effective that the ship was completely hidden from the bomber aircraft when they managed to reach the location. His committee was challenged to develop an artificial fog as effective as the one used to hide the Bismarck. The only additional information advanced about the natural appearance of the smoke screen was that it was possibly made from a material produced by a paper pulp plant. That was the extent of our information.

The Boss raised the question with me as to whether or not a tiny smoke generator I had constructed earlier in the year to test some Canadian smoke filters could be redesigned to be a hundred or a thousand times larger. I was dubious about that possibility, since the smoke generator I had made seemed to be at the upper limit of its capabilities as it was. However, I told the Boss that I would try it despite my misgivings. I built a unit ten times larger, only to find my doubts were justified and the newer design wouldn't work. With this discouragement I was stymied as to what to do when a serendipitous event occurred. As I was returning to the lab after a visit to the Glass Blowers Shop down the hall, I glanced into a room and saw a Langmuir-type mercury vapor pump in operation. This is based on a glass device that heats mercury to its boiling point, and then used the high velocity stream of mercury vapor to draw air from the system that is being evacuated. The mercury vapor is then condensed back to metallic vapor and returned to its supply reservoir.

It occurred to me in a flash that it might be possible to vaporize a high boiling point oil, such as I had used with my smaller generator, and then let the vapor exhaust into the atmosphere where it would condense to form a cloud of oil droplets.

I hurried to my lab, and as I went to my bench my eyes glanced at an ordinary oil can such as is used to lubricate machinery. It consisted of a half-rounded cylindrical reservoir holding 50 cc or so of lubricating oil, topped with a tapered spout having a hole diameter of 1/16" or so.

I placed this oil can on the stand of a Bunsen burner, lit the gas and placed the flame under it. In a few minutes the oil began to boil and spurt hot oil droplets into the air. I then took the burner and heated the tapered spout of the oil can, continuing at the same time to play the flame on the base containing the boiling oil. In a few seconds a high velocity stream of vapor emerged from the spout of the oil can instead of the hot oil droplets.

The vapor immediately condensed to form a white turbulent stream of smoke that looked like the steam jet of a locomotive. As the vapor condensed to form the beautiful white cloud of submicroscopic oil droplets, within a few minutes it filled my laboratory room with what appeared to be a highly stable and persistent smoke.

The success of this experiment was quite exciting to me, since it looked like there was no limit to the size that could be further developed. Unfortunately the Boss was away on one of his frequent trips, so I had to contain my enthusiasm. While the work on the project was not classified as ye,t I felt that it would be desirable to keep from talking about it at least until Langmuir evaluated its possibilities.

Meanwhile I quickly designed a bigger "oil can." This involved designing a much larger reservoir that could be heated with an electric coil and having the orifice for the escape of the vapor as in integral part of the reservoir. I again made the orifice quite small.

I had the new smoke generator ready for testing within a few days and took it to the roof of the laboratory.

The test was a complete success, with the smoke cloud quickly engulfing the entire area below the parapet on the flat roof of the laboratory. In my enthusiasm, I let the unit continue to produce smoke. The smoke was so dense that I could see only a few feet.

All of a sudden I heard a shout and out of the smoke came a group of firemen lugging a hose and looking for a fire! When they saw me and the little device, from which was emerging a high velocity jet of white smoke, I heard some choice expletives.

Fortunately the Fire Chief was an old friend of mine who, when I was an apprentice boy, had served as one of my mentors. Charlie Goethe looked at me in exasperation as he slowly regained his breath and his composure. (They had been forced in their heavy clothing and rubber boots to carry their fire-fighting equipment all the way up the even stories of stairways of the lab to reach the roof and my location.)

He then told me that when someone in the adjoining building had told him that the top of the laboratory was on fire, his first reaction was that it was only some experiment in the laboratory, but when they said they could see flames, he decided he'd better investigate. There, of course, were no flames of any kind, since I was using electrical resistance coils to heat the oil.

I assured him that henceforth I would alert him as to my activities!

Shortly after I discovered the way to make small particle smoke on a large scale, Langmuir told me that we needed to establish the optimum pressure and orifice size to get the greatest quantity of smoke for a given amount of oil.

I decided to start using an orifice with a 3/16" diameter hole. Since my generator was fitted with a pressure gauge, I played around with the oil flow volume until I satisfied myself that the smoke looked best when the pressure gauge showed 10 psi (pounds per square inch).

After I satisfied myself that the 3/16" and 10 psi was close to the optimum values, I returned to my laboratory and noticed that the Boss was in. As I entered the room he announced that he had calculated the optimum operating condition for the generators.

Without giving me a chance to tell of my findings he said "We need an orifice diameter of 0.187 inches and a pressure of 9.4 pounds!" When I told him that my field run had indicated 3/16" (0.187") and 10 psi, he grinned and said, "Good!"

It was then that he told me that the escape velocity of the pressurized vapor jet would automatically produce sonic velocity. This incident reflects the reason why my practical field approach and his calculated theoretical approach made us such an effective team.

When Langmuir returned from his trip I showed him the results of my new generator with a demonstration. He became very excited and immediately started designing a much larger unit that utilized a flash boiler fed by a stream of oil from a pressurized reservoir. Between the two of us we designed a generator that was quickly built by Harold Nelson of the Lab Machine Shop.

Since the new generator was much larger than my glorified oil can, I notified the fire department that I planned to test the new unit down on "the Farm" a region free of buildings near the base of the Bellevue Bluffs. It was an area familiar to me since I had served much of my apprenticeship in Building 101 that was near the end of the Works Avenue and a part of the old Van Guysling Farm.

The testing proceeded well; we tried out a number of different conditions including differing orifices. In order to explore the possibility of producing larger particles, the Boss had suggested making a cavity that would prolong the condensation of the super heated vapor. (We had by this time established that the generator worked most efficiently when the flash boiler had an internal pressure of about 10 psi.) This cavity was built and had three orifices.

It was apparent as soon as the cavity unit was installed and three jets emerged from it that the device was inefficient, since some of the hot oil condensed in the cavity and was spewed out by the jets along with the finer particles of condensed vapor. Then all of a sudden it happened! A ball of fire erupted in the dense smoke cloud about twenty feet away and propagated rapidly towards the generator. Reaching the area of the jets, the fire was extinguished by the high-velocity jets. This was not an explosion, but still a rather frightening phenomenon.

A dense cloud formed again, and again became ignited. After the third sequence I shut the generator down, somewhat shaken by the experience. In seeking an answer to the cause of the ignition sequence, I discovered that the ignition was probably due to an electric spark generated by the charge separation that was produced when the hot oil droplets were entrained by the jets and moved past the much finer condensed droplets. Eventually I discovered that Faraday had studied a similar phenomenon that occurred with a drooling steam jet, in which he was able to generate a voltage difference of 50,000 volts using a cavity very much like the one we built. Thus it appeared that we had, in fact, developed a miniature lightning generator, but instead of developing only a spark discharge, we had produced something akin to a dust explosion as occurs from time to time in grain elevators. Of more recent interest are the explosions that rip apart the tanks of giant oil tankers when they are being cleaned by steam jets. It is likely that these effects are related. Despite the possibility that these phenomena may have a close affinity to the separation of electric charge that produces thunder and lightning in natural convective storms, and similar effects in volcanic eruptions, we never got around to a further study of the drooling nozzle and its electrical effects, nor have I managed to get others to use this mechanism to model thunderstorms!

After making a few runs on GE property, I felt it necessary to run the generator under conditions where I could obtain photographs of the extent and nature of the screening smoke. Accordingly I obtained permission from the Cushing Stone Quarry to make my next runs at the bottom of a large gravel pit west of Scotia (New York) where I had worked one winter some five years before. We made several runs at the pit, and I obtained some fairly good photographs of the operation.

One day, while working in the dense smoke near the generator where I was measuring smoke particle size using a small settling chamber I had built, I happened to look at the sun and was amazed to see its brilliant color. This led to further experiments in controlling particle size by altering the diameter of the orifice of the generator and the pressure in the flash boiler. I found it possible to vary the diameter of the particle form 0.3 to 1.0 micrometers (microns), with color changes in the sun from a deep red to a brilliant blue. When I told this to Dr. Langmuir he worked out the mathematical relationships and had Dr. Blodgett build a filter gauge that consisted of size-colored discs that were mounted in a small wooden frame. When held at arms length, while standing in the dense smoke and looking at the sun, the diameter of the colored discs subtended that of the sun so that a direct comparison could be made. We had found with my smoke box that the best hiding power of the smoke occurred when the particle size of the smoke was 0.6 micron diameter. If the generator operator saw a red or a blue sun he knew that an adjustment was needed in the pressure of oil flow controls.

With the coming of spring we faced the project of needing a test and demonstration site on a much larger scale than thus far utilized in order to obtain quantitative data on screening power and other pertinent details.

I proposed to the Boss that we make use of the Schoharie Valley about 20 miles west of Schenectady, using the top of Vroman's Nose as an observation site. He was not familiar with the area so I took him out to size up the possibilities. We climbed to the summit of the Nose, which is a spectacular east-west rocky ridge, having a vertical cliff on its south side, and rising abruptly from the flat flood plain bordering the Schoharie Creek.

The edge of the cliff discloses a magnificent panorama of the rich farmland below, called Vromansland, that had been purchased from the local Indians in 1712 by Adam Vroman, a pioneer farmer from Schenectady. The valley flats of this region extend for nearly ten miles in a southwesterly direction, while the land is about a mile wide along that course. Thus we had a test area that was about 50,000 feet long, 5,000 feet wide and more than 600 feet deep. Langmuir was much impressed with the Nose as a vantage point and the great flats as a test area.

While we were on the summit I happened to see an oak apple, a gall that commonly forms on oak trees, of which there are so many on the summit and slopes of the Nose. They are entirely of the chestnut oak species.

I cut one of them in half and showed it to the boss. Apparently he had never noticed them before. I showed him how in the center of the "apple," which is about an inch in diameter, there is a chamber supported by a mass of filaments that grow from the inside of the hard-shelled outer sphere. In this chamber is a grub that transforms into a gall fly, that then injects a growth substance into other branches after laying an egg to perpetuate the species. The tree then responds to the injection by forming another apple. When first formed, it has a greenish color and is soft, but it eventually hardens and turns gray. The Boss was fascinated by the gall and greatly puzzled about its formation and role in the scheme of things.

This sense of wonder on his part was one of his great attributes. I could see it wouldn't take much effort to get him involved in a whole new area of research! It remains a puzzle to me that more attention is not directed at the present time by those involved in cancer research to a thorough study and eventual understanding of the growth mechanisms, especially in their early phases of gall formation and fasciation. These unique developments are likely to lead to some major breakthroughs if carried out by the right individuals. "Doc" Whitney, Director of the Research Laboratory, who was greatly intrigued with the Goldenrod gall that is somewhat similar to the oak apple, originally advanced this thought. I have always been fascinated with these marvelous creations.

During the winter of 1941 I built a particle size testing facility in the laboratory. It consisted of a large box 6 x 6 x 7 feet. The box was made of galvanized steel, mounted on the inside of a wooden frame of two-by-fours. The lower box was 6 x 6 x 4 feet, the upper one 6 x 6 x 3 feet. They were welded, the upper box separated from the lower box by a quarter of an inch, thus assuring that there was no heat conductance between the lower and upper boxes. The upper box was wrapped with heating cable, while the lower box was cooled by circulating cold outdoor air within the box prior to a run. A 4 x 4 foot adjustable target rested in the bottom of the chamber that could be raised and lowered with windlasses until the target became visible. Viewing ports were placed at several elevations and at the top, where projection floodlights were also positioned for illuminating the fog.

Another port was located below the inversion for introducing the smoke. A monodispersed smoke was generated with a midget generator fabricated from an artist's spray gun.

Using the smoke box I was able to determine the optimum particle size for obscuring a target under various conditions of illumination, thus establishing the 0.6 diameter particle was the most effective. I was also able to establish that a smoke layer that was 100% effective in hiding the target was quite ineffective in the infrared. I used infrared film to establish this fact.

In the Spring of 1942 we moved our field testing operation to the Schoharie Valley south of Vroman's Nose, using the Jenks Farm, but the generator produced so much smoke that we had to move our test site farther up the valley to the Teller Farm. The tests conducted on May 7 and 12 were so satisfactory that a plan was developed to make a generator ten times bigger. The Boss was put in touch with engineers of the Esso Laboratories (Standard Oil Development Company of New Jersey). Mr. Stewart Hulse came to Schenectady, witnessed a field run and, before returning to New Jersey, promised to have a working unit within a month. Early in June he notified us that the unit was ready for a field demonstration. The unit was brought to Schenectady. It was a spectacular and fearsome thing! Stewart told us he had found the parts of the unit on the scrap heap of the Laboratory. It looked like an ancient wood-fired donkey engine of the middle 1800s. Powered with an oil burner, the flash boiler was fitted with a large manifold bearing ten 3/16" diameter holes.

The preliminary runs with this unit were conducted on June 21 and 23 of 1942, using the George Wilber farm in the middle of the Flats south of Vroman's Nose.

On June 24 a full dress demonstration of artificial fog was planned. When the thirty key persons arrived in Schenectady, they assembled at the Van Curler Hotel, and after dinner were given a brief plan for the next day by Dr. Langmuir, who announced that everyone should be ready to leave the hotel at 3 AM. Despite some mild protests, everyone agreed to be ready.

Everything worked out as planned. The group arrived at the base of the Nose in about an hour. Some of us started walking up to the summit since it was the quickest way to get there. A number of army jeeps were soon loaded with personnel and slowly moved up the old wood road to a terrace within about fifty vertical feet of the summit. By this time dawn was brightening the eastern sky, and the rocky trail to the summit could be easily followed. Arriving at the summit, everyone had a light breakfast, and then watched the action. Before the sun rose a patch of smoke appeared at the generator site and quickly enlarged into a massive smoke screen which rose a hundred feet or so to the top of the night time inversion and spread northerly, flowing down the valley in the drainage wind. It was opaque and quite a sight to see! Everything had worked exactly as Langmuir had predicted. The smoke screen slowly lifted as the rising sun began to create small thermals, but the smoke screen continued to expand and remain intact.

After we had radioed the valley crew to turn off the big generator, several other demonstrations were carried out and then a second breakfast was made available. Shortly before leaving the summit of the Nose at about 10 o'clock, the original smoke screen returned in the up-valley flow, and spread across the valley, thus demonstrating the persistence of the smoke particles.

Everyone was then taken to the generator site and had an opportunity to see it in action. Thus ended a highly successful operation that eventually led to the construction of many thousands of the M-1 Generators manufactured by the Heil Company of Wisconsin, and eventually by Todd Shipbuilding. They were used in many places during WWII, including the Anzio beachhead in Italy, the North African campaign, the crossing of the Rhine, and to ward off Kamikaze attacks in the South Pacific. Although we obtained a joint patent on the smoke generator, the General Electric Company never exploited its use.

The Problem of Precipitation Static

Early in 1943, as the war tempo was increasing, the Boss returned from one of his NDR (National Defense Research Committee) meetings with news that he had been asked to explore certain aspects of airplanes and snowstorms. It was said that with the rapid buildup in plane numbers such as the B-17 and B-24, the number of plane losses were mounting at an alarming rate. It appeared that the combination of young pilots having little experience flying in bad weather and the incidence of snowstorms was a deadly combination. The impact of snow on the plane wings caused a buildup of radio static to the point where electronic navigation devices became inoperative.

After discussing the situation the Boss suggested that we might do some field research at Mount Washington in New Hampshire. Langmuir was familiar with the mountain's severe weather, having climbed it in the winter with John Apperson some years previously.

He made a date with Dr. Charles Brooks, Director of the Blue Hill Observatory south of Boston, to discuss the possibility of mounting a research program at the Mount Washington Observatory that served as a satellite outpost of Blue Hill.

The Boss and I went by car to Blue Hill and met Dr. Brooks at the summit observatory. I found Dr. Brooks to be a fascinating scientist– full of ideas and energy and highly enthusiastic about the possibility of helping us with our project on "the Mountain." He called Joe Dodge, the local manager of the Observatory, and made arrangements for several of us to visit the Observatory in the near future.

On September 28, 1943 I left Schenectady with Dr. Elliott Lawton, an electronics engineer and scientist, and Hubert Tanis, a lab technician, and headed for Mount Washington in a laboratory station wagon. We reached Pinkham Notch Lodge, the main headquarters of the Appalachian Mountain Club, late in the afternoon. Checking in, we discovered the somewhat primitive nature of the accommodations, wherein we had to make our own beds in wooden bunks in semi-dormitory living. The food was plain but in great quantity and well prepared, sufficient to satisfy the appetites of the mountain climbers who used these facilities.

After getting settled we went to an adjacent building where Joe Dodge lived with his wife and children. This was a remarkable experience; Joe was in frequent scheduled contact with the Observatory by radio. After receiving the extended weather report that was also sent to the Blue Hill Observatory, he announced our arrival and arranged for us to obtain the data we needed in order to construct the equipment which we hoped would help us understand precipitation static.

Early the next morning, after a hearty breakfast at the lodge, we picked up food, mail and other item,s and headed for the summit in our car by way of the eight mile Carriage Road. The weather was good and we reached the summit mid-morning. After introducing ourselves and meeting the crew, we set about to determine the measurements of pipes and other structures in the tower of the Observatory to facilitate the installation of equipment for measuring precipitation static.

After lunch we left the observatory with the promise of returning in mid-October to start our experiments. Returning to Schenectady we mounted an intensive construction project to build a set of electrodes, and a voltage detection device that we expected would provide us with the data we hoped to obtain. We returned to the summit sixteen days later, and drove directly to the summit after checking in with Joe Dodge. Arriving there we started our installation, hurrying as much as we could, since there was promise of an intense storm within the next 24 hours. We completed the installation and checkout with our measuring device and were ready for the promised storm. The next morning the storm was upon us and we hastened to obtain measurements. Imagine our dismay to find that within a few minutes our electrodes were covered with rime ice! Thus, instead of being in a position to measure the buildup of static electricity from the breakage of ice crystals on an aluminum surface (that was the objective of our research endeavors), we were confronted with a situation where not only were we troubled with leakage problems between our electrodes and ground, but if a voltage developed it would be measuring the effect of snow on an ice layer– not at all what we had anticipated!

Unfortunately, this turned out to typify our experience at the summit of the mountain. The air accompanying a snowstorm was so moist and rose so rapidly as it encountered the slope of the mountain that in all cases rime ice would form, no matter how intense the snow became.

After spending a good part of the morning trying to obtain data, we finally gave up as the storm intensified and threatened to snow us in for the winter!

Since Ray Falconer of the Mount Washington crew seemed to know the most about electronics (he was a radio ham), we checked him out on the operation of our device and instructed him to continue attempts to get data.

We left the summit with the road rapidly disappearing in the first real snow of the season. Once we left the summit the road rapidly improved so that we were out of snow as we reached the Glen House at the base of the mountain. Incidentally the trip down was the last one of the season!

Returning to Schenectady I reported our lack of success to the Boss and the problem we encountered. He was surprised, but then agreed that it was a factor that might have been anticipated. I suggested that it might be worth trying to develop a facility on the top of our laboratory building using a high velocity blower operated during snowstorms. This worked beautifully and we were soon obtaining excellent quantitative data of the effect of snow crystals shattered upon impacting aluminum surfaces during the frequent snowstorms that passed across the Mohawk Valley. The irony of the situation is that instead of spending half a day of car travel and half a day of climbing to the top of Mount Washington, the summit of our building provided us with exactly the data we were seeking.

As we began to "close in" on the precipitation static problem, we were suddenly informed that we should concentrate our Mount Washington studies on the subject of aircraft icing, since the rapid development of FM radio was solving the static noise problem.

Thus, for the next year, our efforts at the mountain concentrated on measuring such things as liquid water content, particle size and particle size distribution. With all of these studies the Boss remained deeply interested, and frequently added ideas and computations that permitted us to make steady progress toward reaching an understanding of the factors leading to the icing of airplane wings and propellers.

Atmospheric Electrical Measurements on Top of the Laboratory

Following our visits to the summit of Mount Washington in the fall of 1943, and the failure of our experiments due to the rime ice formation on our experimental surfaces, I decided to establish a research facility on top of our Building 5 Laboratory at the Schenectady Works of General Electric.

There was an abandoned greenhouse on top of the flat-roofed building, part of which I converted to a facility that permitted me to measure the charge on falling snow crystals. On the parapet of the building I erected a tall mast, on top of which I located a simple electric potential measuring device with the help of Dr. Elliott Lawton, a friend and electrical engineer in the Laboratory, who had accompanied me to the summit of Mount Washington and had devised all of our electronic gear. Elliott was an old-time friend of mine, a skier and outdoorsman, and a very knowledgeable person about all aspects of electricity. His assistance was indispensable. However, he had his own research projects to take up most of his time, but welcomed the opportunity to go to Mount Washington with me. After assembling the equipment and advising me about its fundamental operation, he returned to his own projects.

During the remainder of the winter of 1944, I spent most of my time on top of Building 5 and learned a great deal about the nature of the storms that swept down the Mohawk Valley or came in from the southwest or the northeast.

I found they consisted of two basic types of storms, so far as their atmospheric electricity was concerned. I called them "cirrus types" and "convective types." The former started with a thickening of cirrostratus clouds coming in from a southwesterly direction, in which the sun was gradually covered with a dense veil of snow crystals, producing a "snow sun," looking as though' it was covered with a sheet of ground glass. The snow from the cirrus levels finally reached the ground, appearing as tiny ice crystals of thick hexagonal plates and hexagonal columns. The falling rate of the crystals remained constant, and they continued to have about the same size, were only lightly charged with electricity, and produced a space charge increase having a positive sign. Throughout such storms, which produced about a half-inch accumulation an hour, the electric potential remained constant.

The convective storms were much different. They approached Schenectady from the west, northwest, or northeast, and were frequently accompanied by wind. The sky would have a heavy overcast typical of convective clouds. The snow would consist of large stellar, spatial dendrites, irregular shapes, long needles, or graupel, and would carry high electrical charges. With such storms the potential gradient tended to change sign every twenty or thirty minutes as successive convective systems passed over the observatory. Perhaps the most interesting effect I experienced was observed at the end of such a storm, when the potential gradient continued to alternate for one or two cycles without any visible cloud present. I called these "ghosts" and believed that the electrical charge remained in the air even after the precipitation had fallen out of it, or had sublimed away.

In the early fall of 1947 I published a summary of the results of these studies. I was able to show the types of crystals that typified the two types of storms, their falling velocity, quantity, sign and amount of electrical charge, mass range in quiet air, and the increased electrification that occurred if the crystals were broken upon impact of a simulated aluminum airplane wing.

This paper, "Properties of Particles of Snow and the Electrical Effect They Produce In Storms," when published, was selected as the Outstanding First Paper of the Year published by the Hydrology Section of the American Geophysical Union.[28]

Bibliography of Papers Co-authored with Dr. Irving Langmuir

Composition of Fatty Acid Films on Water Containing Calcium or Barium Salts, *J.A.C.S.*: 58, 284, (1936)

Built up Films of Proteins and Their Properties, *Science*: 85, 76 (1937)

Optical Measurement of the Thickness of a Film Absorbed from a Solution, *J.A.C.S.*: 59, 1406 (1937)

Multilayers of Sterols and the Absorption of Digitonin by Deposited Monolayers, *J.A.C.S.*: 59, 1751 (1937)

Improved Methods of Conditioning Surfaces for Adsorption, *J.A.C.S.*: 59, 1762 (1937)

Monolayers and Multilayers of Chlorophyll, *J.A.C.S.*: 59, 2075 (1937)

The Effect of Dissolved Salts on Insoluble Monolayers, *J.A.C.S.*: 59, 2400 (1937)

Salted-out Protein Films, *J.A.C.S.*: 60, 2803 (1938)

Properties and Structures of Protein Monolayers, *Chem. Rev.*: 24, 181 (1939)

Rates of Evaporation of Water Through Compressed Monolayers on Water, *Jour. Franklin Institute*: 235, 119 (1943)

---

28 Trans. Amer. Geophys. Union: 28, 587 (1947).

Chapter 6
The War Ends as I Discover Cloud Seeding

A Visit to the Mount Washington Observatory with Langmuir

Late in the winter of 1944 the Boss told me he would like to accompany me to the Mount Washington Observatory on my next trip. Accordingly, I made the necessary reservations at Pinkham Notch, and we headed to the mountain in April. While signs of spring were all around us in the Mohawk Valley, they quickly disappeared as we headed across the Green Mountains. Reaching Pinkham Notch by late afternoon, we checked in, fixed our beds, and went to see Joe Dodge, Manager of the Lodge and the Observatory. Joe was an institution. Called by his friends "the Mayor of Porky Gulch," he was an ex Navy man– a submariner– and had picked up the most colorful and vulgar language I have ever encountered. The strange thing about Joe's emanations was that coming from him, they didn't sound profane at all. We had a delightful discussion, and then went for supper and turned in to bed quite early, since we had an arduous trip ahead on the next day.

Rising early and getting our packs adjusted and skis ready we had an early breakfast, picked up a trail lunch and then headed for the summit. Our route was to head for Tuckerman's Ravine, and then to climb the cliffs of Lion's Head on the north side of the Ravine. The trail was so steep that it was useless to try doing it with our skis on our boots, so we carried them with our rucksacks. Reaching Tuckerman, I searched for a route up the face of Lion's Head and found a route that seemed negotiable. Here our skis were a real impediment, since in some areas they interfered with our freedom to turn and twist among the rocks. However, we persisted and finally reached the top of the cliffs.

The weather was favorable, and although the sky was overcast and looked like it might start snowing at any time, we had no trouble. The wind was present but not excessive. After reaching the top of Lion's Head, we worked our way through the snow to the cone of the mountain and, finding shelter from the wind, we ate our lunch.

While eating we carried on an active conversation, mostly related to our environment and the area in general. The Boss suggested doing an interesting thing. He was an advocate of stereo photography and suggested a fascinating project. Tuckerman's Ravine receives a tremendous deposit of snow since it blows into it all winter, finally reaching a depth of a hundred feet or more. The expert skiers who flock to it in the springtime find it a challenging place. Thus in May and June the slopes of the ravine below the headwall are covered with skiers from all parts of country.

Langmuir proposed that a stereo pair be made when skiing was at its peak. These pictures would be taken from two photo stations about fifty feet apart along the north ridge that would be carefully marked. Later in the year, after all the snow had melted, a second set of pictures would be taken. By superimposing the two sets of pictures, as viewed through a stereoscope, the view would show the skiers to be a hundred feet or more in the air!

A few months later I obtained the first set of stereos with a hundred or more skiers on the slopes below the headwall. While in the process of taking them I became intrigued with the drama occurring at the top of the headwall. A skier was trying to build up his courage to make the schuss. He would check his skis, look around to see if others were watching him, approach the lip and– the courage failing– he would back up to a safer place. He went through this maneuver a half dozen times or more. Finally, he decided to go, and did so– but almost immediately lost his stance and went cartwheeling down the long slope to an ignominious termination. So far as I could tell he didn't get hurt, and after a bit, skied away.

I never did get around to making the second pair of stereos! This remains as a set of unfinished business. I'm afraid that I'll have to start over since I doubt if the rock cairns I erected to mark my camera stations are still intact after 41 years!

Another interesting item noted by Langmuir during our luncheon break was the apparent lack of ice nuclei in the clouds scudding overhead. The temperature was well below freezing so that the darkish clouds passing us must be supercooled. However, every so often a single snow particle would fall from the sky. The Boss made a quick estimate of the concentration of ice nuclei and figured it to be about an ice nucleus per thousand cubic meters. Several years later I established a cold chamber at the Observatory to obtain quantitative measurements of the ice nucleus concentrations in air passing over the summit. The lowest values found a -20°C was 60 per cubic meter. It is quite likely that if we had made measurements at the warmer temperature that existed at the time of Langmuir's observation, the concentration might be comparable.

After having our lunch we put on our skis and zig-zagged up to the summit through the sastrugi[29] and drifted snow. As usual, the staff of the Observatory greeted us warmly, and for the next several days had a wonderful time carrying out routine as well as special observations suggested by the Boss.

This was a very special time in the history of the Observatory. All of the crew were highly motivated and I doubt if any scientific group ever had a better time. The Boss was in his element and was brimming over with new ideas. There was little sleep at the Observatory during our visit!

Langmuir devised techniques for using the data from the Multicylinder device to establish the liquid water content, the dominant particle size, and the particle size distribution. The crude curves that he developed were further refined when we returned home by using one of the first analog computers developed in America.

---

29 Surface irregularities in the snow.

On Easter Sunday morning the Boss and I left the summit to do some skiing on the East Slope, an area that was used for recreation by the summit crew. The sun was shining out of a cloudless sky, the wind had died down, and it was just a magnificent day! Reaching the ski slope we had several fine runs, and then started a slow return to the summit, exploring on the way. I obtained some beautiful photographs as we "poked" around. I showed the Boss a fascinating phenomenon wherein small black pebbles moved through the ice and snow when warmed by the sun. These pebbles move at the rate of about a millimeter a minute and leave behind a pattern that looks like worm burrows. Thus I called these little pebbles ice borers. On a steep slope of our ski area we found the burrow of a woodchuck that had worked its way up through several feet of snow to the surface. Its muddy tracks showed an active interest in its surroundings and were a good indication that spring was near! As we headed for the carriage road and the Great Gulf, we passed beautiful examples of sastrugi, the bane of polar explorers, but something I had never seen before close up. They looked like small sand dunes, but were as hard as ice. Despite their solid appearance, sastrugi are continually changing in form when the winds blow hard and the air is filled with snow. At high velocity the airborne snow is momentarily melted upon impact but then immediately freezes to form the scalloped structure that I photographed that beautiful Easter morning.

As we reached the route of the Carriage Road I asked the Boss to wait for a moment while I took his picture. I did so and subsequently found that I had obtained one of the finest pictures I have ever taken of a person.

The next day we headed down to the lowlands and home. I don't remember the details, but I suspect we skied down the Carriage Road and near the Half Way, took the trail that leads to Pinkham Notch. Thus ended one of my finest sojourns with the Boss.

The Thermal Deposition of Metal

In the middle of August 1944 I started to look at the possibility of preparing very thin metallic films deposited on solid surfaces. On August 20, 1944, I built a 4-order interference color thickness film of tin oxide on a sheet of mica. When its electrical properties were measured, I found it to have a resistance of 48 ohms per square and easily dissipated 12 to 40 watts when an electrical current flowed through the film. When heated, the mica sheet expanded considerably, thus raising the possibility that this type of film might have some important light-transmission properties.

Two days later, I placed some germanium oxide powder on a glass slide, heated it in the reducing portion of gas flame, and found that a very interesting silver-colored metallic film with bright interference colors was deposited. Subsequently I rigged up a "boat" of quartz by cutting a 3/8" diameter tube in half lengthwise, mounted it so it was surrounded by the reducing zone of a gas flame, and found that I could "paint" a thin germanium film on a flat surface or a rotating glass tube. It was important that the surface on which the film was to be deposited be cold, since the mechanism for deposition appeared to be a thermal transport phenomenon. I then discovered that I could remove the film from its glass backing by exposing it for a minute in the vapor from the surface of concentrated hydrofluoric acid, and then lowering the glass through a cleaned water surface. The film could then be deposited on a Formvar film or another surface by standard surface chemistry procedures.

Such films, when observed in the electron microscope, were seen to be coherent, without any holes or other evidence of damage.

On August 31, 1944, my notebook contains a sketch of a proposed bolometer for detecting infrared radiation, wherein a sandwich of germanium and silica could be placed on a thin pellicle of Formvar coated with gold foil. Five days later, I suggested in the same sequence that a germanium/silica sandwich, in successive layers, might be used to form a thermoelectric cell. Unfortunately I was diverted from completing the cell, and thus may have missed an important application for the utilization of solar energy.

At that time one of the diversions was a request to develop a detection cell made of a very thin layer of Formvar of only 50Å (0.005 microns), on which was mounted a layer of evaporated tungsten. To prepare the tungsten film, Bill Ruggles mounted a tungsten filament in a glass bulb that was then subjected to a high vacuum. A fine uniform film of tungsten soon coated that portion of the cooled glass bulb. The bulb was then broken. I placed a piece of coated glass in the hydrofluoric acid vapor for one minute, stripped it onto a cleaned water surface, and then deposited it on the frame on which the very thin Formvar film had previously been deposited. This resulted in an extremely sensitive infrared radiation detector.

A Liquid Water Content of Clouds Detector

On November 10, 1944, I built a new device for measuring the liquid water content for clouds. Part of it was based on word I had received about Dr. Bernard Vonnegut of the De-Icing Laboratory of MIT. He had devised a cloud water collector, based on a small porous ceramic plug, that was cemented onto a capillary tube bent into an L-shape, with the plug headed into the wind, and the capillary filled with water. As cloud droplets hit the end of the porous plug they were pulled into the interior of the plug by the suction provided by the column of water.

I made such a device and found it to work very well. To my knowledge Dr. Vonnegut had not developed a method for recording the amount of water collected by the plug, except to measure the movement of an air bubble in the capillary tube.

It occurred to me that a counting device might be developed consisting of a method for counting the drops that formed on the inner end of the capillary tube. This could be done visually, but would be quite tedious during an extensive storm.

I found it possible to place another tube below the capillary fitted with a knife edge device which, when contacted by a water drop, would suddenly pull the pendant drop away from the end of the capillary.

If the capillary was made conductive and insulated from the lower drop collector, and the two tubes made part of an electric circuit, the momentary bridging of water between the two could cause a transient blip in the current that could easily be impressed on a moving tape. This was done and worked beautifully. I then encountered a problem when the cloud became supercooled. As soon as these supercooled drops made contact with the porous collector, they would freeze and no longer collect measurable water. Having very low conductivity, the porous ceramic collector could not be heated effectively.

I then learned that porous metal filters had been devised by a company in Dayton, Ohio for use in filtering gasoline as it entered a carburetor. I discovered that samples were in the possession of Tony Nerad, one of our engineers. A quick trip over the bridge connecting Building 5 and 37, and I had the material I needed. With little delay I had a cloud droplet collector that gave every evidence of handling supercooled clouds.

It is my recollection at the present time that the cloud water collector worked beautifully for short periods of time, but that when an attempt was made to run it throughout the full period of a storm, problems developed with it. In retrospect I believe we were wrong to expect our efforts would develop a unit that would function without trouble in some of the worst weather in the world. We should have been content to make short exposure of five minutes or so every half hour, extrapolating between the recordings unless there were obvious changes in the storm intensity that were easily observed.

As it was, despite our troubles we were able to classify storms into stable and unstable systems. The former had values in liquid water content of clouds of 0.2 to 0.3 grams per cubic meter, while the unstable ones, that instead of stratus clouds typified by considerable convective activity, had liquid content values ranging from 0.6 to as much as 1.2 grams per cubic meter.

The period from 1943 to 1946 was very exciting, with stimulating research results at the Observatory. I wouldn't have missed my occasional contacts with it for anything!

A Cloud Height Sensor for a Radiosonde

On February 26, 1945, I was asked by an engineer in the Commercial Division of the Electronics Department of General Electric at Syracuse whether I had any ideas for getting information about clouds using an airborne radiosonde. Since I had been quite interested in this subject because of our studies of supercooled clouds at Mount Washington, I showed him some of the things I was doing in an attempt to measure the various parameters of atmospheric clouds. I told him that I'd think about the possibilities and if I got any useful ideas I'd let him know. Within a week I had built a very sensitive lightweight device that could detect the presence of a cloud and which might be utilized to get some measurement of the amount of liquid water in it. The device consisted of a length of No. 60 cotton thread, silk thread, or untreated cellophane tape (hygroscopic) that was soaked in a sodium chloride solution and dried. When exposed to cloud droplets and then passed over metal contacts connected to a small battery, an electric current would flow and activate a sensor. This would indicate an encounter with a cloud and thus could be used as a cloud height indicator. If two units were used, one in free air and one at the stagnation point of a 1-inch cylinder, a rough indication of particle size could be deduced.

While these findings were significant and of potential value, there were other possibilities, and during the following six months I tried out a dozen or more possibilities.

After many experiments, I decided that my original idea of using a conducting thread was possibly the best method, with one modification. Instead of a cotton thread, I found it to be better to use a fine copper wire (0.003-0.005") coated with one or two layers of cotton covering, which in turn was coated with a salt that, when dry, was non-conductive, but when wet, passed current. I also discovered that the conductivity of the wire seemed to respond to the liquid water content of the cloud. In my little wind tunnel I found that under the same air velocity the following differences in conductivity occurred as I changed the liquid water content of the cloud 0.05G/M3 = 230,000 ohms and 0.2G/M3 = 60,000 ohms.

This radiosonde device development was disclosed in a patent letter to the Director of GE Research Laboratory on November 20, 1945. However, as often happens, with the ending of World War II, the funding for the GE Stratometer (as the device was to be called) was canceled and nothing more happened.

Strong Structures of Permafil

About mid-April of 1945 I had a mast made of a new material developed in the Research Laboratory called Permafil. This was a substance that at the present time would be called epoxy. I had found that extremely tough, strong and durable remarkably light structures could be made in the following manner: a wooden form shaped like what was wanted for the shape of the finished product was made, and then wrapped in a sheet of cellophane, and on top of this a strip of glass cloth would be placed on top of the cellophane layer. This cloth, before wrapping, was soaked in Permafil resin and wound onto the wooden form until the thickness was increased by about 0.030". While still wet with Permafil, the fabrication was wrapped with a metal ribbon or, as I mentioned in my notebook, I suggested that the glass cloth be woven with fine wire serving as the warp. After the structure had partially solidified, I suggested that the entire structure be coated with a viscous layer of Permafil so that it would have a glossy finish when the solidifying reaction was completed. Once the Permafil hardened the wooden form could easily be withdrawn and the cellophane peeled from the inner surface of the fabricated surface.

The mast I made for a Mount Washington instrument was so strong and permanent (I still have it after forty years) that I was greatly impressed with the simplicity of the process and the almost unlimited range of objects that could be made from it.

As an example I built a decelerator to be used for collecting ice crystals at the nose of one of the Project Cirrus B-17 airplanes. It consisted of four funnel-shaped objects, having holes of one-inch diameter, and shaped like an expansion orifice. The purpose was to "spill" the oncoming snow filled air, which then permitted the crystals to act as projectiles that would enter the cold area of the nose (the former bombardier's position), where the crystal could be collected without danger of breakage.

This device was used successfully on a number of occasions, including one when it was taken up to the cirrus levels of the atmosphere to collect the ice crystals that formed in cirrus clouds. We found them to be the same as was found during our local "positive current" storms when cirrus type crystals reached the ground.

I made a number of structures with this technique, finding it possible to fashion all kinds of complicated forms without the slightest difficulties.

I could see the unique possibilities of this new technique and tried to find someone in the Company who could envision the sales potential. However, with the pressures on all sides due to World War II, and even though it looked like we were winning it, I couldn't find anyone who could sense the multimillion-dollar business that could and would develop. The fiberglass yachts, autos and similar structures were too far away and it was a quarter century before the engineers caught up with its potential and began to use it.

An Electrically Conducting Film

On October 24, 1945 I concocted a formula for producing a conducting film of graphite flakes and powder suspended in a silicone resin, to which was added ethylene dichloride. The ratio of powder-to-flake was 2:1, while the two liquids each had weights equal to the powder.

This material could be applied by dipping, painting or spraying, and produced a tough coherent film which had a conductance of 50 to 100 ohms per square, dependent on the thickness of the applied film. Such a coating had fascinating properties and potentialities. It could be painted or sprayed onto a non-conducting surface with strips of flat conducting electrodes cemented to opposite sides of the coating, and could be heated to fairly high temperature. It could be sprayed onto cylinders and other structures of Permafil and heated to serve as de-icing surfaces at Mount Washington. I carried out a series of developments with my assistant Robert Smith-Johannsen. We made room-heating panels and found them to be very effective.

After working with me for another year or so, Bob Johannsen,[30] who was an excellent colloid chemist, transferred to the Silicone Department of the company and became deeply involved in the development of silicone rubber and other similar products. He then left the Company and formed his own small development business. Before long he decided to spend considerable time with the further development of conducting films of graphite.

---

30 Johannsen was rescued by Langmuir in 1944 when he was trapped in Nazi-occupied Norway. By using the Norwegian underground and the American War Department, he was smuggled out via neutral Sweden. Johannsen, a Canadian, invented ski wax, and Langmuir thought he could help him in designing de-icing techniques for war planes.

The local power company agreed to build a special house for him that would be heated electrically using conductive wallpaper. In some rooms he also had ceiling and floor heaters and even had an electrically heated driveway!

The house "worked" exceedingly well, but was more than twenty-five years ahead of its time.

The nicest part of this type of home heating is that when one enters a room heated radiantly, it feels like the sun is shining. The temperature of the heating surface above ambient is hardly noticeable, and yet the living environment can be adjusted to whatever temperature is desired.

Bob eventually moved to the west coast and sold his business to Chemelex Corporation, where he remained in a special laboratory built for his exclusive use. His electrical heating tape could be wound around pipes carrying chemicals to batch mixers and adjusted precisely so that the chemicals arriving at the point of reaction would be at the optimum temperature. He also devised a waterbed heater that was so efficient and effective that thousands have been sold for this purpose.

With all of this activity Bob continues to have an active interest in the fabrication of ski waxes. His "Jack Rabbit" brand is named after his father, who at the age of more than 100 years was still racing cross-country in Canada!

Bentley's Method of Measuring Rain Drops

After discovering a way to replicate snow crystals and other ice forms, I became aware of the work of Wilson Bentley, the farmer from Jericho, Vermont, who made such beautiful photomicrographs of snow crystals. I bought his book that was sponsored by the American Meteorological Society and used his pictures as a standard for my own studies. In the course of my replica work I made many photomicrographs of actual crystals to compare them with pictures of replicas made from the same crystals. I found it possible to prepare replicas that looked the same as the originals.

In my reading about Bentley, I was much impressed with his versatility and the breadth of his interests in meteorological phenomenon. At one time I drew Duncan Blanchard's[31] attention to this remarkable natural scientist and was particularly pleased to observe Duncan's enthusiastic response. This has persisted and now he probably knows more about Bentley than anyone in the world!

One of the things that Bentley discovered was in the realm of raindrop size distribution. He found that by putting freshly sifted flour in a shallow pan, exposing it to falling rain, and then putting the pan into a hot oven, dough pellets would form related to the size of each raindrop. These small baked flour pellets could then be screened out of the flour and a size distribution of the falling rain would be established.

---

31 Duncan published a book about Bentley called "The Snowflake Man: A Biography of Wilson A. Bentley," McDonald & Woodward Publishing Co., 1998.

I tried Bentley's technique and found it to be highly useful and an elegant method.

I discovered one problem with the method. In measuring the mist-sized droplets that often accompanied rain from violent thunderstorms, I found it difficult to differentiate between crusty aggregates of flour that passed through the fine mesh of the flour sifter and actual pellets formed by the mist. I solved this problem by adding some methylene blue dye to the flour as it was being sifted. It was then an easy matter to separate out the mist particles.

The beauty of Bentley's photomicrographs, and the widespread publication of them as Dover reprints, gives the general public a somewhat biased view of the nature of falling snow. Anyone who observes individual particles of snow from a number of storms and at different periods in the course of a storm soon discovers that a fall of large perfect stellar crystals is a rare event indeed!

I have found after many years (45+) of detailed observation, that less than 1% of falling snow fit that category. The best conditions for finding perfect crystals in the eastern Mohawk Valley is at night or early morning when a very light snowfall is occurring and the temperature is warmer than 20°F. The exception to this generalization is a fall of cirrus crystals that tend to be mostly hexagonal columns. While they are different and beautiful in their own right, they do not have the esthetic appeal of the larger hexagonal plates and stellar crystals.

An Air Mover Formed by Serendipity

On April 23, 1946 I conducted a few experiments in supercooled cloud generation outdoors using a very high-speed motor having a tiny circular saw in the mandrel. By touching it to the surface of water cooled close to 0°C, a dense cloud formed above the water surface, which drifted away into the cold air and evaporated. While doing this I noticed that a strange ice formation began to form on the notched saw, having sharp pointed spikes extending out from each tooth of the circular saw.

I decided to explore this mechanism in more detail, so I clamped the motor so that the circular saw was horizontal and at the top of the assembly. Starting the motor and letting it come up to maximum speed I used a medicine dropper to apply successive drops of cold water to the spinning saw. The outdoor temperature was quite cold and I soon saw a rather large daisy-like ice formation develop on the rotating saw. When it got to be about three inches in diameter I noticed that astrong current of air was rising above the rotating saw. When I stopped the motor I saw a very strange icy structure. Each tooth of the saw initiated a curved ray of ice that apparently was an effective air mover.

After making the structure of frozen water, I tried making it of wax and metal. It looked like both could be successful. However, each substance posed special problems for making a coherent and uniform structure. Since there were other research projects needing more urgent attention I "put it on the shelf" for future attention and haven't yet got back to it!

The Prevention of Frost Damage to Orchards

In the spring of 1946, a friend who owned a large commercial apple orchard north of Schenectady asked me whether or not the smoke generator we had invented during World War II could be used to protect apple buds from frost. I told him that this would be impossible because the smoke particles produced by the generator were smaller than a micron, far too small to prevent the radiative cooling of the orchards that occurred on clear sky nights and led to frost. In order to protect the orchards from the cooling effect of radiation it was necessary to hold back the infrared rays of the spectrum, and this could only be done with particles having a diameter in the 10-15 micron range. Thus it would be necessary to produce particles of that size, and that could most easily be done with a water spray. Unfortunately with most spray devices then available, the 10 micron diameter water particle was about the smallest that could be produced, most of the water spray consisting of much smaller particles. Despite these shortcomings my friend decided to try it out. With a forecast of frost I joined him and his crew one evening about midnight. He planned to use his new insecticide sprayer using only water for making a fog. The sprayer actually did produce a fairly good fog, but most of the particles were so large that the fog had a very brief lifetime and after a frustrating hour of effort we abandoned the attempt. Fortunately the temperature remained close to freezing, and no frost damage occurred.

There is an erroneous idea among many orchardists that the smudge pots used in Florida and California to prevent frost damage are effective because of the black smoke that is produced by the burning of fuel oil. This idea is erroneous, since the black soot particles are also quite small. These orchard heaters are just that– they heat the air in contact with the trees and thus produce a micro-environment that is often effective. If a smudge pot could be made that would be smokeless, it would do a more effective job.

At the present time (1986) there is a device that does produce a stable, highly efficient infrared screening fog. Its inventor is my friend Tom Mee of Altadena, California, who has invented a nozzle that has an orifice so small that when water is sprayed from it at very high pressure, all of the particles are in the range of 10-15 microns. Fogs produced by it are like ordinary clouds and natural fogs and, consequently, are quite effective in preventing radiative cooling, and thus the residual heat of the earth is retained.

With my interest centered for the time being on protecting fruit orchards from frost damage, I began to search for other methods of protection. I had learned that damage occurs when the buds have swelled and fruit blossoms have opened. I began to wonder whether there was any way to hold back this biological process.

I wondered whether it would be possible to delay the onset of the blossoming cycle. Accordingly I decided to coat some dormant buds with a plastic film. I sprayed a thick layer of polyvinyl alcohol, a water-soluble coating, on several branches of a tree that had buds that had not started to open. I left the rest of the tree as a control. Within a couple of weeks the buds on the main tree swelled and the leaves began to develop on all but the imprisoned ones on the branches sprayed. They looked to be completely dormant, and I began to wonder whether I had in fact damaged the branches. However, with the occurrence of a warm rain the imprisoned buds were released, and they opened in a normal fashion.

Thus it is quite possible that frost damage could be prevented by use of such a plastic spray. However– the economics of the process must be given careful consideration. Until we have better two- or three-week weather forecasts, which are reliable, I doubt whether this method would be adopted. It is another discovery that I have "put on the shelf" for future consideration!

My Discovery of Dry Ice Seeding

During the spring of 1946, using a 6 cubic foot GE Cold Chamber, I made a concerted effort to modify a supercooled cloud. I found that the temperature of the chamber varied from -10°C at the top, to -20°C at the bottom, and -15°C near the middle.

I dusted into the chamber dozens of materials, ranging from chemical dusts to natural soils. While some of these formed a few glinting ice crystals, none of them showed any encouraging effects.

I found the cold chamber to be an ideal testing device, once lined with black velvet and illuminated with a concentrated filament lamp that cast a shaft of bright light at a 45° angle across the length of the chamber. A single ice crystal could readily be detected in the super cooled cloud in the chamber. Exhaling one's breath into the chamber and directing it toward the bottom could easily produce such a cloud. The temperature inversion in the chamber was so intense that the turbulence generated by the moist breath would die out in less than 10 seconds. I found it important in supplying moisture with the breath, to exhale rather than blow. The latter entrains warm, drier air from the air surrounding the top of the chamber, and thus tends to warm the supercooled cloud.

By early July I had reached an impasse in my experiments and was trying to figure what to do next. Then suddenly a serendipitous event occurred. The morning of July 12 was humid and quite warm. Checking the temperature of the cold chamber I found it to be warmer than usual. To cool it down I got a large chunk of dry ice from our storage bin to overcome the warmth of the chamber. The instant the dry ice was lowered into the chamber I saw to my delight that the supercooled cloud had been displaced by a strange bluish fog unlike anything heretofore seen. Quickly lifting the dry ice from the chamber I introduced moisture from my breath and gradually decreased the density of the fine particle blue fog until I could see the glinting of incredible numbers of ice crystals. After spending five minutes or more growing these crystals, I finally was able to produce a supercooled fog.

November 13, 1946. Vince demonstrates his famous ice box experiment discovering cloud seeding. Courtesy of Schenectady Museum.

I then put a smaller piece of dry ice in the chamber only to see it again revert to a bluish fog. Repeating my procedure to cause the fallout of the ice crystals, I then held the chunk of dry ice above the chamber, scratched its surface with a nail and saw a magnificent cluster of condensation trails appear in the supercooled cloud through which the tiny grains of dry ice had fallen. Once I had this incredible phenomenon under control I summoned all of my "neighbors" to see the spectacular display. Unfortunately the Boss was on the west coast on a lecture tour, and it was nearly a week before he returned to see it. Needless to say he was very excited when I showed it to him, and he summoned a number of his friends to see it.

He told me I should make plans to conduct some experiments with atmospheric clouds, and I soon arranged for the cooperation of the GE Flight Test Center at the Schenectady County Airport. The Director of the Center, Mr. Curtis Talbot, volunteered to take me up in his Fairchild Cabin plane, so I designed and built a motor-driven CO_2 dispenser that I planned to use for dispensing crushed dry ice as the plane flew over or through the atmospheric cloud.

As tends to happen with most atmospheric experiments, the fall of 1946 was a period of clear skies! Day after day in October and early November I would get up early, look at the sky and feel disappointed. This continued until the morning of November 13, when a series of parallel bands of clouds were in the sky over our area that appeared large enough to be seedable, and high enough to be supercooled. I called Curt Talbot, headed for the local dairy to obtain the dry ice needed, alerted Langmuir, and was able to take off from the Schenectady Airport at 0930.

My Laboratory Notebook Comments on My First Cloud Seeding Flight

The following is a transcription from an entry in my GE Research Notebook #3986, pages 91-93, and dated November 14, 1946:

> – "Yesterday morning at 9:30 AM Curtis G. Talbot of the GE Flight Test Division at the Schenectady Airport piloted a Fairchild cabin plane taking off from the east-west runway. I was in the plane with Curt with camera, 6 pounds of dry ice and plans for attempting the first large scale test of converting a supercooled cloud to ice crystals. As we took off, the ground temperature was 6°C. In the sky were long stratus clouds isolated from each other and at an altitude of what appeared to be about 10,000 feet. We started climbing immediately and continued for more than an hour. As we climbed we passed through a temperature inversion, but at about 8700 feet emerged from it and from then on followed a typical dry adiabatic lapse rate. The following is the time temperature altitude data.

Time (EST)	Altitude (feet)	Temperature (°Celsius)
09:30	300	6.00
09:40	2500	3.50
09:44	3700	-0.50
09:48	4500	-4.50
09:49	5000	-6.00
	6200	-6.00
09:54	7500	-5.00
	8200	-3.00
10:00	8500	-4.30
10:03	8700	-4.20
	9200	-5.20
10:05	9500	-6.00
	9700	-7.00
10:08	10000	-7.50
10:15	10500	-9.00
10:18	11000	-11.00
10:22	12000	-13.00
10:34	13600	
10:37	14000	-17.50

Note: About 1°C heating due to compression was produced so that all of the temperatures should probably be reduced by about 1°C. Thus at 14,000 feet, the temperature was -1.3°F or -18.5°C.

When we reached 10,000 feet, I pointed out a cloud to Curt that looked as though it would be a good one to work on, and asked him if he thought we could get up to it. He nodded in the affirmative and we started to head toward it. As we approached the cloud I took picture #1 as we headed toward the southern end of the cloud. Passing close to the base at 13,600 feet, I looked toward the sun and observed a beautiful colored corona, and saw it gradually disappear as a sharply outlined disc until it completely disappeared. I also noticed brilliant iridescent colors in portions of the cloud edges, showing also that we had here a supercooled cloud ideal for our experiment. Reaching the southern end of the cloud, the better part of the sun being obscured by the lower part of the cloud, I read the temperature of -17.5°C. At this point the temperature bulb showed a light deposit of ice. Swinging around, Curt flew into the cloud and I started the dispenser in operation. We dropped about three pounds and then swung around and headed south.

About this time I looked toward the rear and was thrilled to see long streamers of snow falling from the base of the cloud through which we had just passed. I shouted to Curt to swing around and as we did so we passed through a mass of glinting snow crystals. We then saw a brilliant 22° halo and adjacent parhelic.

These latter were sharply defined and seemed to be a degree outside of the halo. We made another run through a dense portion of the unseeded cloud, during which time I dispensed about three more pounds of crushed dry ice (pellets from 5/16", down to sugar size). This was done by opening the window and letting the suction of passing air remove it. We then swung west of the cloud and observed draperies of snow that seemed to hang 2,000-3,000 feet below, and noted the cloud drying up rapidly very similar to what we observe in the cloud box in the laboratory. I then took picture #2. While still in the cloud, we saw the glinting crystals all over; I turned to Curt and we shook hands as I said, "We did it!" Needless to say we were quite excited.

The rapidity with which the CO_2 dispensed from the window seemed to affect the cloud was amazing. It seemed as though it almost exploded, the effect was so widespread and rapid.

After heading back for Schenectady, we made one more circle during which I took several more pictures.

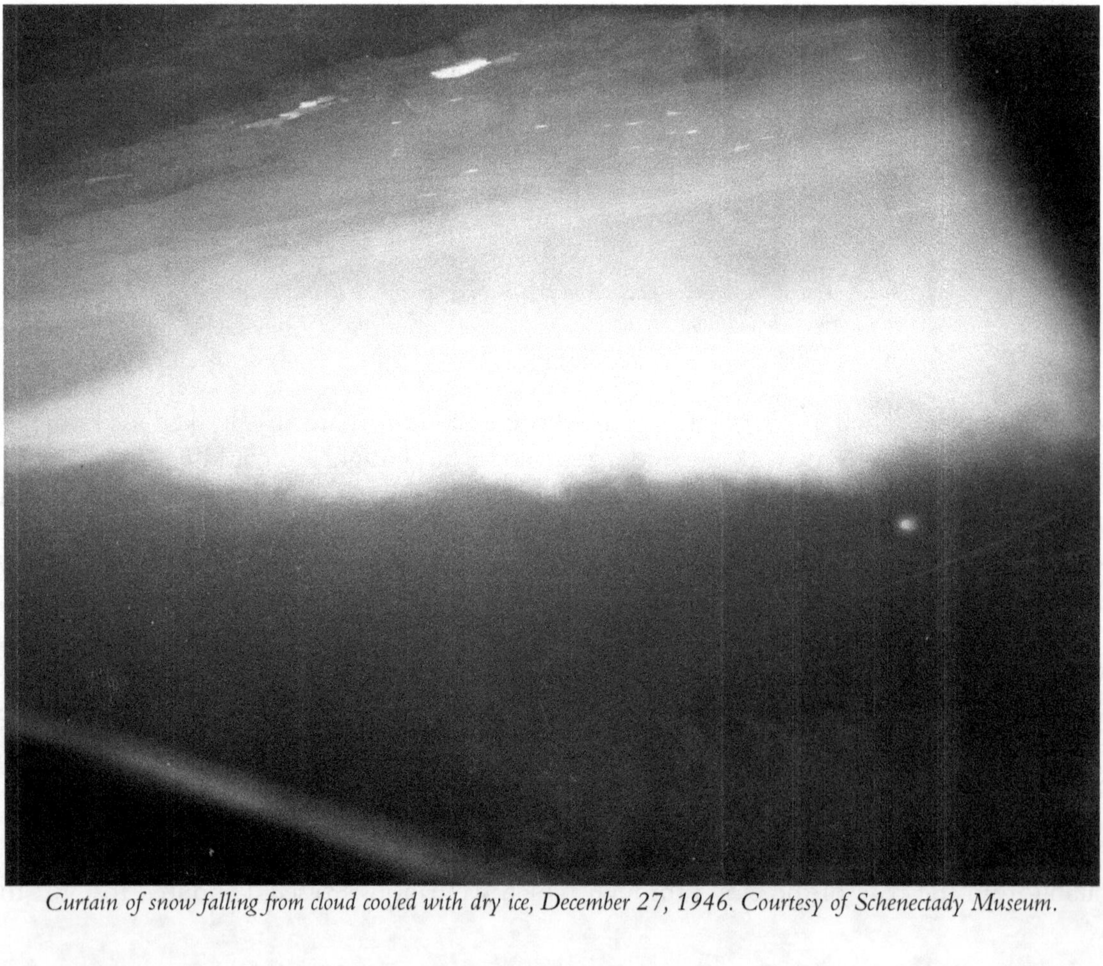

Curtain of snow falling from cloud cooled with dry ice, December 27, 1946. Courtesy of Schenectady Museum.

When we arrived at the port, Dr. Langmuir rushed out enthusiastically, exclaiming what a remarkable view they had of it in the control tower of the GE Lab. He said that in less than two minutes after we radioed that we were starting our run, long draperies appeared from the cloud vicinity.

The above entry in my Laboratory Notebook was signed "Read and Understood R. Smith Johannsen, November 19, 1946." Apparently it was five days before I had Bob witness my entry.

In my Laboratory write-up I failed to record the location where this first cloud seeding flight was made.

As soon as I took my final photos, I asked Curt where we were and he told me we were over western Massachusetts. I then recognized Mount Greylock with its tower, and realized for the first time that our clouds had been traveling quite rapidly from the west. This was the first time I had viewed eastern New York from the air, and was greatly intrigued by the terrain below. On our return flight to Schenectady we made a long glide across the region northeast of Albany, as we flew to avoid the air corridors around the Albany Airport. I found myself searching the terrain for possible Indian sites and examining areas where I had hiked in the past. Thus ended a fascinating experience. While I was elated with the results, it was almost an anti-climax, since I was almost positive the flight would be successful.

I could not envision the countless serendipitous events that lay ahead and are still happening. Just a few days ago I was lecturing to an Elderhostel group at Becket, Massachusetts, almost in the shadow of Mount Greylock, and was asked to recount my observations and reactions to the flight that occurred more than 39 years before.

The Seeding of Supercooled Ground Fog

At early morning on November 21, 1946, as I got up from bed, I glanced down at the Flats below my home and saw that a dense fog had formed during the night. This is a common occurrence on the Great Flats in the fall of the year. Since the temperature outside my window was 32°F (0°C) I knew that the fog over the Flats would be supercooled, since the temperature at the bottom of the hill ranges from 5° to 10° colder than at my house that is on a terrace above the flood plain.

I had been waiting for such an occurrence since my seeding flight of the 13th, so I quickly dressed and headed for the Bellevue Dairy that was my main source for dry ice. To my dismay they had none on hand, so I quickly drove to the only other one in Schenectady– the Sealtest Dairy. Fortunately they had some so I bought a slab (5¢ per pound) and hurried the 2+ miles out the River Road to my home.

Arriving there, I quickly made a small bag out of mosquito screen, tied a rope to it, and headed down the slope to the Flats and the fog. Reaching the edge of the fog I started swinging the dry ice container fastened to the end of the rope above my head as I walked across the field. After going about 50 feet I looked back and saw that I had generated a much denser fog and that, in the early morning light, it had a peculiar appearance. It looked as though I was in a different world. The apple trees nearby had an almost luminous appearance as the pre-sunrise light was reflected from the myriads of tiny crystals now in the air. I continued to walk back and forth, transforming other regions of fog until a large area had this strange appearance. Then all of a sudden the fog was gone as the crystals, that had grown larger than the fog droplets, settled to the ground! As the sun rose above the horizon, tinier crystals were carried aloft in the gentle convective columns that began to form, continuing to grow in the moist air, until finally the air of the entire region along Schermerhorn Road was filled with Diamond Dust that slowly settled to the ground.

As it did so, the twinkling crystals produced a sun streak and a sun pillar toward the rising sun.

Years later, at Yellowstone in the winters of 1960 to 1971, I carried out dozens of similar experiments, but on a much larger scale. We still used the whirling-basket technique, positioning the seeder in the supercooled fog near Old Faithful, and observing the effects and collecting crystals a mile or more down the Firehole River.

One of the nicest features at Old Faithful was that the down-slope westerly drift of foggy air that developed after sundown, and continued until about 10 o'clock the next

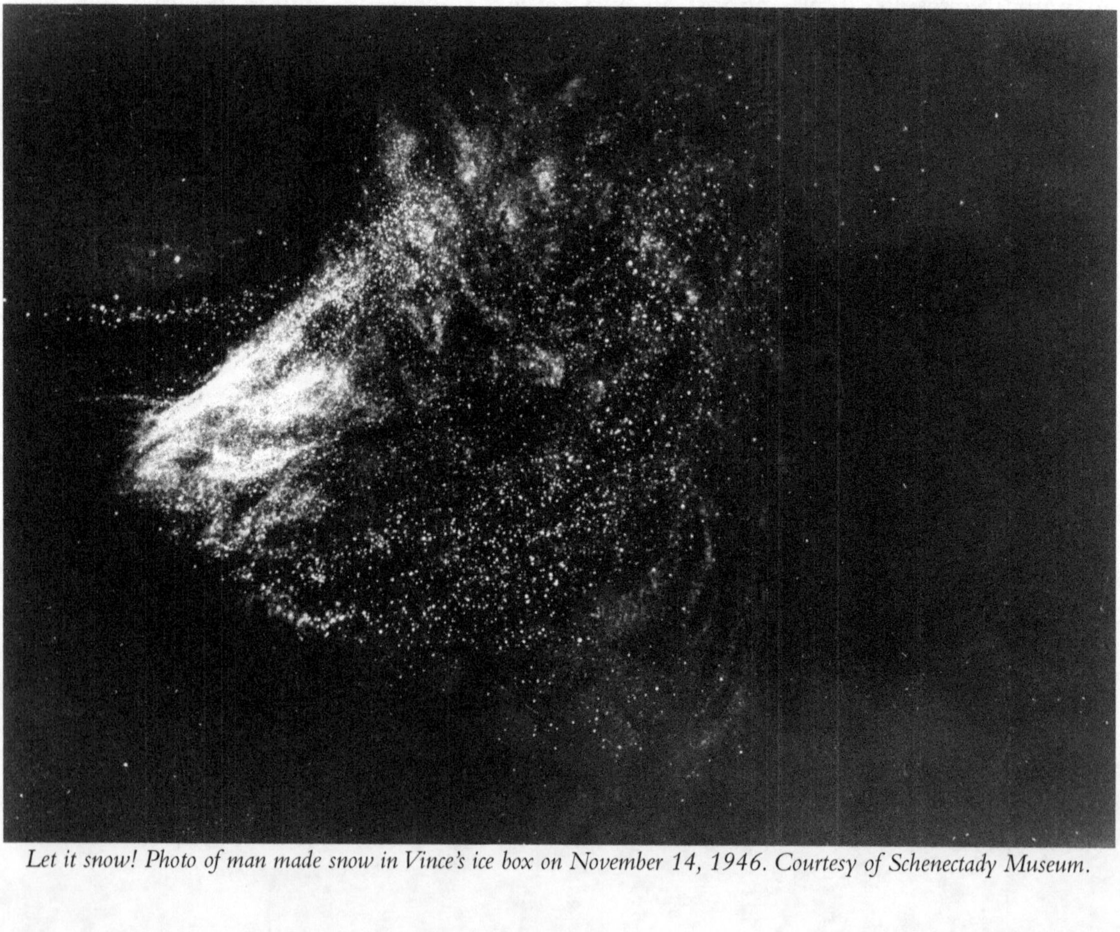

Let it snow! Photo of man made snow in Vince's ice box on November 14, 1946. Courtesy of Schenectady Museum.

17-year-old Kathleen Roan shows Vincent J. Schaefer her snow from a cloud making machine. Courtesy of Schenectady Museum.

morning, would suddenly reverse direction and carry some of the crystals we produced earlier back in an easterly direction, so that the crystals we then collected were quite large.

The first year (1960) I went to Yellowstone, it was necessary for me to fly to Jackson Hole. On the way from Denver in a DC-3 we landed under marginal conditions at Rawlins, Wyoming, and soon after the field was engulfed in fog. My chances for continuing to Jackson Hole looked pretty grim. After being parked in the fog for nearly an hour or more a larger plane loomed out of the fog parked near us and the pilot (who must have been a savvy bush pilot, perhaps from Alaska) put on his brakes and revved his engines at high speed for ten minutes or so. Within a few minutes the fog cleared locally and he took off, followed by us! He must have had a reason for this maneuver, although I never learned whether he realized that his propeller tip vortices were producing myriads of ice embryos, which then grew and for the duration of their falling process, cleared the airport. Reaching Jackson Hole we landed in a snowstorm, but visibility was not restricted by ground fog!

At another time, after finishing one of my Yellowstone Expeditions in the early '60s, we returned to Salt Lake City from Idaho Falls to find the airport in a dense supercooled fog, with all of the in and out flights canceled by all the airlines. To make matters worse, it was the type of fog that sometimes persisted for a week or more. I photographed the flight departure board of the United Airline with every flight marked canceled. Fortunately, by waiting around, there was a momentary period of clearing, during which my flight for Chicago was able to take off.

Vince's experiment reproduced on November 14, 2010 by Don Rittner and Jim Schaefer, during 64th anniversary of cloud seeding invention. Photo by Don Rittner.

Upon returning home I sent a picture of the cancellation legends to my friend Henry Harrison, Chief Meteorologist of United Airlines, and told him that a small airplane loaded with dry ice could quickly dissipate this fog.

Within a short time such a seeding operation was established, and I believe is still used whenever a supercooled fog forms at Salt Lake City and other airports where such a fog causes problems.

The First "Flight" of Project Cirrus

Shortly after the agreement was made between officials of the General Electric Company and the Defense Department that military personnel would become involved in responsibilities for cloud seeding activities in the early winter of 1946-47, a B-26 Hurricane Reconnaissance airplane was sent to Schenectady, with the assignment to provide us with aerial support for some of our preliminary experiments. This followed the order from GE administration, on the recommendations of the Legal Department, that I was to cease all further aerial experiments.

While I was not adequately briefed on the procedures to be followed, I decided it would do no harm to make an exploratory flight to find out the possibilities and limitations that might be expected of a multi-engine airplane. After getting my equipment together, I went to the Schenectady County Airport, fastened on a military parachute, boarded the plane, and was quite fascinated to watch and hear the preliminary checkout of the pilot with the copilot.

After warming up the engines, we were cleared for takeoff. Prior to this point, the flight personnel were loud in their complaints about the cold weather, after having spent several years in the warmth of Florida.

Finishing their flight check, they moved over to the flight runway and, with clearance from the tower, applied power to the twin engines. I braced myself for takeoff. Just as our speed was high enough to lift off, both engines stopped without warning.

Needless to say, there were a few tense moments as the brakes were applied and we slowed down. Fortunately Schenectady Airport is blessed with extra long runways that had been made for B-29 operations. Thus we had runway to spare as we finally stopped.

Both pilots were shaken by this new and unexpected development. Their basic reaction was "Let's get out of this part of the country!" This was the end of our initial contact with the military. Within a month or two we were assigned the use of two B-17s, with appropriate crew, along with the personnel of the US Signal Corps that was charged with working with Langmuir and me as advisors.

From 1947 until our Project Cirrus activities ended five years later, we had no further problems. Excellent cooperation was enjoyed between the military personnel, both officers and technicians, and the civilian staff of the Signal Corps. It was an exciting and fun operation, and fortunately was thenceforth free of any trauma or anxiety on our part.

My Friendship with Bernie Vonnegut

I first met Bernie Vonnegut[32] when he was working on a cloud physics project at the De-icing Laboratory of MIT. I was greatly intrigued with the elegant devices he had developed for measuring the liquid content of atmospheric clouds, and was delighted when he joined the staff of the General Electric Research Laboratory. He initially worked with Dr. Herbert Hollomon on supercooling phenomena.Before long, our mutual interests of ice nucleation and supercooling brought us together, and by mid-1946 he was working closely with our group.

Shortly after I discovered the dry ice effect and demonstrated its effectiveness in seeding clouds, Bernie, after reviewing the crystalline structures of many chemicals, decided that silver iodide should work as a nucleus since its crystalline structure matched that of ice crystals. To his disappointment it did not appear to be effective. Meanwhile I had found that when iodine vapor was introduced into an open cold chamber, large numbers of ice crystals were formed. Bernie didn't give up with silver iodide, however, obtained a purer sample, and discovered it did in fact serve as an excellent nucleus for causing ice crystal formation. At the same time I continued to study the effect of iodine and soon found after

32 Bernie's famous Brother Kurt, the author, also worked for a short time at GE, and as has been found in his books, he did not like his sojourn there.

Bernard (Bernie) Vonnegut, brother of writer Kurt Vonnegut, demonstrates smoke generator in 1950. Bernie discovered that silver iodide made a great material for seeding clouds. Courtesy of Schenectady Museum.

I had isolated it from the air of the laboratory, that its effect diminished and eventually disappeared.

At the time I assumed that there was sufficient sparking of laboratory switches that used a silver alloy as being the source of the metal particles that combined with the iodine to form silver iodide. Some years later in my little laboratory at Fleischmann Hall at the Research Center of the Museum of Northern Arizona I discovered that the lead particles from automobile exhaust were so pervasive that even many miles from automobile roads they were still in sufficient numbers to cause large effects in a cold chamber.

Bernie Vonnegut continued his inventive ability throughout our Project Cirrus activities and developed a number of fine inventions. One of the most interesting was his vortex thermometer. He found that by admitting air into a temperature-sensing device so that it produced a vortex, the errors in velocity measurement, which increased as the plane went faster due to compressive heating, was compensated for when a vortex developed, and his little device gave a true temperature reading. Another device involved the photocell monitoring of a hydrogen flame while sampling ambient air. The normal blue flame produced a yellow color whenever a tiny sodium chloride particle went through the flame. As a result of this device, he could easily identify maritime air at Schenectady even though we were at least 150 miles from the nearest ocean.

Bernie's laboratory space, as well as his desk and even the floor, was cluttered. Every conceivable item having any relationship to his current and past interests could generally be seen within the layers of his equipment. Along with this comment I may usefully interject a study that "Doc" Whitney conducted in his late thirties. He asked Art Westfall, the official photographer of the Laboratory, to take pictures of the workbenches and desks of all the professional staff. Whitney compared the neatness or degree of clutter with the scientific output of each individual. He discovered a close correlation between the ability of the scientist and what seemed to be the "mess" of the desk and workspace. The greater the mess, the higher the scientific achievements. In commenting on this finding, Whitney likened what seemed to be a "mess" to a casual observer as a series of open notebooks that for some reason had not been finally written. By having "notebooks" open, the researcher was constantly being reminded of unfinished business. Thus when a flash of genius occurred everything was in readiness to relate to it.

Bernie's cluttered work area led him finally to leave General Electric. When in the early fifties a new "more efficient" administration began to organize research teams, one of the proponents looked at our laboratory space that I shared with Katy Blodgett and Bernie, and left with raised eyebrows after he viewed Bernie's area. Sometime later when Bernie was on vacation, he returned with a crew and put "order" into Bernie's work area. When Bernie returned to the laboratory he became furious with what had happened and soon afterward he left the Laboratory to take a job as Research Associate at the Arthur D. Little Laboratories in Cambridge, Massachusetts, a very prestigious firm.

Bernie worked for "ADL" until the mid-sixties, when I approached him with an invitation to join me at ASRC. He came without urging and has been with us ever since.

It is an interesting reflection on Doc Whitney's findings that Bernie's workspace now looks like it did in the early days of his sojourn at the Laboratory of GE, and his discoveries and outstanding graduate students continue without pause!

Determination of the Homogeneous Nucleation Temperature of Water

After discovering the effectiveness of dry ice in the nucleating supercooled clouds, I felt it desirable to establish, if possible, the threshold of temperature at which homogeneous nucleation occurred.

I obtained a cold chamber that could produce temperatures as cold as -55°C and carried out a large number of experiments, during which I varied the supercooled cloud temperatures from -35°C to about -55°C, I found that the transition temperature was close to -38° to -40°C. The effects were quite dramatic. It was quite possible to have a supercooled cloud without a trace of ice crystals at -38°C. At -40°C this was quite impossible. The moisture applied to the chamber always produced a dense cloud of ice crystals.

I had considerable difficulty in making sure of the threshold temperature accuracy since I could never be sure that I was measuring the coldest part of the chamber.

It then occurred to me that by using pure substances solidified, placed on the end of a rod and passed through a supercooled cloud at -20°C, the critical temperature might be established by using them at their melting point. In checking the tables of inorganic chemicals, I found that mercury had a melting point of 38.87°C. Since this seemed to be quite fortuitous, I immediately obtained a small drop of mercury, put it at the end of a capillary tube and cooled it by holding it against a chunk of dry ice. When it froze it had a different appearance that in its liquid state.

Once frozen, I passed it slowly through a supercooled cloud and found it would leave a trail of tiny ice crystals in its wake. Continuing to pass it through the cloud as it warmed up, I noticed that as its dull sheen changed to a more silvery appearance, the seeding effect stopped.

It then occurred to me that the use of a mercury thermometer might be the ideal arrangement to tie down the threshold temperature. Accordingly, I obtained a mercury thermometer that was calibrated to -36°F and had a mark that indicated its freezing/melting point at -38°F (-38.89°C). I put the bulb of the thermometer against a chunk of dry ice and watched the thread approach the freezing point mark. It approached the mark but kept right on going! I watched as it reached a region that must have been -41^{0} or -42°F. It suddenly stopped its downward course and then moved back up to the mark where it should have frozen. What I had witnessed was the supercooling of mercury and then its nucleation and return to its proper location.

While the mercury column remained at its solid state mark, the bulb produced vast numbers of ice embryos that quickly grew to a size that scintillated in the light of the illuminator. As soon as the mercury column began to show signs of movement, the nucleation stopped. I repeated this experiment a number of times and concluded that the critical temperature must be close to the melting point of mercury, that is -38.89°C (38°F). For all practical purposes we now say it is -40°C (-40°F).

The Prevention of Ice Formation on an Airship

While in New York City in mid-February of 1947, I presented a paper at a scientific meeting on "Methods of Dissipating Supercooled Clouds." In the audience were Captain Rosendahl and Rear Admiral Settle of the Lakehurst Base of the Lighter Than Airships of the Navy.

Admiral Settle asked me if I thought it would be possible to prevent ice forming on an airship when it flies through supercooled clouds. I told him that I thought it would be feasible and would be glad to offer suggestions as to the approach to be followed. A few days afterward I received a telegram from the Admiral informing me that Ensign Rosendahl would visit my laboratory, at which time I could provide details on the method that I believed could alleviate the icing problem.

When he came to the laboratory, I suggested that the procedure I would use would be to place a boom in front of the airship on which would be mounted an ice nucleus generator. This could be a manifold out of which would stream dry ice, or it could be an assembly of Hilsch Tubes that would put out a concentration of ice embryos, on the order of 100 per cubic centimeter.

To put this in perspective: if the cross section of the airship was 30 meters in diameter and the airship was moving 20 miles per hour and it was necessary to generate 100 ice embryos per cubic centimeter, then the volume needing treatment would be about 1,000 square meters x 10 meters or 103 x 101 x 106 x 102 = 1,012 ice embryos per second. Since dry ice could generate that many embryos in a hundredth of a second per gram of dry ice, the application would require on the order of 1 gram per minute. Thus a pound of dry ice could theoretically protect the airship from icing for at least five hours.

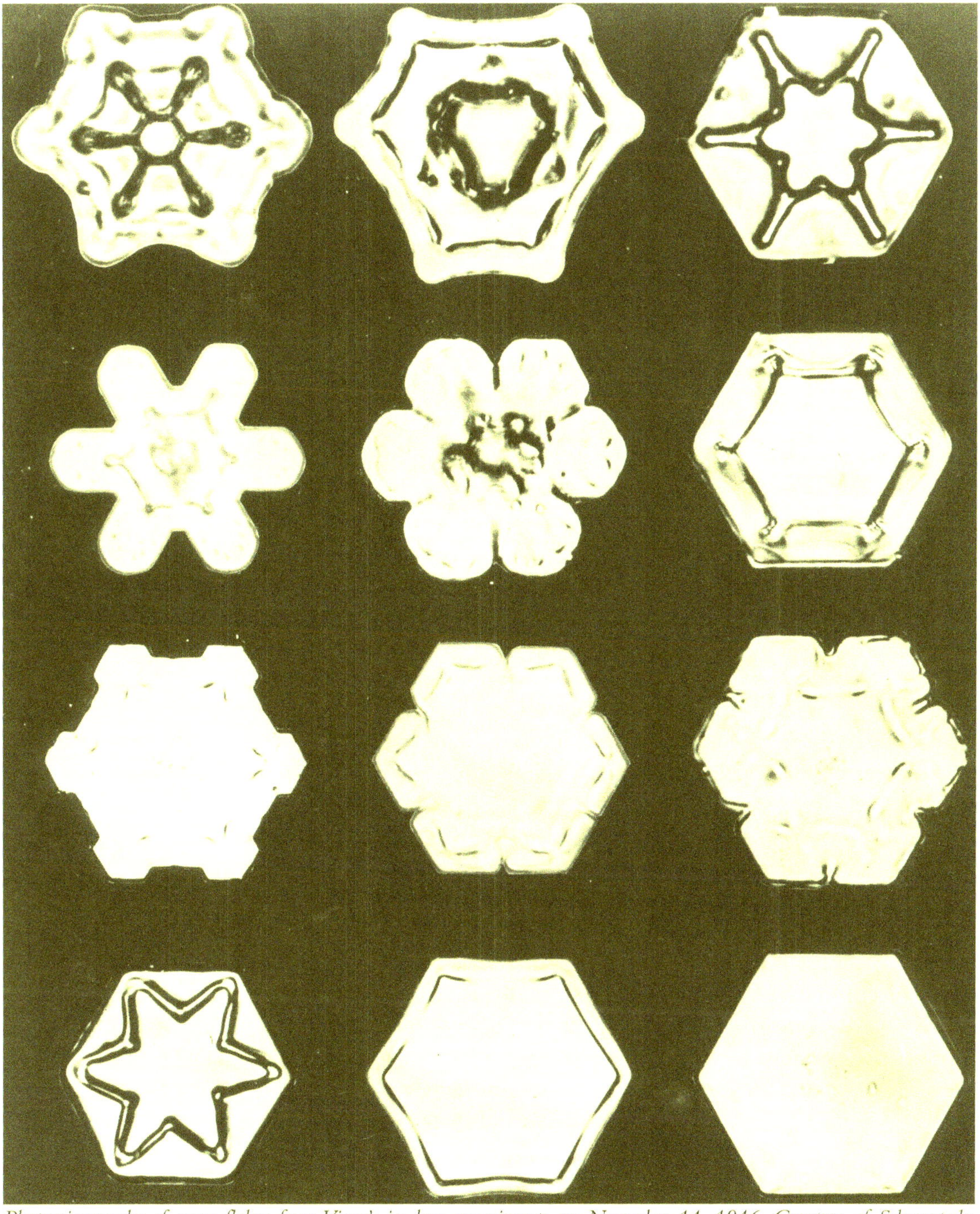

Photomicrographs of snow flakes from Vince's ice box experiments on November 14, 1946. Courtesy of Schenectady Museum.

Whether it could in fact do this would require some field tests, but I believe it would be quite feasible. I suggested that a scale model of an airship be exposed at the summit of Mount Washington, or in the icing tunnel at the NACA facility in Cleveland, to try out such a procedure. Instead of using dry ice, I suggested that this might be a good application of the seeding properties of the Hilsch Tube or a raucous nozzle.

To assure proper distribution of ice embryos, I suggested using a manifold mounted on the boom, having ten or twelve orifices or jets directed forward at an angle of ten degrees or so from the axis of the boom.

I never heard anything further of this project. I am quite sure my suggestions would work!

The Formation of Project Cirrus

After my fourth seeding flight of December 24, 1946, a major snowstorm that produced about two feet of snow, followed on Christmas Day. While I doubt very much if my seeding did anything more than cause a zone of snow that extended to the northeast for a distance of about 30 miles, that reached a depth of about 5 to 6 inches, the subsequent storm was a big one that resulted in strict orders from the GE Legal Department to cease all such activities!

When the Boss was informed about this action he went to Washington and soon was promised active support from the Army, Navy and Air Force! They collectively realized that any degree of weather control was something they should know about. Within a few weeks a program was set up that prospered and continued for the next five years. What is now called a "lead agency" was set up, centered in the US Army Signal Corps, which traditionally was the military unit involved in meteorological activities. In fact it was this agency that pioneered the use of Mount Washington in New Hampshire as a weather outpost, toward the end of the 19th century.

With the Signal Corps supplying the leadership, the US Navy agreed to supply personnel and several twin-engine aircraft with crew, the US Air Force agreed to supply two B-17 four-engine airplanes with maintenance crews, pilots and navigators. The Navy also supplied a meteorologist– Dr. Daniel Rex– who recently obtained his doctorate from the International Meteorological Institute and the University of Stockholm, Sweden, headed by Dr. Carl Gustaf-Rossby.

The coordinated group assembled at the Research Laboratory early in the spring of 1947 and unanimously selected Dr. Daniel Rex as coordinator of the operation. A Steering Committee was set up charged with periodically reviewing the activities of the overall program. I suggested that the project be called Project Cirrus.

Dr. Michael Ferrence, in charge of research programs within the Signal Corps, became Chairman of the Steering Committee. Dr. Langmuir and I were to serve as advisors to the Steering Committee and the Field Operation.

The military was to have final and full responsibility for all field activities, with two of the top civilian photographers of the Signal Corps charged with photographing any and all seeding effects produced by the field operation.

Representatives of the three services met from time to time with Langmuir and me to discuss the type and degree of seeding that we felt would result in the acquisition of useful knowledge.

One of the first things arranged was to get a subsection of the southeastern Adirondacks designated as a test area that would be off limits to other aircraft during cloudy weather. Since this did not interfere with any commercial flight corridors, such designation was quickly granted.

After a few flights had been made, Langmuir and I realized that we had problems! After making a seeding run, the pilots and navigator would quickly lose track of the location of the area seeded. To overcome this problem Langmuir proposed that the Sacandaga Reservoir be used as a visual target and that the pilots and navigator position the aircraft so that photographs of the reservoir could be obtained at an equal distance during a complete circle of the wide waters of the reservoir. When the results of the first flight were analyzed we found that a portion of the flight was more than fifty miles down wind of the target! This highlighted the navigation problems and, after corrections of procedure, within a few more flights we began to get the photographic precision we needed to obtain quantitative measurements of seeding effects.

From then on the pilots and navigators performed beautifully and we soon had a host of excellent photos to measure and evaluate.

Originally it was thought that the B-17s assigned to Project Cirrus could be based at Chamberlin-Wold Field in Pennsylvania. Neither Langmuir nor I favored such an arrangement since we felt that it was highly desirable to have the whole operation nearby. At the Schenectady County Airport, the General Electric Company had built a hangar large enough to accommodate a B-29 aircraft. Since the end of World War II this hangar was no longer in urgent demand, and consequently was available to house the two B-17s. After considerable persuasion, the Steering Committee agreed to recommend a switch to Schenectady, and from then on Project Cirrus operated as a highly coordinated operation.

Seeding experiments were carried out close to Schenectady within a radius of 200 miles, the farthest being the Cape Cod area. Special flights were made to Puerto Rico, Albuquerque, and a seeding flight to Hurricane King east of Florida. A total of nearly 200 flight operations were carried out during the period 1947-1952.

In 1952 our veteran pilots were transferred to become trainers for flight operations in Korea. With this change of personnel we suddenly realized that we would go through a new series of frustrations, due to the arrival of new pilots and navigators. Since Langmuir and I felt that we had obtained sufficient data to answer many of our questions, and that we had a considerable backlog of flight operations to keep us busy for some time, we recommended to the Steering Committee that we close down Project Cirrus. This recommendation was approved and we started to prepare a comprehensive report of our five years of activity.

My report was issued in March 1953 as "Final Report of Project Cirrus." When issued, my report summarized a few of the outstanding flight results, but concentrated on the many laboratory discoveries and applications that occurred over the five-year period.

Langmuir's report summarized the strange periodicities in rainfall that were observed over the Ohio River basin, results which remain an enigma even 35 years later!

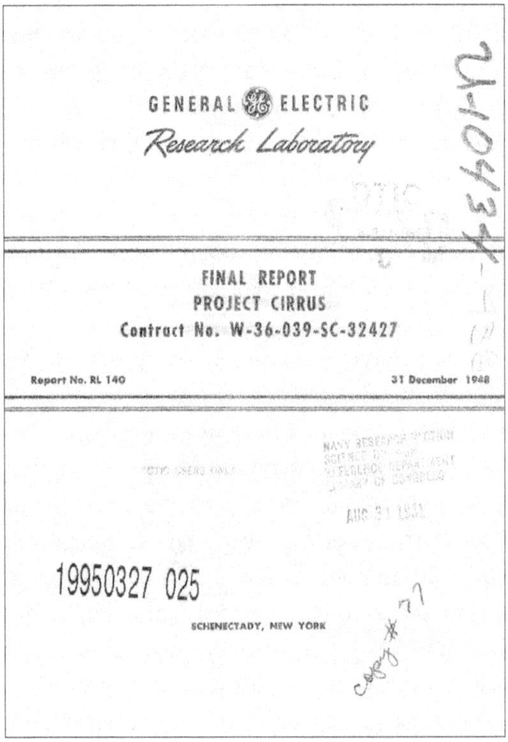
GENERAL ELECTRIC
Research Laboratory

FINAL REPORT
PROJECT CIRRUS
Contract No. W-36-039-SC-32427

Report No. RL 140 31 December 1948

19950327 025

SCHENECTADY, NEW YORK

Chapter 7
Adventures in the Field and in the Laboratory

The Use of the Hilsch Tube for Producing Ice Embryos

On February 28, 1947 I was shown a version of the Hilsch Tube[33] invented in Germany and copied by a member of Tony Nerad's group in our Research Laboratory. This invention was noted for the interesting way in which it took room-temperature pressurized air and generated both heat and cold. One end of the tube became warm while the other end fairly cold. The efficiency of cold generation was not very great, hence the verdict that it was not much more than an engineering curiosity. Many efforts to apply its cooling principles were failures.

Since there was some evidence that a portion of the cold end of the tube might be extremely cold, it prompted me to borrow the copy of the tube made in our Laboratory to see, if in fact, it cooled the ambient pressurized air colder than 40°, the homogeneous nucleation temperature. I was disappointed to find that it was not an effective ice nucleus generator. In examining the device it looked to me as though its design could be simplified. Accordingly, I prepared a design using clear methyl methacrylate (Plexiglass) for the cold end. I hoped the lack of metallic conductivity might produce a colder vortex.

When I tried my new device I found that it produced a tremendous quantity of ice crystal embryos. By means of thermocouples, I found that I could achieve a cooling drop of -39°C. Since I knew from other experiments that the transient cold temperature could best be determined by finding out whether it dropped a portion of the air colder than -40°, its performance indicated that I had achieved a considerable improvement over the Hilsch design.

The most interesting thing I discovered was that, by measuring the general temperature difference using two thermocouples, the maximum difference occurred at the exact point where the tube suddenly emitted a raucous sound. I found that at that maximum difference in temperature between the two thermocouples and the inception of the raucous sound, the tube was most effective in generating ice embryos. It was so effective that the brief introduction of the cold end of the tube into a supercooled cloud produced so many embryos that a bluish fog appeared. With repeated slugs of moisture, the particles compris-

33 This tube cools and heats air at the same time, with no moving parts or electricity. The tube was invented in 1928 by George Ranque, a French physicist. German physicist Rudolph Hilsch made it popular and is named after him. It is sometimes called the Ranque-Hilsch Vortex Tube.

ing such a blue fog eventually grew large enough to show that they were indeed ice crystals.

Since this blue fog phenomenon seemed to be characteristic of a noisy jet of air, I prepared an ordinary airline with a nozzle that could be controlled from a valve from zero to maximum flow. I tried the ice crystal embryo production of an ordinary raucous sounding air jet. It also produced a blue fog of ice crystals! Thus it appears that any noisy gas jet, either of air or some other gas, can generate ice crystal embryos which, if put into a supercooled cloud, will persist and grow.

It is unlikely that the present snow makers used to augment the snow at ski resorts produces much snow because of this phenomenon, a few of the myriads of nuclei causing the atomized water droplets to freeze by contact nucleation!

UFOs and Hurricane King

One of the first things I learned from the Boss was to avoid the danger of "wishful thinking"– that is the procedure when carrying out a research study to favor the data that seems to confirm a preconceived theory, and to discard items that tend to contradict it. He pointed out that everyone who does research is confronted with such conditions from time to time, and can easily fall into such a pattern quite innocently.

He emphasized the importance of being completely honest in reporting research results and that there was nothing wrong in making mistakes so long as they were freely admitted and that one learned something in the process. He was rather ruthless in pointing out mistakes and errors in judgment and, in a number of instances, his own previous mistaken ideas were victims of his remarks!

He frequently cited examples of faulty research and the consequences. One of his examples related to UFOs. On our way to a New York meeting by train he brought up the question of Unidentified Flying Objects and told me that he had been asked by the US Air Force to review the "Blue Book" that contained a review and description of the best examples of UFOs that had been documented to date. He told me that he had it with him and if I were interested, to come to his room after we had checked into the Hotel Commodore.

I did so, and for several hours read accounts of experiences recounted in the Blue Book.

There wasn't a single one that could not be explained by natural phenomena, reflections in windshields, high-flying jet contrails, or a misinterpretation of a transient glimpse of something.

The report that had the clearest picture had a photograph of what could have been a piece of tar paper fluttering down from the sky. It was taken on a windy day and possible could have been carried into the air by a whirlwind.

I have seen five effects that could have been cited. The first of these was observed in eastern Washington State.

The Boss had been asked to go to Hanford, Washington during the early part of WWII to look at the "Canyon" unit of what was to become the Atomic Energy Reservation in the arid lands north of the town. We went by train to Spokane and there transferred to a local line that took us near or to Hanford. At the station we were met by a large limousine and taken for many miles across the desert area. Sometime after leaving our rendezvous, I suddenly saw what appeared to be a group of fast moving lights on objects that seemed to be traveling at a great rate of speed. They suddenly disappeared. Within a few minutes they appeared again and as I was about to draw attention to what I was seeing, I suddenly realized that I was seeing a reflection in the car window of a cluster of lights in the distance, which I saw by reflection. I kept my observations to myself!

A second example was an observation made near Lubbock, Texas. The family and I were heading for a restaurant to get supper. The sun had set, but the sky still glowed in the west. Suddenly my moving eyes picked up a striking object high in the sky that appeared to be a fast moving rocket that was emitting a fiery tail. I drew attention to this phenomenon to the family, and suggested that it was a jet airplane in the high atmosphere putting out a contrail, but the air was so dry up there that the condensed water had a very short lifetime. It appeared to be a flame because its trail was still illuminated by the sun. I mentioned that it would easily be mistaken as a UFO, and in the next morning's newspapers, large headlines proclaimed the fact that a number of excited people had reported the sighting of UFOs the previous evening.

One evening, on the way home from church with my wife, our daughter Susan and her family, one of the boys pointed out a brilliant object in the western sky that appeared to be moving and was too bright to be a satellite. We rushed home; I grabbed my binoculars and Questar telescope and headed far into the Rotterdam hills to a location that had a good view of the western sky and was free from the dazzling lights of the city. Reaching the spot I passed the binoculars to one of the kids and proceeded to set up the telescope on the hood of my car. It wasn't until I had the object in my telescope that I was convinced that it was the planet Venus, and that it was not moving as it appeared to be!

The fourth example was secondhand but attested by a technician from Colorado State University. He gave me a photograph that he had obtained while photographing the CSU installation at Chalk Mountain. He had two photographs of their observatory– one taken at a particular lens opening and a second one taken a few seconds later with the lens opened to the next stop opening. However, when he personally developed the film, the second exposure included a round object that appeared to be made of metal and seemed to be in the sky above the mountain laboratories. It was in sharp focus and was a bewildering development. My friend said that he saw nothing when he took the second picture, although he did it rather automatically, since had already taken the picture he figured would be used. He was quite excited about his picture and was about to publicize it when a graduate student pointed out to him that the UFO looked like a lampshade. He went up to the darkroom area and was chagrined to find that the student was right! In some manner he had inadvertently made a double exposure. Then he remembered that when he had arrived at the dark room it was occupied, and he sat down near the shade to wait his turn. Although he couldn't remember doing anything that would trigger the camera a second time, it must have happened since there was no question but that the image was there! I am sure that if the picture had been published it would have been world famous!

The fifth example I had with a UFO occurred on a flight from Schenectady to Mobile, Alabama. It was during the Project Cirrus days. We had received word that a hurricane was building in the Caribbean, and we were anxious to examine one at close range to determine whether it contained supercooled clouds and thus might be modified by cloud seeding. Our flight took us over West Virginia en route to McDill Field. I was in my favorite seat in the front gunner's position in the plastic nose of the B-17. I was alone and thoroughly enjoying the scenery as we were just skimming the top of a thin layer of altostratus clouds.

Suddenly ahead I saw a white circular object approaching at a great rate of speed on what appeared to be a collision course. It then disappeared just as I expected to experience a collision. I was scared and speechless. Suddenly it appeared again, but this time was on the port side of the airplane, and it again sped by at a very high velocity. I was deeply puzzled by my observation but decided to remain quiet as I puzzled over a reasonable explanation of what I had seen. The Boss was not with us on this flight. He would have been the only one with whom I would discuss my experience. I resolved to tell him about it on our return.

We landed at McDill, learned that the hurricane had passed over southern Florida and was headed for the Atlantic. It was questionable whether there was anything of an organized nature that could be seen. We went to bed with plans for early take off. A few hours later we arose, got dressed and assembled at the base cafeteria. There we ordered the usual bacon and eggs and got the worst, greasiest assemblage of food I had ever encountered. The cook must have been something special to spoil the eggs to the degree to which he had managed! While trying to down the eggs, I noticed one of our pilots, who apparently was familiar with the food at this particular base and had elected to have a bowl of dry cereal. As he was finishing his bowl, I saw him suddenly stare at the nearly empty bowl with dismay as he saw it was crawling with ants! I'm not sure which of us had the best breakfast!

As dawn was breaking we took off to look at the hurricane. I had shifted to a B-29 that had been made available to us, and which would fly at a higher altitude than the seeding and photographic plane. We would also do some photographing, and I planned to take a movie.

We approached the hurricane (King) and saw that it certainly looked like a massive, well-organized storm. A line of huge cumulus congestus clouds spiraled in towards its eye, and some smaller supercooled clouds appeared along the edge of the main spiral that we planned to seed with some 300 pounds of crushed dry ice.

The seeding run was completed, and a great mass of ice crystals replaced the clouds structures that had been there before. We anticipated that the ice crystals would be sucked into the larger clouds at a level where they would create a major modification.

Vince and Bernie Vonnegut examining new equipment. Courtesy of Schenectady Museum.

We then headed back to the base in Alabama since our gas supply was dwindling. Thus we were unable to see whether our strategy did anything to the massive storm.

After having supper and a short rest we assembled our gear and headed back to Schenectady in our B-17s.

We followed the flight corridor back north that we had followed down. As we passed over West Virginia, the matter of the fast moving discs that had scared and puzzled me on the way down came vividly to mind. I was again enjoying my favorite station, alone, and enjoying the lights of the myriads of towns. Then suddenly I saw "it." The sky was cloudless this time, but on the edge of a small town there was a carnival or similar type of operation. At one edge of it was a powerful searchlight that was pointed skyward and was in rotation. I could see the ray of light coming up toward us as it swung around, but this time there were no clouds to be illuminated, and to thus scare the daylights out of an observer who didn't believe in UFOs!

Scientific Adventures—A Fun Project

After participating in the Science Forum once a week for twelve years, and enjoying every program, I was approached by a friend from the GE Publicity Department who wanted to know if I'd be interested in writing a series of bimonthly articles for a new magazine they were planning for the children of the General Electric "family." It was to be called "Adventures Ahead." Again I felt honored to be asked, and immediately agreed to try my hand at it. The first issue was dated September-October 1947 and the final one, 41 issues later, May-June 1954. I thoroughly enjoyed writing this series which I called "Scientific Adventures," since it permitted me to write about a number of subjects that I had found interested youngsters, especially my own!

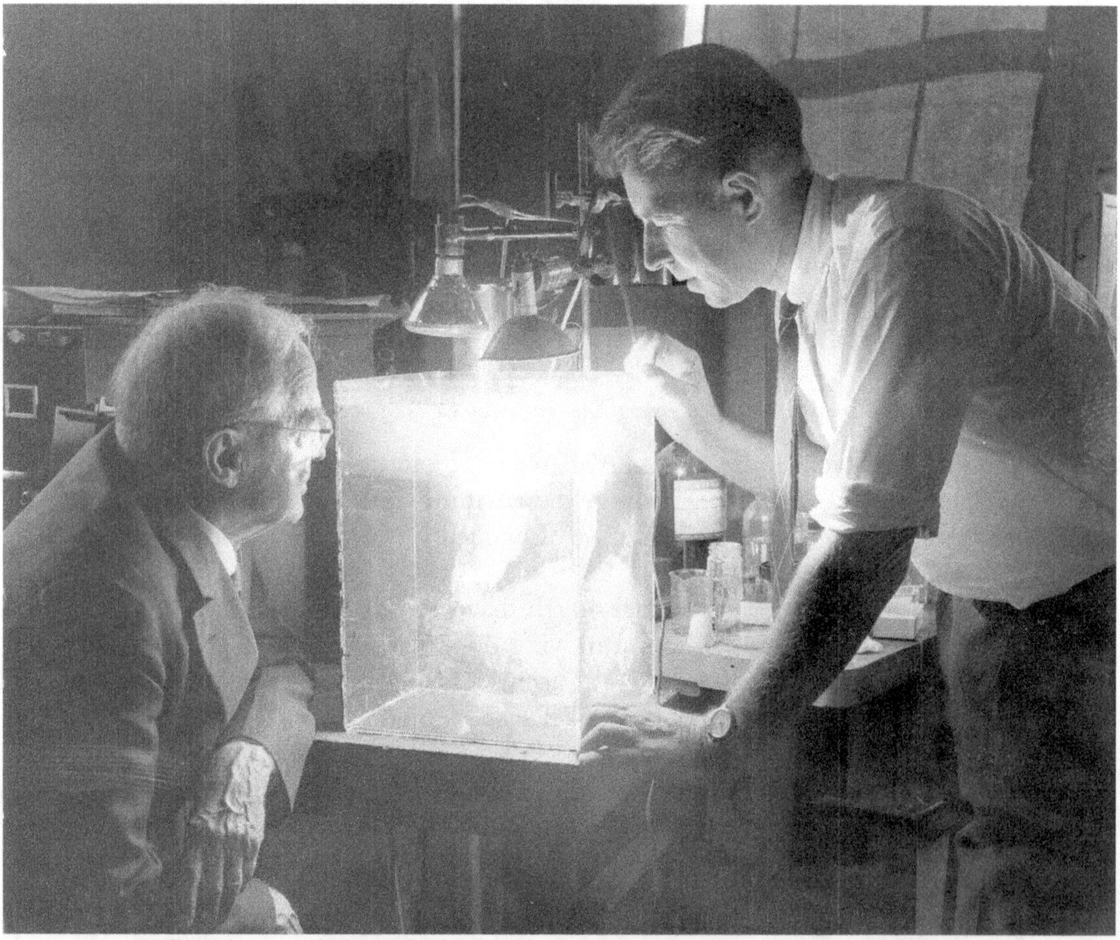

Willis Whitney and Vince Schaefer make cloud in a box. Courtesy of Schenectady Museum.

The Control of the Shape of Ice Crystals

From the time I discovered dry ice seeding in 1946, I have been much intrigued with the differences in ice crystal habit. I quickly found that most of the crystals that developed at -15° to -20° were featureless, hexagonal plates of ice without any further ornamentation. By contrast, the hexagonal crystals that formed when iodine vapor was introduced into the chamber under the same temperature and moisture conditions were incredibly beautiful, with what seemed to be an unlimited variety of intricate detail on the surface of the crystals. When Bernie Vonnegut developed his silver iodide seeding method, I found the resulting crystals to be similar.

For this reason I initially thought that the iodine vapor method was also measuring silver iodide, the silver coming from the sparking of silver alloy contacts within the laboratory. (In later years I discovered that it is quite likely that these effects were due to the presence of submicroscopic lead particles, mostly from automobile exhausts that infiltrated the laboratory air).

Thus when Bernie discovered that a whiff of amyl alcohol would shift the crystalline habit from that of a hexagonal plate to a hexagonal column, I was greatly intrigued with learning more about this phenomenon. I tried introducing vapor from sodium dioctyl sulfosuccinate (OT 100), and found that it also shifted the normal hexagonal plate habit to that of a hexagonal column. I then found that the vapor of silicone Drifilm (a GE trade name) would do the same thing. On the other hand, the vapor of nitric acid and of hydrofluoric acid, while they didn't shift the crystal habit from plates to columns, they did increase the number of trigonal crystals among the hexagonal plates. While this whole subject is fascinating, and quite possibly is controlled by the adsorption of monolayers of foreign molecules on certain growth surfaces, we were far from understanding the mechanism controlling the crystalline growth habits.

I had a visit from my friend Ukichiro Nakaya[34] from Japan, who originated the concept of the change of habit of ice crystals as a function of temperature and moisture, and when I showed him the effect that silicone vapor, amyl alcohol and OT 100 had on their crystalline habit, he was greatly puzzled. Later on, after I had left the Lab and joined the Munitalp Foundation, I was able to arrange a grant to him and a few of his colleagues to occupy the summit of Mauna Loa on the island of Hawaii to conduct research. In an excellent paper resulting from his studies, he described experiments with crystal form when he encountered crystals that did not fit into his classic pattern. He found strong evidence that this anomaly occurred with a weather system that would channel polluted air from Japan to Mauna Loa. Thus in 1956 he reported evidence of the distant transport of polluted air over very large distances. The question of the mechanism of crystal habit control by traces of certain foreign molecules is still an enigma.

---

34 Nakaya became well know for his work in glaciology and low temperature science, and is credited with making the first artificial snowflake. He died in 1962.

A Fun Time in Puerto Rico

The Boss and I had a marvelous time in Puerto Rico. We met in New York City on February 2, 1949. We had dinner, visited his brother Dean, and then went to La Guardia Field for late evening take off for the island. We were delayed several hours and, after a nighttime flight, landed in Bermuda to pick up stranded passengers. Eventually reaching Puerto Rico, we were met by Mr. Lindstrom, the local manager of the International General Electric who showed us around San Juan. I wandered through the old town taking pictures and having a fine time. In late afternoon we took the mail plane to the extreme western end of the island and landed at Ramey Field, where our Project Cirrus flight operations were to take place.

At Ramey we met the Project Cirrus crew and were soon assigned excellent rooms in the Officers Club. For the next ten days we explored from ground, as well as air, the nature of the local trade wind clouds, and learned a great deal about them. We found them to be entirely different than the clouds of the continental United States– the major feature being that in most instances they were warm clouds (that is, no part of them supercooled) and yet they rained easily. It became obvious that this was due to the purity of the air and the deficiency of cold condensation nuclei. With the lack of nuclei the cloud droplets had little competition for the rich moisture source of the sea, and thus grew large very quickly. Langmuir was surprised and quite stimulated by these findings since they agreed so well with his ideas on the formation of warm rain.

We made a number of flights, looking at clouds over the land and over the sea. We found that the only clouds that developed supercooling were over and downwind of the Sugar Centrales, the mills that crushed sugar cane and produced molasses and sugar. The large plumes extruded by these mills overseeded the clouds with nuclei so that they grew much larger and produced supercooled clouds and ice crystals. These large clouds apparently were responsible for producing what the natives called "hog wallow gushers."

One day, when flying was not scheduled, a group of us went by Army jeep to explore the mountain ridge area in the center of the island. We headed for Lares, a mountain town in a beautiful setting. Reaching the center of town, the Boss saw a grocery store and asked the jeep driver to park. He leaped out of the car and disappeared into the shop, and soon emerged carrying a green fruit that, to me, looked like a pumpkin. I could see that he was anticipating a new taste treat. In the area south of Lares, we found some winding roads that took us toward the top of the ridge.

On the way up this road we observed some of the native women crushing corn with a primitive mortar and pestle– the pestle fastened with a rope to the top of a small supple tree that bore much of the weight of the pestle, thus reducing a bit of the labor of raising the pestle. A bit further along we passed a boy carrying what appeared to be a bottle of beer and a blowtorch. We were much puzzled in seeing this combination of objects, and Langmuir was especially intrigued. We found a marvelous site for observing the behavior of orographic clouds along the rocky ridge of the island. After a while the boy appeared and Langmuir immediately started talking to him. He was a bright youngster and talked

English quite well. He explained that he was on his way to repair a television set! The bottle contained fuel for his blowtorch that he would need for soldering wires! He told us that his ambition was to go to New York City to become an electronics repairman. I have often wondered what happened to him. My guess is that he probably owns one of the successful service stores in the big city!

Before the young repair man continued up the mountain, Langmuir went to the jeep and brought out the purchase he had made in Lares. When he asked our young friend to identify the fruit, the boy looked at it and said, "Pumpkin!" Langmuir took our laughs in good grace and quietly discarded the pumpkin.

After we had watched the cloud developments for several hours, we headed back to base going south through Poncé and then around the southwestern coast to Ramey Field.

Near the Officers Club was a walled, open platform projecting out over the cliff edge that served as an ideal spot for observing clouds and taking time-lapse movies. We had two 16mm movie cameras that we placed several hundred feet apart, with the hope of obtaining three dimensional movies of the clouds that we discovered formed, grew, and precipitated every afternoon in a cloud street that extended to the northwest from the tip of the island. We obtained some excellent footage. Several weeks later after the films were developed we set up two projectors fitted with polarizing filters one at right angles to the other. We viewed them with polarizing glasses and found the effects to be spectacular.

One day, as we were watching clouds, I drew the Boss's attention to the sensitive plants that grew like weeds in the grass in front of the Officers Club. He had never been introduced to them before and became intrigued with their behavior. He was on his knees for an hour or more conducting a series of experiments on their sensitivity, their endurance, recovery pattern and other physical properties of the plant.

After we had been on the base a few days, we learned that much better breakfasts could be obtained at the Enlisted Men's cafeteria. I quickly discovered the delicious orange juice obtainable there, and for several mornings had a huge glass to start off the day. One morning, when I asked for orange juice, the waiter said, "No juice!" This puzzled me since there was a huge pile of oranges behind the counter. When I pointed this out he immediately explained the problem when he said, "No ice!" When I told him that I didn't need ice; that I preferred the juice just as it was squeezed from the oranges his face lit up and he immediately produced a huge glass undiluted with melting ice! From then on I was a favorite of his and he drew a rich glass for me when he saw me coming.

One day after our observations, I joined Bernie Vonnegut and Kiah Maynard of our Lab in wandering the beach that stretched for miles along the north coast. We had a wonderful time collecting shells and anything else that was collectible or was a photographic subject. Among the things I found was a stone tube identical in appearance to the tubes made by the caddis worms in our cold streams, but more than five times larger. It had obviously been tossed by the waves for a long time, since the edges were worn smooth. One has to marvel on the structure building ability of the sea creature who searched for and found tiny pebbles that could be fashioned and cemented with a water proof adhesive that would "set" underwater, and when discarded would continue to persist under the constant pounding of the surf.

One interesting project that Bernie asked us to help him with was to find a right-handed curled seashell among the thousands that were strewn along the beach. Despite a diligent search only one was found and that by Bernie.

One evening our whole crew was invited to partake of a fish barbecue. We had a great time! The *pièce de résistance* was deep fried fish– tiny fish that were very much like our sun-fish or "punkin seed." They were fried scales, internals, heads and all until they were crisp and savory. Despite our initial skepticism we soon discovered them to be delicious. The only precaution I discovered in eating them was the need to chew the spiny dorsal fin and the tails so they would not stick in our throat while swallowing.

We soon discovered that the Boss loved to eat tropical fruit. This led to an amusing situation that developed after we had ended the field project and were headed home.

It was necessary for the two of us to travel commercial air, due to space limitations on the plane. We took the mail plane back to San Juan and, after meeting Mr. Lindstrom of IGE, we went to the Hotel Cindedo that appeared to be the fanciest in the city.

After a restful night we met for breakfast. I ordered a pineapple while he had a papaya. As we were finishing our fruit, I saw him eyeing my generous slab of pineapple (which was delicious). He called the waiter over and told him that he'd like to have a piece of pineapple. The waiter looked at the menu and told him that he already had his fruit. Slightly exasperated the Boss told him that he was quite willing to pay extra for the pineapple. The waiter was adamant and insisted that it was not possible to have a second serving of fruit. The Boss sought help from the *maître d*. After listening to Langmuir's explanation that he liked fruit and would be glad to pay for the extra piece, looking at the menu, the *maître d* shrugged his shoulders, shook his head in the negative and left. Langmuir did not get his second helping!

The Measurement of Condensation Nuclei

While walking the beach in the northwestern tip of Puerto Rico, Bernie Vonnegut, in a few minutes made a very useful cloud condensation nucleus detector. He took a beer can washed up on the beach, put a small hole opposite the opening that had been made by the former user, put a small amount of sea water in the can, held a finger over the small hole he had made, created a low vacuum in the can by sucking air out of the larger hole, and then suddenly released the vacuum. By orienting the can so a beam of sunlight illuminated the interior shining through the small hole, and looking through the large hole, a cloud of droplets could be seen. Estimating the number of cloud droplets in a line over a distance of 5 millimeters, doubling that number and then cubing the result, it was possible to put a rough estimate on the number of cloud condensation nuclei per cubic centimeter. It turned out that this number was less than 200 per cc, the number that is characteristic of clean maritime air. This was a very clever and useful device in the absence of the standard Gardner Counter.

Upon my return to Schenectady from Puerto Rico, I decided to make a "pocket" condensation nuclei counter based on Bernie's beer can unit, but somewhat more sophisticated. I made a brass box a little larger than a Prince Albert tobacco can. On one side of the narrow dimension of the box, I made a slit nearly as wide and as long as the vertical dimension of the box, and on one of its narrow sides. Through this slit I lined the sides and bottom of the box's interior with black velvet. On the top of the box I made two holes, one to accommodate a penlight bulb for illuminating the interior, the other for a vacuum gauge. On the side opposite the viewing slit I soldered a 1/4" tube to which a vacuum generating bulb was connected.

After the velvet liner was installed, I cemented a glass window over the viewing slit. With the evacuation bulb I found it possible to produce a vacuum of 20 inches of mercury that was sufficient, when released, of activating all of the particles in the air.

In use, I found several ways to obtain semi-quantitative data on the relative number of condensation nuclei in the air of a specific area.

When the particles were in such high quantities as to make it impossible to estimate the number of particles in a 5 millimeter distance, I measured the time required for sedimentation to occur. It was also possible to observe the presence or absence of High Order Tyndall Spectra that developed, generally in areas far removed from man-made pollution.

While this device was light, simple and highly reliable, I soon found that I could use the Gardner Counter in a similar manner and it had the advantage of providing quantitative measurements. While not as compact, it was quite portable, and I soon found it to be indispensable. Thus, I was soon taking it with me on trips wherever I went, thus beginning a long-range project that culminated in my two-volume report, *Air Quality on the Global Scale*, which I finished in 1976.

The Soap Bubble as an Ice Nucleus Detector

Several years after I published my paper describing the use of a cold chamber for studying supercooled and ice crystal clouds, several useful techniques were developed for measuring ice nuclei in the air of the chamber (in addition to their visual appearance) as they grew into an ice crystal in a supercooled cloud. These were primarily developed with the cold chambers that could not be viewed as with my simple chamber. One method developed was to prepare a supercooled sugar solution, into which an ice nucleus or a tiny ice crystal could fall and continue to grow in the tray holding the solution. Then, when they grew large enough, they could be counted. The main problems with this technique were that is was a messy, sticky mess that, if spilled, posed a cleanup problem, and the slowness of the growth of the ice crystals in the solution.

It occurred to me that this could be improved upon by using a soap bubble made with a soap dissolved in water. I found that the type that could be bought in the five-and-dime store was quite suitable, since it readily formed bubbles, and since the main component of the bubble solution was water, the bubbles readily and quickly supercooled.

I found that if one formed bubbles in the free cold atmosphere, a measurement of the concentration of ice nuclei in the surrounding air could be easily and quickly determined. Not only was this a very fortunate discovery when tried, I discovered that the bubbles generally broke before they became completely frozen and the number of nuclei that had seeded the bubble could be estimated by counting the number of ice fragments that fluttered down from the spot where the bubble broke.

A bubble film was also found to be highly useful when spread on a wire frame and lowered into a chamber. If the crystal habit of the ice embryo would, when further grown, show the familiar six-sided stellar form, it would also produce a six-sided stellar dendrite in the soap film. If the crystal would develop into a needle or a hexagonal column, it would produce a symmetrical four-pronged crystal in the solution. If, on the other hand, the ice nucleus was a soil particle or other asymmetric structure, the crystal that formed in the bubble would be fern-like, much like the type of frost crystals commonly formed on car windows or other solid surfaces.

I then found that I could coat a thin film of plastic or even a piece of black cardboard with the bubble solution and produce similar crystallization effects. The next step was to use a plastic, water-soluble resin, such as polyvinyl alcohol, in a concentration of 3%, to coat the plastic film or the black cardboard.

After the crystals had grown and caused all of the water to be frozen, I discovered that by leaving the frozen film in the cold chamber or in a well-ventilated atmosphere that at no time warmed above 32°F (0°C), the water would evaporate and leave behind a perfect replica of the crystals which formed. This happened because, in the process of the ice crystal freezing, it repels all of the dissolved impurities in the water (the resin or soap) so that they finally end up at the interface between adjacent crystals. These substances then remained at those sites when the ice sublimed away.

In experimenting with bubbles I discovered that at times, when the ice nucleus concentration of the environment is particularly low, a single nucleus will seed the supercooled bubble and grow all the way around the bubble. This is particularly beautiful if the sun is shining or if a strong projection light source illuminates the colored bubble. At times the bubble does not break and the frozen sphere falls and remains in an intact form until sublimation causes it to collapse.

In carrying out my experiments, I also discovered that if I formed a large bubble in the air above an open cold chamber and it fell into the opening it would suddenly stop in mid air as it contacted the cold air of the chamber. Then it would bounce several times before slowly sinking into the very cold air. Generally before reaching the floor of the chamber it would have burst or it would be completely frozen and then remain intact until sublimation caused its collapse.

Speaking about bubbles...I might mention another type of bubble solution that I concocted that is made with very little water! It consists of equal parts of glycerin, a strong detergent such as Joy or Wisk, and white corn syrup. When blended and poured into a large shallow disk, a wire coat hanger bent into a circle is dipped into the liquid, immersed and then lifted. If it comes up with a flat film coating the ring, it is removed and swung through the air! Bubbles of several feet in diameter can be made from this solution. However, such bubble forming activities should be conducted out of doors since the residue of a collapsed film makes a large, sticky mess! Bubbles 8-10 feet long can be formed in this manner, or sheets of film can be lifted from a trough that can be a meter (yard) square!

I only wish that Langmuir knew about these large, tough bubbles. I can see him doing many experiments at his camp on Crown Island. It would have been fun!

My First Honorary Degree—From Notre Dame

Early in the spring of 1948 I received a letter from the President of the University of Notre Dame, Father John Cavanaugh, that I had been selected, along with Cardinal Cushing of Boston, to receive an honorary degree at the June Convocation, and hoping that I could be there to receive it. Needless to say, I was greatly surprised, not having an inkling that this would happen.

I later learned that an old friend from the GE News Bureau had something to do with the nomination, but to this day I do not have any details.

I was extremely pleased that Notre Dame officials had seen fit to confer this honor upon me since of all the Universities in the United States it is the one I would have liked to attend if I had been able to do so. Several of their accomplishments attracted me– the most important being its high regard of morals and ethics. I was also familiar with the germ-free animal program of Professor Reyniers and the remarkable results obtained in visualizing flow patterns in wind tunnels by Professor Brown.

In June, Lois and I took the train to South Bend and, at the University, were given excellent accommodations and treated like royalty. Sunday morning Lois and I had the rare privilege of attending a Mass celebrated by Cardinal Cushing. Father James Joyce served as our host during our visit and things couldn't have been nicer.

At the Convocation a very large class of graduates received their degrees and it made me feel good to realize that I had been admitted to a modest degree to share their traditions and memories.

Lightning Storms in the Northern Rocky Mountains

Late in the winter of 1948, on February 4, I had a visit from Harry Gisborne, Chief of Fire Research of the US Forest Service, Region 1, located at Missoula, Montana. He came to Schenectady to discuss the possibilities of modifying the thunderstorms that often wreck havoc with the northern forests of the United States. His description of the nature of the worst storms did not jibe with the pattern of the frontal or air mass type thunderstorm of the western prairies or the eastern United States, and I became curious about them.

He invited me, with my family, to come to the Northwest. He described his research station at the Priest River Experiment Station and told me that once there, the Forest Service would provide food and lodging. Since this invitation fit into our schedule quite well, as the Boss was going with the Project Cirrus planes to Albuquerque to study the big air mass storms of the southwest, I felt that between the two of us we could roughly evaluate the nature of the western storms both north and south. Preliminary descriptions suggested they were very different.

Accordingly, on July 6, 1948, my family and I left Schenectady in a Laboratory station wagon heading for the northwest, while Langmuir and the Project Cirrus flight crew left in two B-17s shortly afterward.

It took us five days to reach Priest River, Idaho. Since this was the first time I had driven west, it was quite an adventure. I had arranged to include General Electric related activities as part of the trip and consequently stopped at the Case Institute in Cleveland to give a talk and meet with administration and faculty. We went next to Indianapolis to have a session with Dr. William Davis of the Eli Lilly Laboratories. Bill had been one of my "pupils" in the late thirties when I guided him into the techniques in surface chemistry for studying insulin and related proteins.

We continued westward, spending the night at motels. Our quickly-established procedure was to stop at a motel late in the afternoon, wash, eat and to explore the countryside until dusk. Rising early we would pack and travel for an hour or so before breakfast. We generally had lunch while traveling; eating fruit, sandwiches, milk, cookies etc., purchased the evening before. In this manner we reached Denver after five days. We then headed through Rocky Mountain National Park to the Grand Tetons and thence to Yellowstone, where arrangements had been made by the Forest Service to show us some of the Park. We stayed at Mammoth the first night and Old Faithful the second. At Old Faithful we decided to camp out and erected our tent in a very dusty campground. When I checked with the local Ranger, and he learned that we planned to sleep in a tent, he immediately suggested that he'd prefer that we occupied some Park personnel cabins nearby, since bears had been causing problems. Accordingly, much to Lois' peace of mind, we left our dusty tent site and moved into two small cabins. The next morning, I learned from the Chief Ranger, who was looking after us, that a grizzly had bitten five people, tents were slashed and that there had been chaos at the campsite where we had pitched our tent. After breakfast we headed for West Yellowstone, and on the night of July 16, reached Missoula, Montana, where we spent the night at a motel on the bank of Rattlesnake Creek. After checking in with the Forrest Service, the latter ten miles over an extremely dusty road, we were greeted cordially by Harry Gisborne and shown to our house where we would live for a month, while I studied the lightning storms of the Northern Rocky Mountains. On Sunday afternoon, Harry showed me the facilities of the Experiment Station, ending the tour at a very tall steel tower located in the woods near headquarters. It was a very tall tower, and Harry asked me if I would mind climbing it to check some instruments on top of it.

Since I had quite a bit of experience in the high trees when working for the Davey Tree Expert Company I told him I'd be glad to do so. In a few minutes I reached the top of the tower 150 feet up and found it to be ideal for mounting my atmospheric electrical mast. Checking his equipment, I soon reached the ground and showed high enthusiasm for the tower. "Gis," as everyone called him, was apparently very pleased with my reaction. I later learned that it was a standard practice of Gisborne to request a new man to climb the tower as soon as he could manage to do so. Based on the reaction, he apparently evaluated the new man on his performance. Apparently I had passed with flying colors– from then on I had nothing but the finest cooperation from Gis in every respect.

Vince climbing a tree canopy survey tower at Priest River Experiment Station, c. 1950. Courtesy of Jim Schaefer.

After climbing the tower I decided to install my atmospheric electrical equipment without delay, so I installed my mast, radioactive point, shielded cable and photoelectric recorder on the tower. It was by far the best exposure I had ever had. No sooner had I checked its performance with Gis and Jack at the recorder, than I saw a storm moving toward us that looked like it might have lightning in it. I hardly had time to get the recorder in a rain-sheltered box before rain began, with occasional nearby lightning. The three of us stayed at our stations, me on top of the tower, with Gis part way down and Jack monitoring the recorder. The storm didn't last long but it provided us with an excellent trial run and exhibited cloud-to-ground strikes of both positive and negative signs.

With this auspicious beginning, I felt very optimistic about the observational program. Our first storm was not a dry one nor did it move rapidly or produce any fires. However, except for two others that occurred on August 1 and August 4, it was the limit of lightning observed and measured during our three week stay at Priest River.

The August 1 storm was a large and complex one that occurred in the early morning, with the first strike at 9:13. Gis and I went to the summit of Looking Glass Mountain (named after a famous Indian Chief). We spent about six hours looking at the convective clouds that were building over most of the mountain ranges, and concluded that this was a complex air mass type of storm.

On August 4th however, we did have a unique storm of the type described to me by Harry Gisborne when he visited me in Schenectady. It was observed to move 20 miles in 45 minutes, and during this period it produced 77 cloud-to-ground lightning strikes. This particular storm originated in the valley near the Village of Priest River, swept up the valley and headed toward Priest Lake, and then into the Selkirk Mountains of southern Canada, where it was said to have killed several persons climbing in the Bugaboos.

A few years later our Project Skyfire studies gave some evidence that these fast-moving storms originate at middle levels of the troposphere at about 20,000 feet and developed in alto cumulus clouds. Their fast-moving feature seemed in some way to be related to the jet stream. On June 16, 1949, while I was involved in Project Cirrus activities in Schenectady and Albuquerque, New Mexico, "Gis" organized a trial seeding run that turned out to be a hilarious operation. Everything went wrong that could go wrong, including a mix-up in the oxygen valves, a loss of communication with the pilots, and the failure of the planes to dispense the CO_2 seeding agent. Fortunately "Gis" prepared a report on the adventure for the files but sent a copy of it to me. It was published in the *Journal of Weather Modification*, 9: 1, 1977.

To my great sorrow Gis died while in the field investigating the disastrous Mann Gulch Fire in November of that same year. This was his last project as Chief of Fire Research of Region 1. He died as he would have wished, with his boots on.

This calamity put a "hold" on lightning research by the Forest Service. It was reestablished five years later, after I left the GE Research Laboratory and became Director of Research of the Munitalp Foundation. It became one of our major research activities in cooperation with Jack Barrows, a protégé of Harry Gisborne, who became head of Fire Research and developed the Northern Forest Fire Laboratory of the Forest Service of Missoula. Involved in Project Skyfire, and working with me for the Munitalp Foundation, were Dr. Paul MacCready and Donald Fuquay, the latter eventually becoming the leader of Skyfire when he joined the Forest Service.

The Rate of Fall of Fine Particles under Experimental Conditions

There is an interesting demonstration of the effects that sometimes control the settling rate of fine particles. If finely divided, alumina or grains of sand 35 microns in diameter, having a concentration of 1%, is placed in water contained in a glass tube a foot long, shaken up and placed in a vertical position, the fine particles settle to the bottom of the tube in 4 minutes. However, if the tube is tipped from the vertical by 15 degrees the particles fall to bottom in half the time. If the particle size is doubled (to 70 microns) the falling rate is five times faster.

On September 24, 1948 I went to Washington, D.C. to attend a meeting of the Steering Committee of Project Cirrus. As usual the General Electric Company secured accommodations for me in the Carlton Hotel, which at the time (and even now) is one of the swankiest hotels in Washington. I was becoming quite frustrated with the diversions that were cutting into my research activities. Committee meetings, visitors, correspondence, report preparation, talks and scientific meetings were taking up so much time that the number of hours available to carry out research had dwindled to almost zero.

I had been shown this fascinating phenomenon of particle settling rates and was very anxious to explore the phenomena, since I felt that it might relate to the falling rate of snow crystals in the free atmosphere.

Before leaving the laboratory for Washington, I prepared five glass tubes containing measured quantities and five different concentrations of alumina powder of five sizes, and a device to hold the tubes in a vertical position, as well as at varying angles.

Reaching the Carlton I checked-in, then had dinner and returned to my room about 10 PM. I had a big room furnished with a large flat-topped bureau that could serve admirably as a laboratory bench. Filling the tubes with water I started measuring fall rates in different concentrations, sizes and at nine different angles ranging from 0° to 40° from the vertical. I would shake the tube to be sure that the particles were uniformly distributed and then suddenly place the tube in a definite position with respect to the vertical, and with a stopwatch, measure the rate of complete fall out.

I found the experiments to be quite fascinating and completed the various sequences twenty-two minutes after midnight. I had found that some of the most interesting runs involved particle concentrations of 1%.

As a series of final runs, I measured the rate of particle fall in a vertical position of particles having an average size of 50 microns as a function of water temperature. With hot water the rate of fall was 59 seconds, at room temperature it was 83 seconds, and in ice water is was 112 seconds.

The results of these observations have considerable relevance to the free atmosphere. It demonstrates the importance of knowing air motion within stormy skies as well as under certain types of stable air.

The pendules that fall or rise from contrails, the mammato cumulus that form in the under surface of the anvil top of a large convective storm, and the cold air that falls from the base of high-level convective storms are all evidence of instabilities that occur in the atmosphere. The long streamers of crystals, that often accompany jet streams, are another example of this mechanism. It is a very effective method of transporting particles and moisture in the atmosphere at rates greater than can occur by sedimentation or diffusion. By the time it was necessary to leave the hotel for our meeting the next morning, I had all of my data plotted and entered into my notebook. It was a fun activity, making me realize that I had better organize my time better so I could do more!

Forest Fire Smoke and Ice Nuclei

On June 14, 1948, our measurements of ice-forming nuclei at the Mount Washington Observatory showed a sudden increase in concentration by a million fold that continued for half a day, and then dropped back to more normal levels. This tremendous increase coincided with the passage of a dense plume of smoke that reduced the atmospheric visibility from greater than 100 miles to less than 3 miles when nucleus concentration reached its peak. By good fortune I had flown to the mountain in a Project Cirrus plane and visited the summit during this episode. Thus I was able to check the reality of this unusual situation.

In less than two months time I was at the Priest River Experiment Station of the US Forest Service in northern Idaho, where I had a chance to collect sap, bark, wood, needles and cones of the common evergreens that comprised the western forests. I had in mind to test the smoke when these materials burned to see whether it possesses any of the ice nucleating properties that we had encountered at the Observatory.

Among the trees sampled were Ponderosa Pine, White Bark Pine, Western Larch, Douglas Fir, Western White Pine, Alpine Fir, Western Hemlock, and Western Great Cedar. I also had samples of Old Mans Beard– a common lichen, and Bear Grass.

All samples were burned and the smoke tested in a cold chamber at a temperature of -35°C. Of all these materials the only substance that showed any ice nucleation effect was the pitch of the Douglas Fir. While it showed some activity, I estimated that less than 1 smoke particle in a thousand showed this property. Thus it appears unlikely that the high concentration of ice nuclei observed within the plume of the big smoke event observed on June 14 was due to the smoke from the trees.

Even today we are still puzzling over the sources of the natural nuclei that appear in natural clouds. Soil particles from dust storms, certain forms of bacteria, lead iodide (the lead from gasoline reacting with iodine from the sea), special smoke particles from smelters and iron foundries, settling cirrus crystals, and frost fragments are all possibilities. Only on rare occasions, do they eliminate the development of supercooled clouds and the hailstorms, lightning, and massive storms that are triggered by the presence of supercooled clouds in the atmosphere. It is indeed fortunate that we don't have all the answers!

The Cutting of Holes in Solid Decks of Supercooled Stratus

The first seeding flight of Project Cirrus was conducted on March 7, 1947 and consisted of a line of flight fifteen miles long, made just above a solid layer of stratus clouds. A month later, the first of a series of L-shaped seeding flights were made in similar clouds dispensing crushed dry ice at a rate of 1 pound per mile. The results were quite spectacular, and provided us with excellent data about the nature of clouds and permitted us to develop methods to obtain quantitative measurements of seeding effects. After reducing the data and presenting at a scientific meeting, a critic made the statement that he was quite skeptical of the reported effects, since he had seen similar holes in clouds that he was sure were natural.

For the next year and a half, the Project Cirrus aircraft and pilots were given various suggestions for improving observational skills and accurate navigation.

During the summer of 1948, while I was in the Northern Rockies observing thunderstorms, the Project Cirrus aircraft were at Albuquerque, New Mexico carrying out seedings of cumulus clouds, and further developing analytical techniques. After I returned from the Northwest, and our planes from the Southwest, we began to plan for some further quantitative measurements of seeding effects in solid stratus clouds decks.

On November 24, 1948, a solid cloud deck formed over New York State, and we made plans for a dry ice seeding flight. This time, instead of our earlier plan to make an L-pattern, the Boss suggested making a giant Greek letter *gamma*, with the straight legs 24 and 29 miles long, and the circular loop having a radius of 2 1/2 miles.

More than a hundred excellent photographs were obtained of the spectacular effects that ended with the complete dissipation of the stratus clouds in a gamma pattern. The clouds were 1,300 feet thick, with the temperature on top being -5°C, and at the bottom of the deck, -3.5°C. The ice crystals generated from 0.7 pounds of crushed dry ice per mile spread on either side of the seeding line, at a rate of 1.3 meters per second during the first 35 minutes.

At the end of the flight the flight controller suggested that conditions were excellent for further studies, so I suggested to the standby crew that they go to the vicinity of the Griffiss Air Force Base at Rome, New York, upwind of our previous operation. Accordingly, in the afternoon another flight operation was carried out that took the form of a race track, with the straightaways 17 miles long, the parallel legs 4 miles apart, and joined by circular arcs. This seeding was done again with crushed dry ice dispensed at 1.7 pounds per mile.

Observers at Griffiss Air Force Base reported seeding to parallel strips with blue sky visible and virga falling from each side of the strips. After evaluating the excellent photos obtained from this flight, it showed that the rate of spread of the hole was 1 meter per second on each side. Appearance of the seeding effect suggested that the racetrack pattern showed evidence of over-seeding.

When the Boss finished his quantitative measurements and delivered them at a scientific meeting a few months later (at which time our critic was present), he challenged everyone to show us similar patterns that occurred naturally.

He ended his talk by stating that if anyone in the audience was still skeptical of our ability to cut holes in a solid deck of supercooled clouds, the next time we did it we would use the GE Monogram as our pattern!

The Skating of Water Droplets on a Water Surface

On November 21, 1948, I discovered a method to cause water droplets to float on water, and to produce bounce effects and other phenomena that could be studied in detail.

Every observant person has probably seen the phenomena of one or more water drops skittering across a water surface. In washing the metal top of an automobile it is a common experience to witness many drops streaking across the wet surface. In rowing a boat or in swimming it is possible to splash water in such a way that such water drops can be seen for a brief moment or two. Some public water fountains are good sources for seeing such effects.

One of the nicest ways to study the skating of water drops on water is to use a large Langmuir Trough. This, as I have mentioned earlier, is a shallow trough of water whose top edges are made hydrophobic, with the trough over-filled with water so that its surface is higher than the edges of the trough. With such a condition, the surface of the water can be easily cleaned by sweeping it with a bar of metal that is chromium plated, and also hydrophobic.

If such a water surface is freshly cleaned and a small water jet projected across the surface and close to it, a great many drops will be seen to skate across the surface. If a petri dish is submerged near the center of the trough and just below water surface, the drops that skate toward it are captured and remain above the submerged dish. When additional drops come in contact they may bounce away or, more frequently, merge with the floating drop. Many drops may do this before the capturing drop suddenly disappears.

If this "target" drop is looked at from above, it will be seen to be separated from the water on which it is resting by a very thin air film. This film has a thickness of about a quarter wavelength, or about 1,200Å, or 0.12 microns. As such, it displays a bright interference color. As this very thin air film becomes thinner and loses its color, the drop merges with the water below.

By projecting droplets from opposing jets located at opposite sides of the trough, the floating droplets bounce against each other, sometimes merging, but more often bouncing away from each other. This phenomenon occurs frequently in the free atmosphere and is one of the factors that hold back the onset of coalescent rain that is generally responsible for the formation of "warm" rain; i.e., precipitation that isn't dependent on ice crystal formation to produce precipitation. It is this type of rain that commonly occurs in the so-called trade wind regions of the oceans, such as around Hawaii and Puerto Rico. By carrying on this study using a Langmuir Trough, I was able to observe the effects of single layers of molecules, as well as visible oil films on the behavior of the skating drops.

Such films play a major role in the formation and stability of the floaters and some films, even though they are only 10Å (0.0010 microns) thick, may prevent them from performing.

Some day I'd like to make a public fountain that is dominated by water droplets floating on water.

The Concentration of Ice Nuclei in the Air Passing the Summit of Mount Washington

Our first experience at the Mount Washington Observatory in the early fall of 1943 quickly established that ice nuclei in the air passing the summit were in short supply. While a few measurements were made of such nuclei during the next few years, they were quite sporadic and qualitative at best.

It wasn't until the beginning of the new year of 1948 that I decided to obtain routine measurements of ice nucleus concentrations. On January 1, 1948, after transporting a cold chamber to the summit in the fall, we began routine observations every three hours, when other weather measurements were made.

A sample of outside air was brought into the 4 cubic foot chamber in the basement of the observatory. Using a temperature of -20°C, the air was permitted to stabilize.

Then a supercooled cloud was formed and the glistening ice crystals were counted, as illuminated by the light beam that covered a small fraction of the volume of the chamber. In this manner the measurements were made. Over a 21-month period, some 4,540 measurements were made.

When evaluated, the data showed that the concentration of ice-forming nuclei in the air passing Mount Washington varied by a factor of 100,000. Even the highest concentration, that was 10 particles per cc, was five times less than necessary to convert a supercooled cloud to one of ice crystals.

A further study of this data showed that the high concentrations occurred more frequently with an air flow from the west to the northwest, and that the weather tended to be related to a high pressure system involving Continental air of Polar origin, having a trajectory from southwestern Canada, across the Great Plains, and thence into New England. Thus it is likely that the nuclei were mostly soil particles.

One singular occurrence involved a 24-hour period when a spectacular increase in nuclei concentration accompanied a great mass of forest fire smoke that originated in the northwestern part of the Province of Ontario, Canada. This smoke was so massive that a few days later it was observed in Scotland, accompanied by a blue sun.

It is a strange and rather disconcerting fact that these routine, three-hourly measurements, made nearly fifty years ago, are still the only intensive continuous measurements of ice nucleus concentrations that, to my knowledge, have been made.

The Simulation of Clouds in a Water Bath

On October 7, 1948, I had an idea for forming a miniature cloud in a container of water. I made a box of 1/4" thick sheets of clear methyl methacrylate plastic (Plexiglas) that I cemented together. It had the dimensions of 16" x 16" x 20" high. When filled with tap water, it was ready for experiments. I took a 1/4" diameter glass rod and bent it so that it would reach the bottom of the chamber, and extend to the midpoint of the chamber floor, where it was bent upward for a distance of about an inch. I then made an orifice of a polystyrene 1/4" tube that was heated and pulled so as to have a constricted hole of about 1/16" diameter. This was fastened to the bottom of the glass tube with a short length of polyvinyl chloride tubing (Tygon). A pint of distilled water was then placed in a container above the edge of the plastic box containing the tap water. A few silver nitrate crystals, weighing about 100mg, was dissolved in the distilled water. A pinch clamp was placed on the plastic Tygon tube connecting the distilled water source and the upper end of the glass tube descending into the tap water in the plastic box. A projection floodlight was placed above the box to illuminate the water. When turbulence had died out in the water of the box, the pinch clamp permitting the distilled water to flow out of the submerged jet was opened a bit, and a stream of distilled water was seen to rise toward the top of the water bath. Reaching the water surface, the distilled water spreads towards the edge of the plastic box. As soon as the diluted silver nitrate solution came in contact with the tap water, it reacted with the sodium ions to produce silver chloride– a white colloidal precipitate. With a very small flow, a spherical tube-like stream rose with a laminar flow.

With a slightly large flow, the laminar flow begins to wobble, and with still more flow, the wobbles begin to peel away from the vertical stream to form vortices. These vortices then form secondary and tertiary units of the same basic form. With still more flow, the laminar features disappear and turbulence develops, akin to the structure of a cauliflower-like convective mass.

After observing the patterns just described, I discovered ways to produce strong inversions, unstable environments, precipitation, chemical reactions, and other things that have counterparts in the atmosphere.

In more recent years I have found it possible to produce all of these effects in an ordinary aquarium, thus making this exciting cloud simulator available to most anyone.

Movies of the Grand Tetons

The first time I drove west it was by way of the Wind River Mountains, the Wind River, Dubois, and Togwater Pass to Moran. We spent the night close to Jackson Lake. In the early morning, I arose early and drove to the top of Signal Mountain that is the "grandstand seat" of the Tetons. I set up my time-lapse movie camera and obtained an early morning cloud sequence. After breakfast, I returned to the top of Signal Mountain to take pictures of the beginning of convective activity in the vicinity of the high mountains.

Signal Mountain is an excellent location to obtain time-lapse movies, so that a few years later, when we had the Munitalp Cloud Atlas project going, I arranged with the Forest Service to have a camera at the summit so the Fire Observer could get us some cloud sequences.

During one of these times we spent the night near Moose, Wyoming, at a ranch. A spectacular series of jet stream clouds were moving across the area from the northwest. Taking a chance that they would still be present in the early morning, I arose before daybreak, set up my camera at a vantage point, and was favored with an alpine glow on the summit of the Grand Tetons and its neighboring peaks at sunrise. I continued taking pictures from time to time during the day and, at sunset, was favored with a spectacular sequence of clouds, with the Grand Tetons silhouetted against the fast-moving clouds.

This sequence was so good that Walt Disney invited me to his studio and purchased the footage, using it in one of his full-length nature movies.

The visit to Disney's studio was an interesting adventure. I was ushered into his office, passing through a rather narrow hallway lined with cases in which his Oscars and other awards in great numbers were ensconced. I found him to be a fascinating fellow, very friendly, and full of energy and fun. We talked about a variety of things. He mentioned that, at the moment, he had a large company in the field that required sunshine and dry weather. He mentioned with a wry shrug that his forecaster was promising rain!

The sequence of time-lapse movies obtained at the Grand Tetons is one of the best in my collection, and has been used many times in my lectures about the jet stream.

Soil Samples from the West

As a result of my ice nuclei studies that started in the mid-forties, before my discovery of the dry ice method of seeding clouds, I developed a continuing interest in the effectiveness of natural soils in ice-nucleating abilities. While I realized that such studies were primarily of academic interest, since I couldn't see any way that heterogeneous nucleation could possibly supplant homogeneous nucleation, with the possible inception of load and silver iodide. Nature was often dependent on the former type of nucleation. Consequently, on the yearly trips west, I made it a practice of stopping every several hundred miles to gather soil samples. In addition, I had solicited samples from friends in other parts of the world, ranging from Mexican volcanoes to the "outback" of Australia.

While I never got around to testing all of the samples I had assembled, I did manage to evaluate several dozen of them. Interestingly, the most active of those sampled, I gathered less than ten miles from my home. It consisted of very finely divided rock flour, in the form of a glacial lake deposit from post-glacial Lake Albany, in Guilderland. Of all the western samples, I found that the most active nuclei were from weathered volcanic soils gathered in the field north of our Arizona Field Station, at the Research Center of the Museum of Northern Arizona, near Fleischmann Hall. Its threshold of nucleating ability was at 0°F, while that of the clay deposit near Guilderland was active at about +10°F.

I had an amusing confrontation with an agricultural officer near the Oregon-California border, as I was entering California by car with my family. When stopped, he asked me if we were transporting any fruit or other materials that might be damaging to California agriculture. When I told him that a particular box contained soil samples, his interest perked up and he told me that he'd probably have to confiscate them. I protested such action and told him the bottles were sealed and I had no intention of opening them while in California. He remained adamant until I pointed out to him the layers of thick dusty soils on the hubcaps and other parts of the wheels of the car. These had accumulated all the way from the Mid-West where, after getting established in a motel and having dinner, we frequently explored the local countryside, often traveling on dusty dirt roads.

After this was pointed out, he figuratively threw up his hands and left my samples alone!

The Old Ranch of the School of Mines

During the summer of 1949, I spent a fascinating time at the ranch of the New Mexico Institute of Mining and Technology. This was a vast acreage that extended for several miles into the foothills of the Manzano Mountains, directly east of Albuquerque, New Mexico. During World War II, the School of Mines, headed by Dr. E.V. (Jack) Workman, was intensively involved in the development and testing of the Proximity Fuse. This was a detonating fuse used in shells to cause them to explode when they came near a metal object. There was a huge, all-wooden tower located at the test range, and the western slope of the ridge that constituted the boundary of the ranch was literally covered with exploded projectiles that were used in the testing procedures.

With the enthusiastic support of Jack Workman, the Project Cirrus operation was given temporary quarters at the Ranch, where we used his radar and expansive terrain as a site for testing cloud seeding using silver iodide ground generators and B-17 aircraft, the latter being based at the Albuquerque airport.

The Boss loved the desert and its vast reaches, and he was in his glory that summer. Our planes and ground generators seeded the large cumulus that formed over the Manzano Mountains virtually every day during the summertime. Using zero lift (balanced) balloons that were released from the ground generator site, we were able to estimate the direction and altitude of the silver iodide plume, while at other times, using radio communications with our B-17s, we watched and photographed the effects which occurred when large cumulus congestus clouds were seeded with dry ice. A 3cm radar operated throughout the seeding periods, and with the excellent radio communications we had with the seeding and photographic planes, we carried out a wide variety of interesting experiments.

During part of the operation I manned a station many miles to the south of the Ranch, where I obtained some excellent photographs of seeded cumulus which, later back home, showed the relationship of the seeding to the growth of the radar echoes, and provided excellent proof of a successful operation.

One day, Jack Workman reactivated one of his artillery guns, fitted up a hollow shell with liquid CO_2, and programmed the projectile to explode within the supercooled cloud. It worked beautifully, but was only of academic interest since, it would be impractical to use this method in routine seeding operations.

A strange and amusing happening occurred while we were at the Ranch. One day a group of National Security personnel approached us during an operational day, and discovered that our B-17s were operating over the Manzano Mountains, taking pictures of clouds and the general area. They requested that all such photos be sent to them for review. This was done and, to our surprise, when the photos were returned we found that in all of the photos taken to the north of the Ranch, a rectangular portion of the picture had been neatly cut out. This was a puzzling action from our standpoint. We then realized that the cut-out portion was a view of a relatively small, isolated mountainous hill that had previously mildly puzzled us, since it was surrounded by several tiers of high metal fences. While we were never sure, we strongly suspected that tunnels had been excavated in this hill for storage of the early stockpile of atomic bombs, made up the river at Los Alamos.

If the security folks hadn't bothered us, their secret would have remained intact. By this senseless use of a razor blade they had pinpointed an area that they had hoped would remain a secret. Even more ridiculous– they had doctored only a print– we still had intact negatives that could have been used to prepare as many enlargements as we had wanted. As it was, we didn't bother to do so, since there was nothing of Project Cirrus interest in any of the prints so treated.

The 1949 season at the Ranch was the final year of our storm seeding research in the vicinity of Albuquerque. The following year we shifted all of our New Mexico research activities some 80 miles to the south, and on the west side of the Rio Grande, at the new facilities of the School of Mines at Soccoro, New Mexico.

At both places the Boss was in fine fettle. His sequences of black and white photographs, taken with his trusty 35mm Leica, and his meticulous notes of everything that transpired, are beautiful examples of the care and detail he devoted to everything he did. I can still see him with the wind blowing his hair, his tie-less shirt open at the neck, his bronzed face, and the look of one who is having a wonderful time.

The Localized Seeding of a Line of Towering Cumulus

One afternoon, during the cloud seeding activities of Project Cirrus at the Ranch of the School of Mines east of Albuquerque, the mission had ended and our radio contact indicated that the B-17s were about to head for the airport. A line of five towering cumulus were visible to the north, in a region outside of the day's activities. I asked the head of operations if the seeding plane still had a supply of dry ice, and if they could see the cumulus visible to us. Both answers were affirmative, so I suggested that the seeding plane head for the line of cumulus and seed the second tower from the west.

The suggestion was accepted and the plane immediately headed for the cloud, which was seeded with a few pounds of dry ice. Within a few minutes after we were informed that the seeding had occurred, that cumulus began to show signs of modification and, in less than thirty minutes, the clouds had completely precipitated, leaving a strange-looking line, with a big hole existing in the line of cumulus.

It was a spectacular demonstration of pinpoint modification that added further confidence to our ability to transform suitable supercooled clouds.

Chapter 8
Research in a Variety of Areas

The Nucleation of Steam in a Turbine

On February 28, 1949, I had a visit with Dr. John Fisher, who was with Herb Hollomon. He was interested in ideas for nucleation in steam in a turbine. Apparently there was a lack of sufficient nuclei for water droplet formation in the high-velocity movement of steam in such a device and, consequently, large drops formed, which caused cavitation and other mechanical wear on the turbine blades. Presumably, with more condensation nuclei present, the condensed steam droplets would be smaller and less erosive.

I told John that in the fall of 1946, a similar question was raised with me by Mr. Glenn Warren, Superintendent of the Turbine Department, and Mr. W. L. Fleischmann of his staff. I told them that I would recommend that they either arrange to have heated wires of nichrome either in the boiler, or along the entry tubes to the turbine, since such wire, when heated to a dull red, produces very large numbers of nuclei. An alternative would be to have a series of electrodes that could be activated by a Tesla Coil to ionize the air. Either of these methods would produce very large numbers of condensation nuclei and, in fact, were commonly used in my laboratory demonstrations to produce spectacular increases in cloud density.

I told John that there are other methods that could be used, but they were not as simple or innocuous as the nichrome wire or Tesla Coil. He was grateful for the ideas and told me that he was planning to make a demonstration unit, and would try out these ideas. I never did hear whether or not he found these methods effective.

His visit and questions were typical of many that occurred quite frequently at the Laboratory. We considered them to be part of our research activities and welcomed the contacts that they provided, since they always broadened our horizon. I am sure that some of our suggestions were useful and solved the immediate problem or, in their case, widened their horizon. We rarely followed up on such visits unless the questioner initiated further contact.

In this manner our professional aid was sought on an every increasing scale, and we had a good time.

An Artificial Snow Storm in Miniature

On May 5, 1949, I prepared a two-page "essay" in my Lab Notebook #3986 on a series of thoughts backed up by simple experiments. The essence of my understanding was that an object colder than the ambient air, when put into a supercooled cloud, quickly frosts up and, within a short time, begins to shed fragments of frost. These fragments then begin to grow at the expense of the supercooled cloud droplets, since the vapor pressure of ice is lower than that of water. The water molecules thus always move toward the ice and soon produce an ice crystal on every one of the frost particles. Sometimes the crystal is symmetrical, but more often it has an irregular shape.

I discovered several very nice arrangements to illustrate this phenomenon. By placing a flat metal plate, having a temperature of -10°C, above a tray of water having a temperature of 1°C and positioned 3/4 of an inch above the water surface, and illuminating it with a parallel beam of light, a veritable snowstorm in miniature can be seen.

A smaller difference in temperature is also effective, but it takes a longer time to see it.

My original observation of this phenomenon occurred in the early forties, when I was using my abandoned greenhouse to produce frost crystals. It was located at the lower terrace of my land along Schermerhorn Road, where an ancient road climbed the hill to the woodlands beyond. I had placed a large bucket of hot water in the greenhouse in the late evening to raise the moisture level of the air so frost would form on the under surface of the glass in the greenhouse. After a cold, clear night, I would find magnificent large frost crystals covering the glass surface.

One morning, I was in the greenhouse just before sunrise and had been examining the frost crystals with a flashlight and a pocket lens. The sky was clear and the air quite cold. Just as the sun rose above the horizon and illuminated the greenhouse, I noticed the air to be suddenly filled with ice particles that were falling from the frost coatings of the greenhouse glass. I was much puzzled with this occurrence, but then concluded that it must have occurred because of the temperature difference between the glass surface, the adjacent crystals, and the still cold air in the greenhouse, that led to a thermal shock. I remembered that most of the glass of the greenhouse was ancient glass, probably at least a hundred years old. It was of uneven thickness, and had a greenish color that I have been told is characteristic of a manganese impurity in the glass. Such glass absorbs infrared quite readily, and thus could have produced a thermal stress on the crystals.

At about the same time, I had occasion to observe another fascinating and very intriguing phenomenon related to the surface of dry ice, or of anything else colder than -40°C (-40°F). When exposed to ordinary air at room temperature, a layer of frost quickly coats its surface. As soon as the frost begins to form, vast number of icy particles stream from the surface of the dry ice. To show that there is nothing peculiar about the fact that carbon dioxide is sublimating from the surface, the dry ice can be placed at the bottom of a closed plastic bag. The stream of ice particles continues to stream away as before. However, there is another fascinating thing to be seen. If the surface of the cold plastic is viewed with a hand lens of 10X magnification, the frost that forms there assumes the shape of tree-like dendrites. The most interesting feature of these particles is the fact that they appear to be made of many separate particles, held together with an electric force. The dendrites are not at all rigid, but frequently twist around as though they have many loose ball joints. Every so often one of these dendrites will suddenly shoot away at high velocity, and sometimes can be seen to break apart. If such a set up is held above a supercooled cloud in an open-top cold chamber, these fragments will readily seed the cloud and grow.

There are many puzzling features about this phenomenon still not understood!

The Production of Snow at Ski Areas with Snow-making Jet Nozzles

On May 25, 1949, I discovered that by taking water at 0°C and spraying it at pressure of 90-100psi, a large number of ice crystals are formed having spherical centers. It seemed likely that the larger spray droplets are seeded by contact nucleation.

Again on August 25 I returned to the same theme, just prior to leaving for the Project Cirrus operation in Albuquerque, New Mexico. On my return I planned to utilize a cold room operating in the temperature range of -10 to -14°C. This method could be used to produce either granular ice particles or fluffy snow that could be controlled by the ambient temperature, the quantity, and the size of the spray particles, and the pressure used on the jet nozzle. Katy Blodgett witnessed both of these lab book entries on May 26 and August 25, 1949.

An Adventure in England and Switzerland

This consists of a copy of my daily log entries which presents a brief picture of the type of activity, meetings, and adventures that occurred quite frequently during my twenty-five year sojourn at the General Electric Research Laboratory.

The main purpose of this particular trip was to participate in the Centenary Celebration of the Royal Meteorological Society held at Oxford and London during the early spring of 1950, at which I had been invited to present a paper on weather modification. While abroad, I also accepted an invitation from its Director to spend a few days doing research at the Snow and Avalanche Research Institute on the Weissfluhjoch at Davos, Switzerland.

The special lecture at the Physical Society of London had not been planned, but developed when Sir Watson-Watts (said to be the primary inventor of radar) heard my presentation during the Royal Society meeting and asked me to present it at a hastily convened meeting of the Society. I had a good time!

Thursday, March 23, 1950

To New York City by train on 0832 with Mr. Thomas of Int. General Electric. Drs. Langmuir, Landsberg and Harker on same train. Arrived N.Y.C. about 1 hour late. Met men from Saudi Arabia and Nicaragua at General Electric headquarters on Lexington Ave. Afterward, to American Museum where I saw Gordon Atwater and John Saunders. With John to see new Ecology exhibit. It is coming along but slow. To flower show in evening. Not so hot! To bed at Hotel Commodore fairly early.

Friday, March 24, 1950

Up at 0715. Met Kiah Maynard and Bernie Vonnegut and had breakfast at Commodore. To GE Building on Lexington Ave. for Project Cirrus Steering Committee meeting. Fairly good meeting in afternoon. New Mexico trip OK if desirable. Got tickets for London and Zurich from Lynch and some English money from GE cashier. Left GE building at 5 pm and met Ed Fogarty (my brother in law) at laboratory (Western Electric) on West Street and went by ferry and train to Rutherford, New Jersey. Had dinner with my sister, Gertrude's, family and spent a nice evening. Had call from Lois and the kids. Talked to my sister, Margaret, in Hawthorne. Left Fogartys about 11:30pm for N.Y.C. In bed at Commodore at 12:30. Letters from Ray Falconer and Lois. Day was partly cloudy though only fair weather cumulus most of the time.

Saturday, March 25, 1950

Up at 7:30. Breakfast at Grand Central. To N.Y. Public Library. Nice exhibit of printing. Light rain occasionally. To St. Patrick's for noon Mass. Back to Grand Cent. Rest for dinner. To air terminal at 2:15. Checked in and then went to Idlewild. T.O. in P.A.A. Flying Cloud. Climbed rapidly and hit base of rugged stratus no turbulence at 4500 and emerged at 8000' -7°C. About 10000' saw thin ice crystal clouds very white and with brilliant pillar and refl. sun dog. Perplexed until I saw that the snow was falling mass of cirrus. We reached successively thicker cirrus as we continued NE and at 23000' with temp at -37°C we were in and below thin cirrus. It was forming in air probably -45 to -50°C halo around the setting sun with portions below. No sun pillar or very little in cirrus. Heavy snow seen near sunset. Down at Gander with rough engine for about 2 hours. Snow flurries.

Sunday, March 26, 1950

0940 GMT 1510 mi from Gander NF 23000' 288 mph S 20 284 mph -25°F -33°C. London temp 50°F. Will contact MizenHead S coast Eire 11:30. Mr. Riches Montclair Newfoundland-Greenwood Lake. Arr. London Cu 1:30 PM. Left for Hotel Savoy by bus and cab and then after unpacking left for a walk around town. Headed along the Thames to Black Friar Bridge and then over to St. Paul's Cathedral, a huge church. The area around was badly damaged with entire blocks completely demolished and much still desolate. Headed back to hotel toward evening. Had fine supper of Filet of Sole, and it was better than in Park Plaza and 12/60. To bed fairly early. Excellent room and bath at 6/18/42 nights. Bed felt excellent and room is very quiet and overlooks Thames.

Monday, March 27, 1950

Left hotel early and started down river. Before this I went to see Haynes at IGE who fixed up airplane ticket to Switz. and return. I then returned to Strand to Charing Cross and thence down hill to river looking for Westminster Bridge. Thence past Big Ben and Houses of Parliament and on to Victoria Bridge. From there I continued in southerly direction back away from river looking for Westminster Cathedral. Could not find it but believe this was due to fact I was too close to river. I then went to Westminster Abbey and spent several hours in this fascinating place. Heard harpsichord played. Sun shining through big rose windows was very beautiful. From here I went to Horse Grounds, St James Park, Admiralty Building, Trafalgar Square thence to hiking shop where I bought several maps and guide books. Before this I went through Covent Garden market. Then up Museum Street to British Museum. Saw library, Mendenhall and Montezuma treasure and much of early church treasures– most beautiful. Left about 5 PM and returned to hotel late in afternoon. Stayed up till midnight reading guide books.

Tuesday, March 28, 1950

Day cloudy all day. Up at 7:00 and left about 8:00 for Paddington Station. Caught train for Didcot. On train had breakfast of beans, sausage, coffee. Changed to Oxford train at Didcot and arrived at 11:15. Took cab to Christ College and found my room TOM-6-2 just opposite registration office. Met Ashford and several others. My room is most interesting– very chilly, plainly furnished in bedroom, nice in living room which has nichrome heaters. Walls about 4' thick. This part of the college was designed by Wren and built under direction of Cardinal Wolsey. It looks like very ancient and has a beautiful yard! The hall where we eat meals is magnificent. Interesting meetings in afternoon. Had bull session at night with Rex, Willet, Rossby, Eaton and ----.

Wednesday, March 29, 1950

Up at 6:30. Very cold but I slept well– warm after putting on socks. Checked over slides and made rough notes for talk. Took walk up to Clarendon Lab by another route. Yesterday, I went up High Street to St. Mary's and checked up on mission for Father Finn. Only writing I could find said DOMINA NOSTRO Our Lady. In morning I presented talk that was well received and enough interest was shown to require an evening meeting. In afternoon I left rather early and walked down Longwall to high up hill and then down by St. Ebbes– thence to gardens below Christ Church thence back where I decided to catch up. Day partly cloudy, sun out early and late.

Thursday, March 30, 1950

Up at 6:15 with brilliant blue sky. Took walk up to St. Mary's, High St., Radcliff Camera, Longwall St., St. Peter's of the East, and other regions for photos. Took quite a few. Then back to breakfast. Had porridge, eggs and coffee. To meeting via road up St. Aldate's. A rather heavy session with Rossby, Petterson, etc. More interesting discussions with Brewers, Palmer, et al. Had many compliments on talk. In afternoon I left for London on 2:05 and arrived at Royal Geographic Society. Met Seligman, Bill Ward and several others and then John Hardy. Good to see him. Made plans to meet him at RAE on Thurs. With Seligman to meeting of Glaciological Society. Small but interesting group. After had dinner at Athenaeum Club– very swank– had sole– not so good. Rode with Sutherby of Bath back to Didcot. Arrived at Town Gate 12:00 PM. Read for awhile and then to bed. Had darndest dessert– toast with gooey mushrooms!

Friday, March 31, 1950

Up 7:45. Was awakened by Englishman who takes care of area who shined my shoes and fixed hot water. To breakfast and then rapid walk to Lab. Rather interesting talks by Sir Johnston, Sir David Blunt, etc. I like English humor! At end of meeting back to Tom Quad and then back to lunch. Thin pie, potatoes and water. On guided tour to some colleges and Bodleian Library. Most interesting since I saw several gardens that I couldn't get into previously. After tour was completed I went to Botanical Garden and then to Christ Church Meadows and thence back to Quad. Dinner of fish, potatoes and a tartlike pie. After went to play "The Man with Load of Mischief." OK. Thence back to room to write and then to bed. Cloudy all day. However, took a few more pictures of buildings, flowers, etc.

Saturday, April 1, 1950

Partly cloudy, cool. At Bedford College which was reached about noon. To exhibits of Met. Services in afternoon and Conversazione in evening. Left Oxford at 10:15 after walking to the station. We certainly had an interesting time at that place. Reaching Paddington, I engaged a cab and shortly after reached the college. My room which is on the 3rd or 4th floor overlooks the gardens of the college in Regents Park. After luncheon, most of us left for Harrow where we saw a most impressive display of meteorological instruments, techniques, etc. We then returned, had dinner, and then left for the rooms of the Society. This was a rather formal affair although I did not dress formally. My movie seemed to make quite a hit and Sir Robert Watson-Watts asked me to present it before the Physical Society on Wednesday.

Sunday, April 2, 1950

Partly cloudy, heavy showers with sleet in morning. To St. James at Spanish Place, Westminster Cathedral, Waterloo, Blakingston, Salisbury Cathedral– Amesbury and Stonehenge. No rain after Salisbury. Arrived located Cathedral from train and headed for it. A very inspiring building– the finest I have seen. After taking quite a few pictures I searched for and found bus to Amesbury. Arrived, and found way from little Irishman. Stopped at Avon Hotel and had best meal so far in England. Roast beef! Started up onto Salisbury Plain and was fascinated by it. Reached Stonehenge and took quite a few photos. Returned to London, retrace route, Stonehenge, Amesbury, etc. Interesting map discussion with Prof. Manley of Bedford College.

Monday, April 3, 1950

Up early, had breakfast at Bedford College and left for Imperial College at London Univ. On way to meeting met Palmer and never did get to the meeting. Left at 11:45 for Cambridge where I arrived at 1:30. Hunted for Cavendish Lab and had quite a time. Finally found it but everything was closed tight. Then to Christ College but Ted Prince was away. I then went over to the Backs and obtained some nice photos (I hope). It is very beautiful from the backs but the town is quite a mess with nothing very picturesque. I have a more kindly feeling toward Oxford. Left on 4:10 train and back to Bedford about 6 PM. To Centenary dinner where we had a grand time– very formal– with toasts and much show. It was interesting. Back to Bedford with Dan Rex.

Tuesday, April 4, 1950

Day mostly cloudy with some rain, clearing in evening. Up early at Bedford College and headed for Aldermaston Court via Reading and Aldermaston. Met at train by driver and arrived at the Court about 10:30. Met Dr. Albelime, Dr. Billig, Dr. Haines and a number of others. They are doing some beautiful work with electron microscope and some exciting new work which may give map. of 1 x 106! This latter was electron diffraction and conversion lenses using wave fronts. Saw work in Cd sulfide, tellurium, selenium, etc. They are working in thin films of germanium. At 5PM left with Dr. Billig to his home for dinner via Windsor Castle, Runnymede, and Hampton Court. Chilly homes! Nice pictures. Billig is fine fellow. His brother is doing some interesting things with reinforced concrete. Home (to Savoy) about 10:30. Checked in, wrote and to bed 12:30 PM. Fine time!

Wednesday, April 5, 1950

Up about 7:30 after fine sleep. Left via Waterloo for Kew Bridge where I crossed to Kew Gardens. Entered at 10 AM and wandered length and breadth of gardens including one greenhouse. Took some photos. Some of the early flowering shrubs were very beautiful. Had bunch of grapes and coffee and 3 sweet during all of day. Left for Nat. Science Museum at 2 and was impressed with mineral exhibits. Zoology N.G. Then to Physical Society meeting where I had a nice audience for a talk on Exp. in Cloudy Skies. Introduced by Sir Robert Watson-Watts who was very complimentary before and after. Met Dr. R.W. Lunt, I.C.I. and had a nice trip with him to Chelsea on the way back to hotel. Am to have dinner with him and daughter. To Victoria and Albert Art Inst. Beautiful stuff. In evening with Lunt to Prospect of Whitly Inn below London Bridge. Very interesting.

Thursday, April 6, 1950

Up early. Left hotel and went to Roman Wall. Took some pictures of wall and diggings. Then toward Tower of London. Blundered into Billingsgate fish market and had most interesting time. Should have some rare pictures! Then to Tower of London and London Bridge which I crossed. Took train to Waterloo. Bought two pounds of grapes and then just managed to get train to Fairborough. Reached R.A.E. and met Hardy and Frith. Had some fine talks with them. Then to Hardy's home for lunch. Met Mrs. Hardy and Mrs. Frith and had excellent dinner including huge gooseberries. Headed back about 3:30. Met Mr. Mearns at Savoy and went to his home. Met Mrs. Mearns and two daughters, Jean and Margaret. Grand people. Back to Charing Cross and to bed about midnight.

Friday, April 7, 1950

Very beautiful, warm cloudless day. At 13,500' -17°C. Smog over London– hazy below 10,000 feet even in Switzerland. Up at 6:45. Packed and left Savoy at 8:00. Arr. Kens. Air Sta. then to Airport. Passed through customs and boarded plane at 9:30. Corvair. After take-off we headed over through cloudless sky Westminster Abbey and down Thames. Took some pictures around London and thence around Dover and Cape Griz Nez. Much evidence of bombing. Took several pictures of intensive farming in northern France. Approaching Switzerland saw snow on high rounded ridges. Dropped down into Zurich about 12:30. Arrive at Zurich city and after some confusion I finally located porter, baggage, ticket and train in plenty of time. After starting had an excellent meal as we went along Lake Zurich. Changed at Landquart and finally reached Davos dorf– it is beautiful with lots of snow. Met Marcel– took walk on promenade– had supper and to bed after writing Lo. Letter from her. To bed fairly early.

Saturday, April 8, 1950

Up at 7 AM. Had breakfast and met Marcel about 9:00 AM. Bought boots, rucksack and rented poles and skis. Boarded the funicular to Weissfluhjoch. It took about 20 min. to reach top station. Went through Avalanche Lab. It is very fine with the very best of facilities and equipment. We then went down to beginning of trail to Wolfgang and for next few hours had some grand skiing. The skiing is simply marvelous with tremendous slopes of every type of terrain and unlimited room, without trees or stones to worry. As we approached lowlands the trail wound through some nice woods. Returned to Davos by private car. Decided to go up again in afternoon and this time went from the summit Weissfluhjoch down the route to Klosters 10.5 Km. Again it was magnificent skiing. Arrived just in time for train. Had fine bath and then to de Quiervain's for a fine dinner. Left shortly after 11 PM.

Sunday, April 9, 1950

A very beautiful cloudless day. Up about 7:00, had breakfast and then to Church. A very beautiful simple church located only a few hundred yards from the hotel. Church was High Mass lasting from 8:30 to 9:45. It was quite inspiring to watch the people and listen to the very beautiful singing. After Church, I walked to Davos Platz and took train to Davos Gloria, a small mountain town to the south. Took a number of pictures of old houses and church. I then started upstream and stopped at Frauenkirche to photograph the "avalanche" church. Had my lunch (and what a lunch) alongside a tiny, clean looking mountain stream. Finally reached Davos Dorf about 12:30. Rested, washed, packed and met Marcel and Rita at 3:30 to leave by train for Pontresina. Boarded train and went through some very beautiful mountain country. Arrived about 6 PM at Hotel Bernina. Very nice in every way.

Monday, April 10, 1950

Arose about 7:30. Breakfast at 8:00 and then to Pontresina station for train to Berninahauser. Beautiful day, nearly cloudless but with a few cirrus. Put on skis shortly after getting off train and started the climb to Diavolezza hut. It was most interesting though strenuous. Climbed for about approximately 3 hours starting at 11t. 2049 m and climbing to 2977 m or 928 meters. After resting for about an hour and having tea and lunch, we started down to Pers Glacier. The scenery was magnificent and I took pictures 35mm and 16mm. The descent from the hut to the glacier was steep but fast but I had very little trouble. We then crossed the glacier and climbed to the SW side of Pers Isle. Here the going was very steep and I took a couple of good tumbles. Finally reached the Morteratsch Glacier. This was a magnificent run– fast but wonderful. Took a number of glacier pictures on the way down. Arrived at foot of glacier at station of Morteratsch just in time to board train. Arrived back at Bernina Hotel about 3:30. Walked through town, had tea, then lunch after walk around town.

Tuesday, April 11, 1950

Left Pontresina (Hotel Bernina) at 10:00 AM. Previously after breakfast we walked up to the edge of the mountain and passed some very nice old mountain houses. Stopped at an ancient church built in about the 10th or 11th century. It had some nice old paintings, several layers in fact. An old castle was nearby. Saw some door hinges and smaller ones quite similar to those I have from Palatine Church. We then walked down to station and boarded train for Semaden, changed to train to Filisur then to train for Davos Platz. The day was overcast with an occasional trace of sun and spit of snow. The route between Filisur and Davos Gloria is very beautiful with fantastic gorges and high mountain villages. It would be fun to explore this area with Lois and the kids. Arrive Davos Platz and Hotel Meirhof about 1:15 PM. Had 3 letters from Lo and then lunch. Will have de Quervains for dinner at hotel. Walked through snow on promenade home.

Wednesday, April 12, 1950

Up at 7:00 AM. Had breakfast and took skis to funicular to Weissfluhjoch. Met Marcel and at 8 started for laboratory. Arrived and immediately started experiments. Met staff shortly after arriving. Nice bunch. Made attempt to produce replicas of lake ice surface and found their 2% solution too dilute. Made it to a 3% and found I could make excellent replicas showing fine optical effects, grain boundaries, etc. Made replicas of spontaneous nuclei in -40°C cold room and found they were identical to those made in cold chamber at Schenectady. Made AgI with Ag sparking and iodine vapor. Made replicas of blowing snow, frost, granular snow. Everything worked fine. Marcel seemed happy about situation. Took lapse time movies of snow showers from lab. Dull day with snow showers most of time. Skied down from Weissfluhjoch to Hohenweg station and then to Meirhof. Had evening with Marcel and Rita and Andre Roch and wife. Fine people.

Thursday, April 13, 1950

Left for Weissfluhjoch at 8:00. Snow in valley with low clouds. Reaching summit found it above most lower clouds with very beautiful cloud effects. Spent nearly an hour on summit above labs taking lapse time movies. Snow came in of graupel and stellar and I went down to lab. Spent most of day working with Marcel getting replicas of glacier ice. Prepared some nice samples using several grades of emery paper, finishing polishing with chamois, velvet and paper. Used sublimation etching and produced some beautiful replicas of grains at -10°C. Nearly missed train down due to enthusiasm. After fine dinner at Meierhof I went to Davos Platz with Marcel and Rita, Andre Roch and wife and gave talk before Davos Nat. Sci. Society. Very well received. Room beautiful– dating to 1664. Before dinner went through Met. Obs. of Dr. Morikofer at Davos Platz. They are doing very nice work with radiation studies.

Friday, April 14, 1950

Left Davos Dorf on 9:44 train after packing, paying bill, etc. Very beautiful with snow sticking to everything. Met Marcel at station and journeyed with him to Zurich via Landquart. Day was cloudy with high summits in cloud. Reached Zurich about 12:30. Had dinner at station and then Marcel got car and took me to Swiss Met Office where I met about 50 scientists including meteorologists, physicians, etc., including Dr. Lugeon, Director of Met Office, and Dr. Wanner, Asst. As soon as I arrived I found that they would like me to give a talk. This is I did and was received with much enthusiasm. Prof. Sanger, a grand gentleman, was most enthusiastic and asked me to visit him at Polytechnic tomorrow. He has accepted leadership of national group established to study hail problem. Dr Lugeon also invited me to visit him on Saturday. After fine dinner had most enjoyable visit with Marcel's sister and his brother and wife. Home 11:30 PM.

Saturday, April 15, 1950

This morning several inches of snow covered the trees and ground from a wet snowfall during the night. Had a fine sleep and good breakfast. Hotel is very comfortable and quiet. Met Marcel about 9:20 and he took me to Polytechnic where I met and had long talk with Prof. Sanger. Gave him my proof of Exp. Met. Incl tables which had been left out. Saw projection television and some very nice scintillation counters that give amplification of 1010. Met Dr. Lugeon at 11:30 who took me on a nice drive through city and surroundings ending at his home where I met his wife and had a fine meal. Then back to hotel where I proofread Part 2 of the paper. Met Marcel and Rita at about 3:30 and explored more of the city and several small villages and woods near Zurich. Summits of mountains in clouds all day with occasional light rain. Damp but pleasant with many early spring flowers covered with patches of snow. Back at hotel at 6 PM. Expect to go on trip with de Quervain tomorrow.

Sunday, April 16, 1950

Up at 6:15 to St. Martin's to Mass, met Rita, Marcel and his brother and wife at 9:00 and we headed into country. A rainy day with clouds covering the mountains. Went by the See of Lug and Lucerne stopping for a little while at the latter. Very interesting region. Before this we went through beautiful limestone cave called Hollgrotten Baar. It had a remarkable botanical exhibit– 2 living roots from spruce extending 20' through the chamber to the water below. Once that took 4 years to reach water about 9 ft. Trees 10' above roof. Many ferns around including Harts Tongue. Stopped for dinner at Brienz on lake. Brienz thence to Interlaken and Grindelwald. Very impressive though peaks in clouds. Had tea and cheese in home of clergyman. Then to Thun and back into very beautiful Bernina farming country. Met an M.D. friend of de Quervain quite by accident and had supper of bread, tea and local cheese. Excellent. Beautiful house at Zazieil and Signeau– tremendous size. Returned about 9:20 in the rain. From Thun to near Zurich weather was bright and clear.

Monday, April 17, 1950

Up at 8:00. Sunshine with some st and cu clouds. Will take walk into town this morning for a few pictures. Down to Lake of Zurich and photographed building cumulus. They would go into towers to about 10,000' then flatten off downwind into stratus. Occasional rain fell from them. Took a number of pictures at 3 min intervals on way down and at lake. Returned to Waldhaus at noon via Dolderbahn. In afternoon walked through beach woods to summit of Dolderberg and then down through pretty ravine to town. Again along lake to different portions of old town. Wandered back toward Waldhaus late in afternoon. At hotel cleaned up and packed for morning. Watched sun sink below mountains and a sail becalmed on See of Zurich. Very beautiful. Bells rang for 5 min. Dinner with de Quervain and wives. Very grand people. I hope I shall see them again.

Tuesday, April 18, 1950

Up at 6:30. Had breakfast, finished packing and by train to Bahnhof. Thence by bus to Zurich Airport. Partly cloudy with .8-1.0 cover not very thick. Flight from Zurich via Basle. High alps had active cumulus (orographic) and were quite beautiful. Thickening clouds toward France– thicker over England 6500' -0°C. Light rime toward London with what looked like snow. Illumination very poor. To IGE in London. Talked with Seligman re glacier replicas, promised note. Light rain. Left airport terminal at 6:20 T.O. PAA 8:00 PM. 10,500 -22°C below cirrus, nearly cloudless over Ireland. To land at Shannon. Approaching Shannon we suddenly turned and started back to London! Halfway there we turned again and finally landed at Shannon with news that one engine was bad and we would transfer to a Constellation of TWA.

Wednesday, April 19, 1950

Spent from 10PM last night to 5:15 this morning trying to get some rest at Shannon. Spent quite a bit of time talking to Capt. Jesse Bird of Baldwin, LI, about 30,000 ton tankers and lumber op near Bernadillo. Took photos of cirrus above us. Finally off at 5:13 and at 10,000' we headed toward Gander. At 0800 we were at 53°15N and 23°W 10,000' +2°C with cloud deck (solid) at 5000', 2500' thick. Nearing Gander we suddenly had break in clouds and could see ice pack with open leads and one large two pointed berg. Took some pictures of sea ice and fog. Landed at Gander and in few minutes took off for New York. Saw on big Cu over Nova Scotia– only one of entire trip. Thickening haze toward Boston with high cirrus thickening. Landed at New York and after two hours getting through customs boarded plane for Albany arriving in about 30'! Great to see family again! To bed early.

Thursday, April 20, 1950

Into Lab in morning, rainy day, rain starting with thunderstorms at 2-3 AM. Waded through mail and had talk and discussions at lab. Home at noon, rested and read in afternoon. Rain stopped about 4 PM with slow clearing trend. Will stay home tomorrow and work up reports. To bed early.

The Production of a Cirrus-Type Overcast

There are times when the trail of an airplane produces a condensation trail. Depending on the degree of moisture at plane level, such trails quickly disappear, persist, or actually grow. Such trails, when produced at temperatures colder than -40°C (-40°F), consist of ice crystals that form by homogeneous nucleation.

Ray Falconer, of our group, found that he could make a small bag of muslin or netting, fill it with dry ice, and send it aloft tied to a weather balloon. Whenever the dry ice passed through a region of moist air colder than 0°C (32°F), it would leave behind a trail of ice crystals. Sometimes such trails appeared at levels in the atmosphere far below the homogeneous temperature level. We found that by dispensing dry ice from an airplane at such middle levels in the troposphere, it was sometimes possible to produce a growing trail of ice crystals and, in fact, in a few instances, such trails eventually formed an overcast made up of ice crystals.

While such an operation might have little, if any, commercial value, we found such experiments extremely interesting, especially since it suggested and demonstrated the mechanism whereby high flying aircraft, especially those powered by jet engines, often produce an ice crystal (false cirrus) overcast. Such massive amounts of ice crystals sometimes settle into the troposphere and seed convective clouds or massive layers of alto stratus cloud formations.

When such seeding occurs, it is sometimes quite noticeable. I have seen alto stratus clouds along flight corridors completely modified by such settling cirrus, while non-modified clouds adjacent to the flight corridor, instead of a gray appearance with an undersun, appeared white with a bright glory.

On one of the flight corridors from San Francisco to Chicago that passes just west of Navajo Mountain, the false cirrus is so pervasive that after many tries, I have not yet been able to get a clear photograph of Navajo Mountain. The same situation appears to exist along the Great Circle route from Anchorage, Alaska to London. An ice crystal haze obscures views of the Brooks Range and other regions of the Canadian Arctic as the corridor approaches Greenland.

In several instances, Project Cirrus flights intentionally produced extensive overcasts of ice crystals by seeding moist air layers with dry ice.

A Cloud Seeding Event at Socorro, New Mexico

In July of 1950, Project Cirrus Experiment #172 was carried out west and north of Socorro, New Mexico, and clearly demonstrated the effectiveness of cloud seeding.

The day was one that was typified by an air mass condition with no fronts or other disturbances in the region. Clouds started forming as soon as the sun heated the land. We alerted our B-17s that were stationed at the Albuquerque airport. They arrived on schedule, having 18 pounds of crushed dry ice in the seeding aircraft. The radar was activated; our cameras prepared both still and time-lapse.

The B-17 was scheduled to seed all large cumulus in a line from northeast to southwest, and to seed at a rate of 2 pounds per mile. About fifteen minutes after seeding began, the radar picked up signals along the seeding line. These echoes grew rapidly and reached a peak of intensity about two hours after the seeding started. Visible precipitation started at the place where the seeding started, and progressed in a line toward the southwest. The first precipitation appeared as line of virga, but this quickly reached the ground as a dense curtain of rain began to appear. This increased in density until it extended almost the whole length of the distant horizon, although there was a definite limit to both ends of the line of precipitation. It was an awesome sight. I had never seen anything like it previously, nor have I seen anything like it since. The heavy rain continued for more than three hours and appeared to be in the uninhabited valley of the Rio Salado that, heretofore, had not had any water since last year.

When the runoff appeared in the river, it produced a wall of water that rushed down its valley, flooding the "dip" that crossed the normally dry riverbed. A large truck and a car were engulfed in the floodwater and carried down the river for a considerable distance. Fortunately there was no loss of life.

The river, which I said had not carried water since last year, was in flood and it ran into the Rio Grande for two days! Everyone was very pleased with this experiment, since everything planned was carried out as scheduled. Our only regret was that the official observer of the Weather Bureau elected to sleep in this day and missed seeing the entire operation!

The time-lapse movie, which I obtained during the operation, was excellent and has been shown many times to audiences interested in weather modification. Such movies are quite rare, since during most of such activities there is a lack of manpower to operate the cameras, or the seeding effects are not in locations that can readily be photographed.

With our field operations over the Manzano Mountains east and southeast of Albuquerque that were carried out the previous year, most of the outstanding pictures were obtained from the B-17 observation plane that was in close contact with the seeding plane, and thus could get into a position suitable for good photography.

With flight 172 the best pictures were obtained from our ground location, since our communication with the plane and our radar operation pinpointed the effects produced, and the precipitation zone was in a perfect location for photography. It was a spectacular show!

We Go Over the Datil Mountains

One of Jack Workman's favorite tales was the story of the Lost Adams Diggings. It seems that there was a lone prospector who discovered an extremely rich source of gold in the Malpis– a large mass of "a-a lava" that had oozed out of rifts in the ground south of Grants, New Mexico, and spread over an area more than 30 miles long, and ranging from 6 to 15 miles wide. It was a forbidding and yet fascinating area of black cindery rock that was so jagged and cracked, that it takes about a day to go a mile cross-country. Here and there the lava, for some reason or other failed to cover the land, so that there were small park-like regions scattered throughout the huge area covered by virtually impassable terrain.

It was in one of these places where Adams had found his trove of gold. We went searching for the diggings along the primitive road that borders the eastern extremities of the region on an earlier trip, but that is another story.

This time Workman proposed that we go southeast of it into the wilderness of the Datil Mountains, north of the road that skirts the Magdalena Mountains. We traveled in a four-wheel drive van– the same one that took us the previous year to Meteor Crater and the Grand Canyon. Its transmission problems had been cured and it had turned out to be a very reliable vehicle. After heading west from Socorro for fifteen or twenty miles, we ran out of roads and were forced to go cross-country. This was slow going, but we found a dry creek bed that provided us with a useful way to go. The traveling was far from ideal, however, and the route so bad that at times we wondered whether we would ever get back to civilized country.

At one place we discovered a cliff dwelling that was apparently still in use in certain seasons of the year. Its occupant had apparently received some exposure to civilization, since on the flat surface of the cliff, the successive years of occupancy were carefully carved into the rock: 1908, 1909, 1910– and ending with 1950. It was obvious that the occupant was not around at the time. We continued up the creek several more miles, and then decided to camp for the night. As soon as the sun disappeared, it began to get cold. We built a big fire, accumulated a pile of driftwood and, after a good supper, fixed our beds for the night. It was cold. I didn't sleep too well since I didn't have my down bag with me. After a fitful sleep I finally got up about day break. Several others of our party had a similar night and were already by the fire boiling coffee. One of the group had stayed by the fire all night. He was the smart one! As sunshine flooded our campsite, I looked at the van and found it to be covered with frost.

After breakfast we talked over our predicament. Workman, who was driving, declared that under no circumstances would he consider backtracking. He said that the way ahead couldn't possibly be any worse than the route we had followed and that, eventually, we were bound to find civilization if we kept on our southerly course.

I volunteered to climb a small rocky mountain near our camp to see if any signs of civilization could be detected. After considerable effort I reached a good lookout, but had to report that there was no evidence whatever of roads or anything except sparsely wooded, undulating hills. It did look like our way south was blocked by fairly high slopes.

Despite this news, Workman decided to continue. Leaving the dwindling streambed, we were climbing, encountering all sorts of obstacles, but surmounting them without undue difficulty. At one point we all had to remove fallen trees and other barriers. In this process Jack had to remove his heavy sheepskin coat as we struggled to remove an obstacle. Proceeding on, it was an hour before Jack suddenly realized that his coat was missing! By this time we had surmounted a high hill and were proceeding down its other side. The sun was sinking in the west, we had no food left, and none of us were looking forward to another cold night. Jack felt all of these problems and decided that, under no circumstances, would he retrace our route to recover the coat.

An hour or so later we encountered a wood road, then a slightly better road and then the highway at Pietown. Saved!

We fortunately found a gasoline station that replenished our nearly empty gas tank, and we were soon headed back to Socorro.

We all agreed that there was little evidence that any of the country we had traversed could be related to the Adams Diggings, so that the next expedition, if the opportunity developed, should go farther west to the Malpais.

An Adventure at the Fernandez Ranch in New Mexico

On July 19, 1950, after spending a week following the termination of our Project Cirrus program at Socorro, in visiting the Bellota Ranch in the mountain east of Tucson, Arizona, and then the Grand Canyon, the family and I arrived at the Fernandez Ranch near San Mateo, New Mexico, on the north side of Mount Taylor. This large cattle ranch owned by Floyd Lee occupies one of the ancient Spanish land grants. Its headquarters is a massive adobe ranch house that is very old and beautifully preserved.

Floyd was much interested in using cloud seeding to enhance the precipitation over his ranch lands and asked me to visit him to look at the clouds, and to advise him on the usefulness of a new type of silver iodide generator.

During the 3 1/2 days we were at the ranch we had a fascinating time. We went to the summit of Mount Taylor, one of the sacred mountains of the Navajo, participated in an old-time round up, had a steak dinner out on the range, visited some ancient Indian ruins, and evaluated Floyd's new silver iodide generator.

This latter episode had a humorous twist. The "generator" consisted of a common road flare of the type which has a massive wick, which is normally immersed in a reservoir of crude oil and, when ignited, burns with a smoky flame.

When Floyd showed this device to me, I volunteered to calibrate it for him. He had a large storeroom that, among many other things, contained a large cold chamber in which frozen foods were stored. I found that it had at least four cubic feet of vacant space. Finding a chunk of black cloth (to serve as a liner) and a strong flashlight, we were ready for a test. I lit the wick (which burned a silver iodide-sodium iodide solution in acetone) and made sure that it was burning well. I then let it burn for a short period in the closed storeroom. In order to spread the invisible iodide smoke throughout the volume of the storage building, I suggested to Floyd that he walk up and down the aisle with a large flat sheet of solid material, waving so as to mix the air. Only after he had completed the maneuver did I notice that he had used a large oil painting of one of his ancestors! It would have made quite a movie.

After the air was mixed, I made measurements of the concentration of nuclei produced by the generator. While it did produce some particles, the values were quite low and I recommended that he not bother to try to modify clouds with that device!

Floyd showed me the ruins of a very large multi-room pueblo, not far from headquarters. It appeared to be in untouched condition, the ground strewn with potsherds, beads, and other artifacts. Since archeology is one of my main hobbies, I had a wonderful time exploring the ruins. The next morning I arose early and again went over to the site. As I was searching the surface of the site, I suddenly saw something that looked interesting. As I went to pick it up, I realized that it was a rattlesnake! Fortunately, because of the cold of the early morning, it was in a comatose state, so I passed on!

Later in the day I was in the headquarters ranch and was greatly impressed with the design of the building. It had an inside court exposed to the sky, with the roofs all sloping inward and extending over a portion of the court on the four sides. The design of the roof made it an excellent reflector toward outer space so that, under clear skies at night, the roof cooled, the cold air draining into the enclosed court and displacing the cooler air in it.

The four-foot thick adobe walls of the house had another unique feature, in addition to its thermal stability. I saw Florence Lee take an axe and several large boards into one of the rooms. Seeing my questioning look, she explained that she needed more bookshelves. Wielding the axe, she quickly chopped a depression into the 4-foot thick adobe wall, cut notches for the boards, and then, with some adobe mud, filled in the irregularities. As soon as the adobe dried it was whitewashed, and looked like it had been there since the construction of the hacienda.

While at the ranch, horses were provided to our youngsters and they were soon having the time of their lives. Our older daughter, Susan, had a love of horses from her earliest days, and was in her glory. Her favorite at the ranch was Judge; a white horse that had served the Lee's for many years and was retired from active use. He was still frisky enough for Sue's needs, and she had a marvelous time. When the time arrived for departure, Floyd told Sue that she could have Judge if she could get him home! She was realistic enough to realize that this wasn't possible, but for many years her desire to have a horse of her own was partially satisfied with the realization that she did have a cow horse at the Fernandez Ranch near San Mateo, New Mexico.

The Development of Portable Cold Chambers

With an active interest in measuring nuclei for ice crystal formation, I tried developing a number of small portable units that could be taken into the field. None of them were an outstanding success.

The first one was the size of a tobacco can. The inside of the can was lined with black velvet, the illuminator was a penlight, and the coolant was dry ice. I had no trouble forming a supercooled cloud, but the air volume was so small, and the nucleus concentration so low that, after a number of attempts, I decided that a somewhat larger size was needed.

I then built a unit having an outside dimension of 18" x 20" x 4", with an inner chamber made of copper, and the coolant– exposure to the winter atmosphere of the Northeast. The chamber, when left outdoors assumed the ambient temperature; a supercooled cloud could be formed and observations made. Again the illuminant was a penlight. The amount of light was inadequate, so a concentrated-beam flashlight was utilized. This worked better, but was limited to the temperature of the ambient atmosphere and was seasonal, and only useful in the wintertime.

Eventually, Meteorology Research Inc. of Pasadena, California, at my urging, developed a fine, portable chamber which, by a clever cooling system using alcohol cooled with dry ice and circulated with the CO_2 gas evolved from the dry ice, provided a temperature range of -10 to -25°C. This proved to be an ideal combination. It had a coordinated optical illumination system that provided excellent viewing conditions, and proved to be ideal for surveys from small aircraft and river rafts.

We sponsored six aerial survey flights from coast to coast, and another one on the Green River of Utah, to find out if areas long distances from auto roads still would show the presence of submicroscopic lead particles in the ambient atmosphere. By using the iodine reaction, we found airborne lead even in the depths of Desolation Canyon on the Green River.

The Development of the Metal Multicell

On August 1, 1950, Dr. Suits,[35] Director of the Laboratory, asked me to meet with Clifford Fick, an electronics engineer, who was interested in developing a color television tube. As the problem was discussed, I learned that the ideal structure that was envisioned consisted of a mass of tiny tubes sixteen inches in diameter, each tube being 0.4 inches long and 0.020" in diameter. Thus there would be about 400,000 of them. In addition, they all must be focused at a distance of 18 inches! This appeared to be a challenging assignment!

I made a number of quick moves to explore the possibilities. Within a week I had decided to take a half-inch roll of 0.001" thickness Nickel and pass it through a pair of fine-tooth gears. This produced a spiral, using a similar strip of un-crimped metal foil as a spacer. In this manner a circular coil could be fashioned, having the requisite sized holes (dependent on the pitch of the gears). The resulting fabrication resembled a coil of corrugated cardboard, except that it was of metal and had much tinier holes!

With the production of the large number of tubes solved, it was necessary to find some means of focusing them at a fixed distance. A number of techniques had been tried, but all resulted in failure.

By early October however, I was having some success in focusing multicells. I coated them with a dilute solution of an adhesive and, after drying, pressed them into a curved mold. This produced some degree of focusing, but resulted in a six-sided pattern of "faults," which developed as the thin walls of the multicells collapsed.

On November 4, 1950 I awoke from a sound sleep in the early morning with and idea for the collapsing wall problem. I would fill the holes with water and freeze it before compression took place. Prior to this date, I had found it possible with a steel multicell to heat it in a hydrogen atmosphere at which the surfaces sintered together. I then froze water in the tubes. Upon pressing the structure I found that the ice supported the walls of all of the tubes and this distributed the deformation in a uniform manner. The focus was beautiful.

I turned the focusing multicell over to Dr. Lewis Koller, who proceeded to coat the walls of the multicells with the three fluorescent powders that produced red, yellow, and blue colors when activated with an electron beam.

My notebook contains many ideas for exploration of the multicell structure, ranging from air and water heaters, light collimators, toasters, x-ray collimators and a host of other applications. While I had no desire to be diverted into a manufacturing direction, I called attention of these possibilities to friends in General Engineering, the Works Lab and other parts of the Company.

By the latter part of 1950 I became much interested in the development and use of a continuous diffusion cloud chamber for exploring the properties of particle free air and homogeneous nucleation. Accordingly, I returned to do further studies of the multicell and

---

35 Chauncey Guy Suits (1905-1991) was the head of GE Research from 1945-1965. He was a founding member of the National Academy of Engineering.

found it could be used as an elegant source of water vapor when the tubes were filled with water.

Production of Smoke on a Small Scale

At times there is need for a method of producing a dense screening smoke in a relatively short period of time. With the 10 gallons and 100 gallons per hour units which we invented and developed in 1941-42, and which were used in very large numbers during World War II, I made a very little one which dispensed about one thousandth of the amount of smoke generated by our "Gopher" unit.

The unit I made and tested was a tiny version of my first test model that was an oilcan. I took a small 1 gram ampule containing the material to be used to make the smoke– sulfur, diesel lubricating oil, stearic acid, etc., etc., and after heating one end of the glass, pulled it quickly to form a fine capillary. This was then broken at a place where the exit hole was about .010" diameter. This was then placed in the end of a soldering heater of the type which has a hole normally occupied by the soldering tip, but is deep enough to accommodate the full ampule and open tip.

Shortly after the substance reaches vaporization temperature, a dense smoke emerges from the fine tip of the ampule at high velocity, looking like a small steam jet. The particles in such smokes tend to be smaller than 1 micron, and thus has very little falling velocity and moves with the air.

The Accumulation of Water by Evergreen Trees

On April 27, 1951, I met Eugene Bollay for the first time, and the next day accompanied him to his Lake Arrowhead camp. While there we met with Stuart Cundiff and a Mr. Cage of the California Electric Company to discuss plans for establishing a network of ice nuclei observation stations. I learned that three weeks previously a group of local persons had met to discuss forming an organization to represent the groups concerned about protecting the interests of cloud seeders in California.

Out of this organizing effort developed the Weather Modification Association that now represents that business activity throughout the United States. While enroute to Lake Arrowhead, Gene and I discussed the role played by evergreen trees intercepting cloud droplets. Underneath every evergreen tree the ground was wet with water that had been collected by the needles of the trees. Some trees actually had tiny streams of water running away from the tree trunks that were soaked with the collected cloud water.

On May 1, 1951, as soon as I returned home, I remarked in my notebook that this was an important phenomenon and should be studied. It wasn't until early July that I was able to make some field measurements on the summit of Mount Washington. I took branches of six different evergreen trees, weighed them when dry, and then exposed them for a minute to a passing cloud having a liquid water content of 0.2G per M3, in a wind of 10 miles per hour, and a drop diameter of 10 microns. After weighing for a second time, I was able to determine the effectiveness of the evergreen needles in their ability to collect cloud water.

Now, twenty-five years later, the subject of cloud water collection by evergreen needles is again of considerable interest. Since it has been found that the acidity of clouds is considerably greater than that of rain, the effect on the welfare of the evergreens on the mountains of the Northeast is getting considerable attention.

The Development of the Diffusion Cloud Chamber

On November 13, 1950, I constructed a cloud chamber in which the effects of cosmic rays and other sources of ionization could be observed as they nucleated a supersaturated vapor of methyl or ethyl alcohol, and produced a line of tiny droplets that formed in the wake of the ionizing radiation.

I then decided to substitute water for the alcohol, and to explore the nucleating characteristics of a warm and a cold atmosphere. By surrounding the lower part of the chamber with chunks of dry ice, I found it possible to generate a temperature in the lower part of the chamber to -15°C. Once all of the particles in the air enclosed in the chamber had rained out, and the air became supersaturated with water vapor, I found that by introducing a cluster of silver iodide smoke particles in to the warm zone in the chamber, they would serve as cloud condensation nuclei. These condensed water droplets grew fast and fell with increasing speed into the cold part of the chamber. As they passed through the -4°C level, they were seen to flash suddenly into ice crystals. Replicating these crystals, I found that they were not frozen droplets, but had crystallized to form either plates or hexagonal columns.

Further research on this chamber was delayed by fieldwork that took me in May to Socorro, New Mexico, and in August to the Northwest.

Late in August I was back in the laboratory at the Knolls, involved with further study and development of the continuous diffusion cold chamber.

My new design, which I constructed in a few days, consisted of a tall glass chamber 8 inches square and 18 inches high, which rested on a copper box which, in turn, rested on a large block of copper that served as a cold sink, since it was completely covered with chunks of dry ice. With this unit I was able to get temperatures as cold as -55°C. With a fine thermocouple probe I ran a temperature profile and found that the top of the chamber had a temperature of +15°C, the 0°C level was about halfway down the chamber, while the -40°C was about 5 inches from the bottom. This chamber was an elegant device for studying both heterogeneous and homogeneous nucleation for ice crystal formation.

On October 23, 1951, I built the final version of my chamber that included a much larger copper box for the base. The thick upper walls of the copper box established the -40°C level of the chamber air so it was easier to observe phenomena at the homogeneous nucleation temperature. In addition, I combined a 7 inch telephoto lens with a 7X ocular, so that the homogenous nucleation area was magnified by ten fold. What I saw was fascinating! After all of the nuclei in the cold chamber had rained out, I could introduce a small cluster of cloud condensation nuclei of sodium chloride. They would grow very rapidly in the supersaturated air, and continue falling through the -40°C level under observation. Passing through this zone, no visible change occurred to the liquid droplet. However, in the wake of the fallen drop, a bright starlike object would appear, which would rise a millimeter or two and then start falling. However, as it reached its maximum elevation, this particle would separate into many much smaller ones, so small that they exhibited Brownian motion. These trembling particles would then fall as a sort of cascade toward the bottom of the chamber.

While I never thoroughly understood the phenomenon, I describe I had a feeling that a group of submicroscopic crystals formed spontaneously in the rich moisture wake behind the falling cloud droplets formed on the sodium chloride nucleated droplets. As soon as the ice crystals formed, they released their heat of crystallization that caused them to rise in the highly stable air inversion. Once the growth stopped, Brownian motion separated them from their original close proximity, and they cascaded toward the bottom of the chamber. After carrying out these and many other fascinating experiments, I then dried out the multicell which I was using as a water supply at the top of the cloud chamber, and replaced the water with Benzene. All of the phenomena observed with water were repeated except for homogeneous nucleation. If it occurs it must be quite a bit colder than -40°C.

Dropping tiny grams of dry ice through supersaturated benzene vapor generates a condensation trail of very beautiful benzene crystals so thin as to disclose brilliant interference colors.

I never had a chance to do any further experiments with benzene, but had a feeling that some fascinating experiments were possible using the diffusion chamber.

Attempts to Control the Local Environment of Trees

It was in 1952 that I became involved in a big squabble in the Northwestern fruit producing areas of eastern Washington, related to cloud seeding. Dr. Irving Krick was a pioneer commercial cloud seeder who had been employed by Judge Horrigan to attempt to increase precipitation on the extensive wheat fields in the Horse Heaven Hills. The fruit growers of the Yakima-Prosser area claimed that rain falling on the ripe cherry crop would devastate the cherries by causing them to split.

I attended a public hearing on the matter and was dismayed at the near riot atmosphere of the crowd that attended.

It seemed that common sense was needed to solve the problem. Since the critical period of harvesting the crop extended for only a few weeks in June (and this was a period when climatic records showed a minimum of cloudiness), it seemed to me that the logical solution to the problem was to declare a moratorium on cloud seeding activity during the sensitive harvesting period.

While discussing the problem with Leo Horrigan, son of the judge, I came up with the idea that it might be worthwhile to try enclosing the tree in a plastic bag. Upon returning to Schenectady I wrote a patent letter dated March 25, 1952 to Dr. Suits, Director of the Laboratory, detailing my thoughts. Polyethylene plastic film was just being marketed for the first time and, while quite expensive, I managed to get a small supply for experimental use.

The growing season was still ahead and the trees were still dormant, so I made some bags and proceeded to encase branches of a lilac and plum tree in a transparent plastic bag. As the sun climbed higher in early spring, I kept a close watch of the branches and found that the localized environment was greatly modified by the presence of the bags. When the bags were removed about three weeks later, the buds on the rest of the shrub and tree were just beginning to open, while the protected branches were nearly in full leaf and flowers were about to unfold.

This striking demonstration showed that some degree of growth and environmental control could be exerted with a plastic covering.

My basic thought in this matter was that if a means could be found to inexpensively enclose a fruit tree in a plastic tent link cover which could protect it from rain and, perhaps, serve as retainer for a fumigant for insect control, there might be some important advantages. It would be necessary that such a device could be easily removed to permit pollination, although its environmental control might even be utilized for this purpose. I also devised a method for making tiny holes in the plastic that would permit ventilation, but would exclude raindrops.

I finally obtained a very large plastic bag; large enough to encompass a medium-sized tree. I decided to enclose a symmetrical maple tree that was growing back of my garage.

On what seemed to be a calm morning, I erected a high step ladder along side of the tree and proceeded to try to put the bag over the tree.

What a disaster!

I found that despite the appearance of calm air, the vagrant breezes that move through relatively calm air played havoc with the plastic sheeting. Every attempt to deploy the bag over the tree ended in complete failure. As soon as I had a small portion of the bag located where I wanted it, a flow of air would quickly move it away! Within a few minutes the folds of the bag were tangled and moved to areas where they became hopelessly collapsed in disarray. It became increasingly apparent that my ideas had not adequately considered the fact that what I was doing was to try to construct a device that would never stand the buffeting of the wind.

This experiment was a valuable one– one that once again stressed the importance of field-testing.

It reminded me of failures experienced at the summit of Mount Washington in stormy weather. More than once, in the calm of the laboratory at Schenectady, I would devise and assemble a device that looked to be eminently practical and well-designed. I would pack it to the summit, prepare it for mounting in the storm, and only as I climbed the inside ladder to the parapet and felt the vibrations of the tower caused by the buffeting of the hurricane-plus-force winds, would I realize that the device couldn't work under actual atmospheric conditions!

My plastic bag disaster ended my active work on the bagging of trees. It was an intriguing idea but apparently an impractical one!

Our Contact with Solar Energy

Sometime prior to the fall of 1951, we were approached by Dr. Maria Telkas[36] of MIT, where she was involved in the development of a heat storage substance for use with a solar house. A hydrated chemical called Glauber's Salt was being used as a heat storage substance. The heat from solar panels was used to melt the Glauber's Salt crystals. Then, when heat was needed in the house, a liquid circulating system would extract the heat from the molten salt. When it reached its melting point and began to crystallize, the heat of crystallization would continue to supply heat until all the crystals had solidified. The problem which had developed was that the Glauber's Salt supercooled, and thus the heat of crystallization feature could not be utilized. Dr. Telkes approached us since she had learned of our work with crystallization and supercooling. I obtained samples of the hydrated salt and quickly found that, once melted, they supercooled as a regular feature of the melt. I also discovered another disturbing feature of this material. Sometimes, when it finally crystallized, the normal combination with water, which required 10 molecules of water, sometimes produced a different type of crystal that required only seven molecules. Unfortunately, when this crystal formed its ability to melt again, when heated, failed to occur. When this happened, the material was no longer useful.

The only solution to the supercooling or crystal modification problems that occurred to me seemed to require that a means must be provided to prevent the melting of all of the crystals. If such a feature could be provided, then in the cooling cycle, the right type of crystals would develop at the proper temperature. Meanwhile, I had given some thought to developing a heat storage material that would be relatively inexpensive and would behave better than the salt hydrates. After checking through the melting points of a variety of chemicals, I found that stearic acid came close to the ideal temperature. After trying it, I found that molten stearic acid never supercooled and had most of the good features of an ideal heat storage material. Subsequently I discovered that varied amounts of oleaic acid and diphenyl ether provided a considerable range of crystallization temperatures. While much of this effort was directed toward developing a heat storage substance for the new GE Heat Pump (the Weathertron), the economics were such that fuel savings was not as important as it became twenty years later!

---

36 Maria Telkas (1901-1995), dubbed the "Sun Queen," developed a lightweight water distillation device using the sun, during World War II. Her "Dover Sun House" project began in 1948. This complete solar house used Glauber's Salt to store the sun's energy. In 1953 she developed a solar oven for use in impoverished countries, and was funded by the Ford Foundation. It reached a temperature of over 350°.

Spontaneous Electrification of Pentaerythritol Crystals

On December 29, 1953, I had a call from Dr. Hillig, who was working with Drs. David Turnbull and Jack Newkirk, asking me whether or not I was familiar with an electrical effect noticed when pentaerythritol crystals formed in a supersaturated water solution. He told me that, if true, he couldn't understand the phenomenon, but admitted he had not tried to see whether or not it was a reality. The next morning I obtained some crystals of the substance, made a saturated solution in water, and placed a few drops on a microscope slide that I cooled down in my cold chamber so that the solution became supersaturated. Almost immediately lines of pentaerythritol crystals formed on the water surface, floating on it and positioning each other at distances of four or five dimensions from its neighbors. Meanwhile, they were growing and trembling on the surface as their repellent forces caused a continuing distance adjustment. It was a fascinating phenomenon.

After seeing this initial effect I secured a sheet of glass about 10" x 12" in area, prepared a saturated solution in hot water, and poured it onto the glass surface. Within a few seconds, a dense stream of floating crystals began to flow across the water, until they covered the entire surface. I had used just enough water so that, by the time the water had cooled and the crystals reached size equilibrium, they were still isolated from each other as the water dried. It remains a beautiful remembrance of this exciting experiment.

I continued studying the electrical effects of these crystals. By placing a tiny wire electrode close to the surface of the growing crystals, I found that it took +5,000 volts to cause them to move away from the wire the same distance they were repelling each other.

Some years later, while preparing a saturated solution of this chemical in hot water, the solution boiled over and the overflow melted and formed a very finely divided aerosol. I happened to have a cold chamber in operation about twenty feet away. Imagine my surprise when I found the chamber filled with ice crystals!

It then occurred to me that a finely divided smoke of pentaerythritol– like silver iodide– was an excellent heterogeneous ice nucleus. I quickly prepared a Formvar replica solution, replicated a slide of crystals, and found the slide covered with two types of beautiful ice crystals.

Those that were symmetrical had some of the finest and most complicated patterns I had ever seen, quite on par with those of silver iodide. The other were quite asymmetric, consisting of four rays on one plane, with the other two of the hexagon as sort of wings that projected upward at angles of 60 degrees from the horizontal. They were identical with others I had become familiar with which had formed under the influence of a strong electric field. Thus the apparent reason for their strange shape suggested that this growth was due to the electrification phenomenon that developed with this strange crystal.

A search for a reason for these highly unusual phenomena might be related to the molecular form of the crystals, that were highly symmetrical in one plane, but very asymmetric at right angles to it.

Far from answering these puzzling facts, the situation with respect to pentaerythritol is still cloaked in mystery and, after more than thirty years, poses many fascinating questions!

The Critical Concentration of Ice Crystals to Quickly Seed Supercooled Clouds

In the course of my studies of ice crystal effects and my efforts to understand homogeneous nucleation, I prepared hundreds, if not thousands, of supercooled clouds in my laboratory cold chamber. Quite frequently, when producing high concentrations such as a blue fog generated with a Hilsch Tube, a raucous nozzle, a jet of liquid CO_2, propane, Freon 22, heavy use of liquid nitrogen, or of dry ice, the initial concentrations were in the excess of 105 per cubic centimeter. This could be established by feeding the crystals with a continuous supply of water, and then plotting the concentrations on a logarithmic scale as a function of time.

When this was done, there was found a sharp break in the linear relationship that could be correlated with the sudden appearance of the coexistence of ice crystals and supercooled cloud droplets. This was a very striking effect, and quickly became an extremely important number.

After many experiments, during which the length of time for the break in the linear relationship ranged from 120 to 540 seconds depending on the type of homogeneous nucleation utilized, this critical concentration for the coexistence of supercooled droplets and ice crystals was established as 50 crystals per cc. Any concentration higher than 50 quickly used up the supercooled cloud, while anything lower than that number never used up the supercooled clouds. In all experiments the cold chamber temperature was held at 14°C, the temperature where the largest difference exists between the vapor pressure of water and that of ice.

This finding means that in order to most efficiently convert a supercooled cloud to ice crystals at 14°C, it is necessary to provide a concentration in the air of 50 ice embryos or crystals per cubic centimeter. Larger crystals will probably require lower concentrations. In fact, at Yellowstone, crystals of 100 microns in cross-section appear to be about 100 per liter (i.e., 0.1 per cc).

Since dry ice is capable of producing at least 1,015 ice embryos per gram of material when effectively distributed, one cubic kilometer could be seeded in an optimum fashion with 50 grams of dry ice. Thus, a pound of dry ice could affect 10 km^3.

The Measurement of Atmospheric Electricity

Starting in 1943, when I became involved with studying precipitation static on airplanes, I learned a bit about atmospheric electricity and have been intrigued with the subject ever since. I have never had the time or inclination to become deeply involved with the subject that would require the development and application of field mills and other more sophisticated measurements. Rather, I concentrated my observations on the variations encountered and sensed with nothing more elaborate than a radioactive point, well insulated and connected by shielded cable to an ultra-sensitive micro-ammeter that, in turn, was connected to ground.

Many of my observations, especially those related to snowstorms, jet streams, and lightning, were made at the weather observatory I fashioned on top of Building #5 at the main plant of General Electric. However, I have carried and used my telescopic mast, radioactive point, and micro-ammeter (or a similar combination) made by Meteorology Research, Inc., at such places as the geysers of Yellowstone, the waterfalls of the Little Colorado and Niagara Rivers, the surf and volcanoes of Hawaii, the head of a wildfire in Arizona, and at the top of a 150-foot tower for measuring lightning in northern Idaho. I also found some extremely interesting relationships by traveling cross-country in my car with the insulated point stuck out of my car window.

I have described the equipment and some of my findings in a paper published in the *Transactions of the American Geophysical Union*. It consisted of a brass point coated with gold foil impregnated with radium sulphate. This was connected to a wire-shielded, single-conductor cable which led to an RCS ultra-sensitive micro-ammeter. The point was insulated with a knob of Bakelite that was mounted on a cylindrical skirt of the same material. Even in a driving rainstorm the insulation remained intact. A 35-foot collapsible and light (but rugged) aluminum mast completed my field equipment. I didn't use a recorder, since I found it quite feasible to read the needle fluctuations on the meter and record the data, finding I could obtain points every two seconds. Such a time interval was quite sufficient to enable me to subsequently graph the data and come up with an excellent portrayal of the nature of the electrical behavior of the atmosphere.

The MRI portable electrometer provided similar measurements with the data recorded on a Rustrak Recorder. However, despite its convenience, I found that my simpler measuring device to be more adaptable to better note-taking and higher sensitivity.

In measuring the electrical nature of the dense smoke in an actively "rolling" wildfire in Arizona, I found a strongly positive current– so strong that my colleague, Jack Dietrich, received an electrical shock when he connected the meter to the shielded cable. On the other hand, there was a strong negative current of ions in the vicinity of the canyon's edge at both the Grand Falls of the Little Colorado and Niagara Falls on both sides of the river. The potential was higher at the Grand Falls than at Niagara, presumably because of the much larger spray particles coming out of the canyon in Arizona. The potential, on the other hand, during the geyser eruptions of Old Faithful at Yellowstone in the wintertime, were very much smaller than I had anticipated, but I suspect this was in some way related to the unusual purity of the entrained air. At Grand Falls the spray particles were mostly mud pellets!

An interesting measurement at Yellowstone was made when a pint or so of hot water was thrown into the air when the temperature was 45°C below zero. A sudden pulse of positive current developed as a dense cloud of ice crystals formed from the hot water thrown into the air, and a strange "whispering" sound occurred in the process. Similarly, a positive current was observed in the crashing of the surf on the edge of the sea of Hawaii, the current fluctuating with the intensity of the spray from the waves. At a spot near the summit of Mauna Kea, at about 12,000 feet above the sea, the positive current was so uniform, that I was wondering whether there was something wrong with the meter. In the misty rain along the Saddle Road east of the col[37] between Mauna Loa and Mauna Kea, the electrical current fluctuated with the intensity of the precipitation, and even showed evidence of mild electrical discharges that were probably occurring in the warm cloud trade wind convective towers that were forming over the coast line near Hilo.

At the Priest River Experiment Station of the US Forest Service, my point collector responded quite actively to the lightning storms that swept across the Priest River Valley, after forming over the mountain ridges to the west. The lightning discharges had both positive and negative signs, while some of them fluctuated from positive to negative, and others negative to positive.

During my earliest experience with this operation, in 1943, I measured and then identified three basic characteristics of the potential gradient that occurs in unsettled weather. In what I called a "Cirrus Storm," the snow was preceded by a thickening cirrus, in which the edge of the sun's disc was very fuzzy, with the sun gradually disappearing as the cirrus crystals moved toward the ground, until finally the snow reached the ground. The crystals consisted of hexagonal plates. The air-to-ground current with such storms always had a positive current, and one which was not much greater than the fair-weather field which is frequently about 0.02 micro amps.

The other major storm, that I called a "Cross Current storm," occurred with large convective clouds associated with it. The flow of electrical current alternated between positive and negative, with alternations occurring every fifteen or twenty minutes. Lightning frequently occurred, both cloud-to-ground and ground-to-cloud. The ice crystals that accompanied such storms in the wintertime ranged from long needles, graupel, large wet snowflakes, large stellar crystals, spatial dendrites, and sometimes, sleet. The peak of the electrical current often exceeded 1 microampere.

The last major effect observed at my Laboratory Station at Schenectady was what I called a "Ghost Storm," wherein after a precipitating storm had passed and blue sky dominated the scene with all visible clouds gone, strong manifestations continued for an hour or so in the wake of the storminess. It was my conclusion that the precipitation particles that had formed evaporated, leaving the air parcel still electrified by either positive or negative ions.

There were other times when electrification occurred which seemed to be related to the onset of a jet stream. These manifestations were not easy to correlate, although my measurements, with the radioactive point projecting out of my car, provided a number of examples where an anomalous departure from the common levels observed occurred as my car passed under such clouds. There were so many other ground effects noticeable that I could never be sure that I knew the cause that produced the current flow.

On March 31, 1953, I headed for Mount Washington by way of Burlington, Vermont, returning south to West Lynn, Massachusetts, and thence back to Schenectady. I had my ra-

---

37 A pass between two mountain peaks.

dioactive point projecting out of the window of my car with the micrometer where it could be easily seen. Much of the time the current flow ranged from +0.01 to +0.1. Crossing rivers and running along rivers and lakes, the current invariably swung negative and would go as high as 0.4 microamps. At Pinkham Notch, I went to the base of Glen Ellis Falls on the Cutler River and measured negative currents as high as 1 microamp, with the highest current appearing in the densest spray at the base of the falls.

The only other major effect observed on the trip was a considerable change in positive current as we passed from clear skies to cloudy skies and rain. However, the degree of current flow was quite variable, ranging from +0.15 to +1.1 microamps when we were passing under jet stream cirrus clouds.

Other measurements of a similar nature were made using my car. The most notable one was a very large negative current observed between Schenectady and Binghamton, when we passed a large power plant. It is quite possible this high current was caused by the large electrical precipitator that was in operation at the time.

At the present time (1986) there is a revival in interest in such measurements, particularly because of the planned construction of high voltage direct current power lines. It would be very desirable to get a series of measurements, under all sorts of weather conditions, along the corridors where such power lines are scheduled for construction.

Irving Langmuir in flight gear on board a B-17, inspecting camera equipment for a Project Cirrus hurricane cloud seeding mission, c.1948. Courtesy of the CECOM LCMC Staff Historian's Office, Aberdeen Proving Ground, Maryland.

III

Reflections on the Langmuir Years

Chapter 9
A Retrospective of My "University" Years

The Effect of Jet Streams on Television Signal Propagation

On December 9, 1953, I had a visit from Dr. R. L. Shue and Leonard Abraham, who were interested in determining whether the jet stream was in any way responsible for causing anomalous propagation of television signals. They were monitoring signals from a Binghamton station that, from time to time, produced a strong signal.

I suggested that Dr. Abraham check their records to see whether or not they had obtained a good signal on December 8, since my records showed that we had an excellent display of jet stream clouds coming from the west, and that the Weather Bureau records showed the stream's velocity to be more than 100 knots.

He checked his log and became quite excited when he found that the first good picture signals were picked up on that day!

On December 11, at sunrise, we had an excellent display of jet stream clouds between Schenectady and Binghamton. I planned to call Dr. Abraham as soon as I reached the laboratory. Arriving there I found a note asking me to call him. When I did so and told him that there was an excellent display of clouds in a southerly direction, he told me that they had an excellent 200-megacycle television picture signal, the first early morning signal they had observed. I suggested that they would do well to put a recorder on their receiver, and we would do the same with our atmospheric electrical detector.

Within a few weeks we were able to make comparisons and found that there was a definite correlation with some events, but not in every instance. It would have been more significant, I believe, if we had a field station at the southern end of the Schoharie Valley for measuring atmospheric potential, since it was in that region where the reflection would occur.

Further studies and observations eventually showed that the best occurrences of anomalous propagation were found when the lower atmosphere was stable, and there were lenticular wave clouds in the middle atmosphere.

My Use of Photography, 1920 –1986

From the time my father gave me his box camera, when I was about 16 years old, until the present, I have been interested in photography. I initially became involved in preserving a record of beauty spots encountered, but soon began to branch out into many other activities.

I found photography to be an essential tool of the experimental scientist, recording phenomena that occurred during the course of my research. Photographs and photomicrographs I found to be of tremendous value in recording transient phenomena and providing illustrations for scientific papers. A picture is far better than a word description, since it provides an element of reality that no combination of words can provide.

As our house was being planned, I decided to prepare a visual record of construction progress and have a complete record of the activity. These include the little Schermerhorn Burial Ground when it was still intact, and then the excavation for the foundation for our house, the framing, roofing, interior work, bricklaying, and then the final result.

As one of my first developments, in our basement I built a dark room and fitted it with sinks, a water supply, shelves, enlarger, safe light, and supplies of chemicals. It has continued to serve as such a facility until the present time. My homemade enlarger has been replaced with a professional one, and other improvements have been made so it is more convenient to use.

Over the years my original Kodak box camera has been supplanted by improved cameras, my Zeiss 3 1/2" x 4 1/4" single film holders and film packs and 35mm Exacta being slowly replaced by better ones– eventually ending up with Nikons for color pictures and a Linhof for 4" x 5" black and white negatives. Throughout this period of more than 50 years I have sold enough pictures to provide the funds for purchasing all my equipment.

My laboratory notebooks and film collection attests to the value of photography to my interests. It consists of more than 50,000 color transparencies stored in archival quality polyethylene slip-sheets that, in turn, are filed in loose-leaf binders. My black and white negatives are safeguarded in similar sheets.

In addition, I have about 5,000 stereo pairs taken over a period of some 35 years. This collection began when I went to the Grand Canyon with the Boss in 1949. While there, he proposed that we take some stereo pairs of the canyon by standing about 100 feet from each other and, at a given signal, taking pictures of identical objects at the middle distance in the canyon. I was so impressed with the results of these photographs that I routinely made stereo pairs from that time on many suitable subjects. I found that in most cases, it was not necessary to take pictures at the identical instant with most subjects, but that it was possible to get excellent results by taking a picture and then walking quickly to a prearranged site. Even if clouds were in the picture, the few second difference made no visible difference.

One evening, while developing some negatives, I inadvertently turned the white light on for a moment before realizing that the development process was not completed. Imagine my surprise when I found that this act had produced a positive rather than a negative image!

This discovery, made on a Friday evening, set off an intensive, nearly round-the-clock series of experiments on my part, which continued throughout the weekend.

When I returned to the Laboratory Monday morning, I told a friend there, who was very knowledgeable about photography, what I had discovered. He told me that this phenomenon had been discovered many years before by Sir John Herschel, noted for his work in photography, and the phenomenon I had discovered is actually called the Herschel Effect. I sought more information about his findings and, to my pleasure, discovered that I had duplicated most of his experimental results!

At about the same time I developed a photographic process that, to me, was extremely interesting. I found that I could mix some potassium dichromate in a water solution of polyvinyl alcohol. A film cast from this mixture, dried, and then used as a photographic film, showed the same kind of effect that occurs when a gelatin film containing a dichromate is exposed to light. Depending on the intensity of the light falling on it, the film becomes progressively insoluble with increased amounts of light.

After finding the selective reaction would occur with the polyvinyl alcohol dichromate film, I then put it in a silvering solution. To my delight I found that the silver was deposited to form a beautiful picture that seemed grainless. I have never done anything further with this intriguing phenomenon. I have recently wondered whether this mechanism wouldn't be worth some exploration with respect to the production of the ultimate resolution with microchips!

GE Building number 37 opened in 1926 the same year Vince began working for the Research Laboratory. Courtesy of Schenectady Museum

My Friends in the Shops of the Research Laboratory

Since I started my association with the Research Laboratory during the fall of 1926 as the lowest paid worker in the entire organization, I had a chance to become acquainted with everyone, since I had nothing to lose. When I left in 1929, it was with the good wishes of the friends I had made. When I returned in the fall of 1929, I resumed my contacts and friendships and, during the next four years, enlarged my circle of friends.

A very good friend was Superintendent Robert Palmer, who headed up all of the Service Shops of the Laboratory. It was Bob Palmer who advised me against investing in a plant nursery business just before the start of the 1929 Depression. One of his managers, that of the Machine Shop, was Nathan (Ginger) Adams. I worked under Ginger during both periods I was in the Shop, and found him to be a very friendly person.

Then there was Bill Ruggles, who ran the Glass Shop. It was in this shop where I had a number of glass tubes fabricated for experiments with ultraviolet light effects, and where all of my evaporated metal films were made, and where my multicells were sintered.

Alan Cooley was head of the Metallurgical Shop. where all sorts of interesting operations were conducted. including the early fabrication of silicon carbide, the drawing of tungsten wire, and other fascinating operations were undertaken. I believe it was under his direction that I was given the challenging job of threading up and weaving a multi-strand tungsten filament braid. consisting of about twenty strands of 0.0005" diameter tungsten wire. The wire was so fine as to be almost invisible. so that I had to use my fingers to locate it and to thread it from the bobbin to the place where it would be braided. I was finally successful but, when finished, was somewhat relieved when I learned that the experiment for which it was made didn't work out as anticipated!

Dr. William Coolidge, shown here in 1967, was the second director of the GE Laboratory in which Vince worked. Coolidge invented the X-Ray tube. Courtesy of Schenectady Museum.

I got to know most of the skilled workmen in the Shop and knew the ones who could do the best job for a particular device we needed. Art Parr and Clarence Nelson were among those I depended on. We had a master welder – Sam Lospinoso. Unlike so many others in that profession, Sam was a philosopher, an avid reader, and a well-educated individual. After Doc Whitney relinquished his Directorship to Dr. Coolidge, he asked Sam to teach him the rudiments of the welding arts.

Another of my good friends was Claude Perkins, a master carpenter. Claude was born in the Adirondacks and often, while he was building something for me, he would tell me of some of his experiences in the mountains. The head of the Carpenter Shop was another friend, Jesse Scrafford. When I needed a special job, such as a picture frame for a farmer who had been highly cooperative in our artificial fog studies in the Schoharie Valley, I prepared a beautiful enlargement of his farm taken from Vroman's Nose, as a token of our appreciation. He couldn't have been more pleased.

After I was invited by Dr. Langmuir to become his assistant, I made it a point to continue all of these friendships as best I could, and continued to live in "two worlds." Most of these friendships lasted throughout my participation at Langmuir University. I still have warm memories of those golden days!

My Special Friends in the Research Laboratory

I have mentioned some of my friends in the other General Electric Laboratories, the corporate organization and in the Service Shops of the Research Laboratory, and now will mention some of my special friends on the professional staff. I got to know a great many, but there were some with whom I had much closer relationships.

I have previously mentioned the Boss, "Doc" Whitney, Katy Blodgett, Bernie Vonnegut, Ed Hennelly, Ray Falconer, and Mary Christie. Dr. William D. Coolidge, who became the second Director of the Laboratory, was a very kind man, and though I never had a close relationship with him, I greatly admired his administration of the laboratory and his cordial treatment of me.

I was much closer to Albert W. Hull, who for a considerable period of time frequently stopped by to see how things were going. When I was trying to decide whether to leave the Laboratory to join the Munitalp Foundation, I discussed the matter with him in some detail. He approved of my decision to leave, but hoped I would retain a tie with the Research Laboratory. Another great person, in my opinion, was Dr. Saul Dushman.[38] It was Saul, more than anyone else, who took it upon himself to keep a close contact with the newly recruited staff of the Laboratory. There was always a sense of hilarity around the large table in the cafeteria where he held forth each noon.

There was a special sense of friendship among those of the Laboratory staff who participated with me in the Science Forum. Frank Norton, chemist; Lewi Tonks, physicist; Winton Patnode, chemist; along with me, constituted the "Original Four." When one of us was unable to participate, we had a highly capable group of replacements, including Bill Cass, Bob Krieble, George Baldwin, Jack Wolfe and Milan Fiske. Special friendships with all of these fellows developed over the twenty years of the Forum, and when we had a reunion a few years ago, we had a fantastic time!

Another person in the Laboratory, who I greatly respected but never got to know well, was Dr. Lawrence "Larry" Hawkins. He was the Staff Engineer and had an office next to Doc Whitney. He was noted for his knowledge of Greek, and it was said that many of the

---

38 Saul Dushman (1883-1954) authored several science textbooks, and his thermionic emission research is known in the form of the Richardson-Dushman equation.

Dr. Albert Hull, inventor of the Dynatron vacuum tube. Courtesy of Schenectady Museum.

"tron" tubes invented by Albert Hull and others were christened by name after consulting Larry. I was also told that he was the one who originated the painting of a white line on the edge of a highway and that the first application was placed along the road from Bolton Landing to Lake George Village, with Larry paying for the paint out of his personal funds. Larry was a very kind and considerate man, and I admired him very much.

Another special person was John Payne. An engineer type and inventor, it was John who developed the first remote control "hands" used to carry out chemical reactions and other tasks during the early days of radioactive studies in this country. John organized an expedition with me and several other lab personnel to visit and explore a purported gold mine in the Paul Creek Valley in the town of Day, north of the Sacandaga River. A very old man guided us to the site that was a deep hole in the ground, obviously excavated by man. He claimed that the gold ore was transported by mule back to Hadley, where it was loaded on a train for transportation to a refinery. We obtained some of the "ore," and Mac Safford of the Laboratory took on the task of evaluating the ore sample. After several days of chemical tests, he unfortunately discarded the critical solution (by mistake) that would have told whether there was gold in the ore. We never let Mac forget about this episode, and to-day the subject still poses a question.

I also had special feelings for Francis Bundy, Herb Strong, and Fred Wentdorf, who, in later years, became noted as the "Diamond Group" since they, along with Tracy Hall, had worked out the mechanism for the production of man-made diamonds.[39]

Over the years, the ranks of the Old Timers have thinned, until now there is only a small group remaining. Frank Norton and Milan Fiske have lunch with me about once a month. We have a great time discussing our current interests, and only occasionally dipping into the past. It is fun!

My Contacts with the General Engineering Laboratory

The General Engineering Laboratory was just across the hall from us in Building #5. There were many informal contacts between the two laboratories. In his own way, Everett Lee, the Director, was a character of the pattern of "Doc" Whitney. His enthusiasms were well known, and he seemed to have a good relationship with his staff.

Over the years I had friendly contacts with about ten of their engineers, and profited a great deal from these relationships. One of the first was Charlie Bachman,[40] a physicist, who was interested in electronic devices, from early television tubes, to the electron microscope. I became quite involved with him in attempts to improve the contrast and brightness of the television image. We cooperated in the development of an improved fluorescent coating for a television tube. I also helped him in preparing test samples for his electron microscope development.

José Malpica was an effervescent Mexican friend who was an engineer in the Laboratory. He was a frequent visitor to my lab, and we had many enjoyable discussions. He eventually returned to Mexico and became a professor at the University of Mexico.

Ted Rich[41] was one of my closest friends, and a brilliant engineer and inventor. One of his more than one hundred patents, one of the most interesting, and the one were we had much common ground, was his highly practical and rugged condensation nuclei meter. It was the commercial version of the Rich Counter, manufactured under license from General Electric by George Gardner, Ted's former supervisor. It was with one of these units that I measured air quality at hundreds of locations, ranging from the stratosphere to the depths of caves, from the center of Los Angeles to the Fiji Islands, the center of the Atlantic Ocean to the top of Mauna Kea, and the woods of the Adirondacks to the deserts of Arizona.

---

39 Hall invented the first reproducible process for making synthetic diamonds in the Laboratory in 1954, creating a multimillion dollar business. Hall was given a $10 savings bond by GE for his invention. Hall left GE after that and created his own company.

40 Charles Bachman developed the first successful database management system in 1961.

41 Rich developed equipment in the 1950s to search out snorkeling diesel submarines and a "people sniffer" during the Vietnam War.

Many thousands of measurements were utilized in preparing my report, "Air Quality on the Global Scale," which was issued as my last major research study printed by the University at Albany, in 1976.

My relationship with the General Engineering Laboratory personnel and the Works Laboratory, where I also had friends, was one of the delightful relationships that, in those days, made it such fun to go to work!

My Finances While Attending Langmuir University

During my final year as an apprentice boy, and when I first entered the GE Research Laboratory, my salary for a 48-hour week was something like $8.64 a week, or 18 cents an hour. As I graduated to become a journeyman, this salary was increased, but I don't have a record of the amount received. My income while with the Davey Institute, and the subsequent year of freelancing, is also unclear. I know it wasn't very much!

When I returned to the Laboratory in 1929 it improved, but again, my records are not definite for some reason. The first firm data I have shows that in the year 1934, the year when we started building our house on Schermerhorn Road, my total income was just about $2,000. Each subsequent year my salary increased about $250– until the war years when it leveled off, but then increased considerably until, in the late forties, it went up in increments of several thousand a year.

During and after the construction of Woestyne South, my income appeared to be adequate to pay for the building supplies, salaries and other costs involved with our construction and the establishment of our home. To my recollection, we never went into debt, but neither were we in a position to save much money! We lived on simple foods and made our own entertainment. In the process we all had a great time!

In retrospect, I now realize I was extremely fortunate in the way things worked out for me. In obtaining my education at Langmuir University my tuition and other costs of getting an education were the opposite of going to a regular university. Instead of paying out, I always had money coming in! This permitted me to buy whatever books were relevant to the immediate activities of the moment, so that I ended up amassing an invaluable collection of technical reference books, monographs, and other materials that continue to be useful to me. That part of my preparatory training in which I learned how to use tools was of tremendous value, since it engendered in me a sense of stability and independence, so that I had no fear of trying new activities. I knew that I had the training and ability that I could fall back on if, for any reason, my new venture didn't work out.

I have a strong feeling that every young person should spend three or four years in their late teens acquiring intensive training in the practical use of hands and brain. Such an effort would bridge the gap between adolescence and maturity, and would be an investment that would be of benefit for the rest of their lives.

My Experiences with the GE News Bureau

One of the facts of life in a large company like General Electric, is that their news staff is continually searching for newsworthy "angles" that might be publishable in magazines and newspapers. Thus, from time to time, we would have visits from one of them to see if we had "anything." It turned out that with our far-ranging activities, a number of interesting stories were developed.

The Boss's award of the Nobel Prize was a major story of the first order, and it tended to focus News Department attention toward our laboratory on the 4th floor of Building #5. Stories about some of our protein studies, of our smoke screen production, non-reflecting glass, and then of my cloud seeding discoveries, were highly newsworthy.

We enjoyed an excellent relationship with the News Bureau. It extended from the head of it, Chester Lang, to their chief reporter, Clyde Wagoner, to Joe Pauze, Roger Hammond, and Joe Morton. It was the latter two whom were most involved with us, especially during Project Cirrus days. When I told Roger that I had decided to leave the Laboratory to become Director of Research at the Munitalp Foundation, he appeared a week before my final day at the Laboratory with his friend and colleague, Joe Morton. With considerable ceremony they presented me with a fancy certificate bearing the following citation:

Institute of Tele-Fortitude

This is to certify that

Vincent J. Schaefer

Having successfully undergone all rigors of tele-control and tele-control procedures, and having the requisite tele-training, is hereby awarded the degree of

Doctor of Loquacity

And is hereby qualified and commissioned to the following privileges of this rank:

- Unguided and unmonitored to answer his own telephone even in cases where the caller is a reporter.
- Unguided and unmonitored to comment on the weather *viz.* "some rain we're having."
- Unguided and unmonitored to indulge in such references as calling a light bulb a light bulb instead of a lamp.
- Guided and monitored to enjoy uncommon success, fame, faith and happiness.

(with best wishes of the undersigned)

Roger P. Hammond

Joseph D. Morton

February 19, 1954

Of the considerable number of such items that have been given me while at Langmuir University and in later years, it is one of my most cherished possessions!

The Backup I Enjoyed from Lois and Our Children

None of the things I managed to accomplish while attending Langmuir University would have been possible without a happy home life. This I enjoyed to the ultimate with Lois, Susan, Katherine and Jim.

No matter what I had to do– the considerable periods when I was on expeditions– that is all of the things that an active scientist is required to do as part of his advancing responsibilities, were approved without question by Lois. I knew that my frequent absences put a double or more burden on her in bringing up our children, but it was never a point of contention. The youngsters on their part were exceptionally mature in helping her or creating a minimum of problems.

I made it a point to take the family west with me every summer while conducting my fieldwork. Whether it was Priest River, Idaho; Missoula, Montana; Socorro, New Mexico; Flagstaff, Arizona; or places in between, we had wonderful times. At Priest River and Flagstaff, the US Forest Service supplied us with housing at the Experiment Stations. At Missoula we rented a house, while at Socorro we occupied student housing rented from the New Mexico School of Mines.

The trips to our western stations were full of adventures. In a couple of instances, Lois drove out alone with the children and met me at a prearranged rendezvous. In these trips, which took five to six days, we had a chance to see our youngsters under varying degrees of stress, and were able to better understand various aspects of their character far better than parent-child relationships at home.

In the course of these adventures, our children developed a self-sufficiency that was to stand them in good stead in later years. Our only regret was that it gave them a degree of wanderlust and a high regard for the West so that two of them, Katherine and Jim, chose to spend their lives in the West. I am sure that it wouldn't take much for Susan to do the same, but her husband's work centered mostly in the Boston area, so that we are fortunate to have them relatively nearby.

The memories of our many trips and adventures in the West, and the friendships that developed in the process, have provided a continuing thread of good feeling and connections that have sustained all of us to a wonderful degree.

During my days with Langmuir, it was great fun for all of us to visit him and Marion at the Crown Island camp. The Boss would take the youngsters for fast trips in his Chris-Craft across the wide waters of Lake George, and it was obvious that everyone was having a super time. Afterward, he would show them tricks and other things attractive to young folks.

Whenever the youngsters and their families get together, it is a joy to see the love and friendship they have for each other, their children, their friends and us.

Family picture, Christmas 1960. Lois and Vince with their daughter Sue Schaefer Sullivan, her baby Michael and husband John W. Sullivan. Courtesy of Jim Schaefer.

My Friends in the GE Organization Beyond the Laboratories

As I became increasingly active in research activities with the Boss, I occasionally encountered others in the General Electric structure whom became my friends. One of my treasured memories involves Mr. J. R. Lovejoy, who was high up in the corporate structure. It was J. R. who accompanied "Doc" Whitney and me on an Indian relic hunting expedition down the Hudson, and who for the next year talked about the great time he had to anyone who would listen! It was also J. R. who gave our fledgling Van Epps-Hartley Chapter of the State Archeological Association funds to acquire the Percy Van Epps collection of Indian artifacts and his substantial library, professional design cases for displaying collections at the Schenectady Museum, and the digging tools and surveying instruments needed when we excavated the Schermerhorn site.

Another friend was Charles E. Eveleth, Manager of the Schenectady Works of GE. I approached him with the idea of preserving the original models of the electron tubes and other devices being invented by Langmuir, Albert Hull, W. D. Coolidge, and others, in both the Research Laboratory and the General Engineering Laboratory. I knew that they would eventually be scrapped and felt that the Company could establish a Museum to preserve them. Everything was proceeding fine with our plans when a major change occurred in the GE organization; our plan was shelved and subsequently abandoned.

During the Great Depression, due to my being Director of the Mohawk Valley Hiking Club, I was approached by E.W. Allen, who had volunteered to head up the regional activities of the Works Progress Administration (WPA).

His high post in General Electric, and his administrative ability as head of engineering for the Company, made him a fine choice for making things happen in our road to recovery after the climactic days following the end of 1929. He asked for my cooperation in establishing a geological pyramid made up of a sequence of the rock strata underlying our region. He also wanted to explore the possibility of establishing a County Forest in the abandoned farm country in the high lands north of Duanesburg. I suggested he get professional guidance for the rock pyramid from Professor Smith of Union College, and that Albert Getz, a fellow member of the hiking club (who at the time was unemployed), could supervise the construction of the pyramid. I also recommended that Al Getz could also supervise the planting of the County Forest, since he was deeply involved with us in planting trees at the Christman Wildlife Sanctuary, where we were helping Mr. W. W. Christman, "the Poet of the Helderhills," plant a number of his hilly acres with seedling trees.

Mr. Allen immediately followed my advice, and both the pyramid and the County Forest were established with the active supervision of my close friend Al Getz. Now, after fifty-five years, the stone pyramid has been moved to the grounds of Union College where it was designed, and the County Forest is now a favorite cross country ski area in the Duanesburg Hills.

Another friend and an engineer was Mr. Phillip Alger, a very dynamic person, who also had a hobby of encouraging good government at the local level. He was responsible for the introduction of the City Manager type of city administration to Schenectady, but was not a particularly savvy politician. He managed to talk the Boss into running for Councilman, that he agreed to do. Unfortunately, the politicians who had been deposed by Alger's action rallied in opposition and, as a result, all of Alger's slate was defeated and the politicians emerged victorious, and apparently carried on as before with a somewhat different organizational plan! I think the Boss was secretly pleased with his defeat, since such service would have taken precious time away from his research activities.

Another friend of mine was Chester Rice, son of E. W. Rice, Jr., who was responsible for getting Dr. Whitney to form the Research Laboratory. While I never got to know Chet very well, our paths crossed frequently. When he died, his widow offered to give me as many books from his rather large library as I might wish to have. Consequently, I obtained a number of fascinating books, including a complete set of Faradays Experiments and a number of other fine books on the history of science.

I became involved with the Boy Scouts of America in 1926 when I became Scoutmaster of Troop 26, the "Poorhouse Troop," so called because our meeting quarters in the Hamilton Hill area of Schenectady was at the County Poor House. In my contacts with the Scout Executives Office, I became acquainted with Ben Fisher and Ralph Carter. Ben was another engineer who was trying to purchase a chunk of land near Rock City Falls on the upper waters of the Kayaderosseras Creek. I became quite involved in his efforts and helped in the establishment of Boyhaven, as it is now called[42]. Ralph Carter was also very actively involved in Scouting. He was a major purchasing agent for General Electric. When he learned that we were interested in buying several bolts of high quality balloon "silk," a long staple of Egyptian cotton made by the Wamsutta Mills, he offered to help us obtain the cloth through the good offices of the General Electric Company. This was accomplished, and we were able to make our bags of the finest quality material available.

There were many others in the corporate structure of the Company that I became acquainted with. In those days there was a sense of neighborliness and a sense of civic responsibility that eroded away when the company headquarters moved to Connecticut. I was indeed fortunate to be with the Company for some years before that happened.

The Professional and Other Organizations that I Joined

In the early days of my assistantship with Irving Langmuir, I was interested in improving my professional status. Toward the latter part of the thirties, I was invited to become a member of the American Chemical Society. My application was approved, and for a number of years I received the Journal and published a number of papers in it.

Subsequently, I joined a number of other scientific organizations, including the American Geophysical Union, the American Meteorological Society, the Ecological Society of America, Society for American Antiquity, American Association for the Advancement of Science, and the National Speleological Society. I was admitted to membership in Sigma Xi, Sigma Pi Sigma, and the American Institute of Chemists. One of my most enjoyable relationships was with the American Geophysical Union. I was asked to become a member of the Committee on Snow and Ice. This brought me into personal contact with Dr. James E. Church, Dr. Francois Mathes, Dr. Ukichiro Nakaya and Dr. Marcel de Quervain. With the latter two, I developed the Snow Classification System that has been used for many years as the basic classification of snow and ice forms.

Except for this committee, I didn't become actively involved in the business of any of the other professional societies except that of the American Meteorological Society. With this organization I was elected to be a member of the Council, and also a member of its Education Committee. As a member of the Council, I became quite involved in the acquisition of 45 Beacon Street and a variety of other activities, which transformed the Society from a group of meteorologists discussing localized problems, to a highly professional organization which, under the long time leadership of Dr. Kenneth Spengler, is one of the best American scientific endeavors.

---

42 Boyhaven is still in operation and run by the Twin Rivers Council.

Honors Received while at Langmuir University

It may seem to be a process of self-serving to list the several honors give to me during this formative period of my life, but for the record– and to illustrate the way in which our cultural pattern in America recognizes certain types and forms of achievement– I will list the various honors I received, placed in chronological order.

- Certificate of Apprenticeship - General Electric (December 6, 1926)
- Young Man of the Year Award – Schenectady Jr. Chamber of Commerce (1940)
- Fellow - Rochester Museum of Science and Art (May 13, 1943)
- Member - Sigma Xi, Union Chapter (April 20, 1948)
- Outstanding First Paper - Div. Hydrology, Amer. Geophys. Union (April 21, 1948)
- Doctor of Science (Honorary) – University of Notre Dame (June 6, 1948)
- Man of the Year Award - Notre Dame Club of Schenectady (1948)
- Robert M. Losey Award – Institute of Aeronautical Sciences (1952)
- Doctor of Loquacity – GE News Bureau (February 19, 1954)
- Fellow – American Association for the Advancement of Science (July 19, 1956)
- Professional Member – American Meteorological Society (1956)
- Outstanding Contribution to Advancement of Applied Meteorology - AMS (1957)
- Advisory Council - New York State Museum (1957)
- Fellowship - Woods Hole Oceanographic Institute (1959)
- Member (Honorary) - Sigma Pi Sigma, Albany Chapter (1960)
- Fellowship - CSIRO, Australia (1960)
- Distinguished Science Lectureship - State University of New York, College of Education at Albany (1960-61)
- Fellow - American Meteorological Society (1967)
- Doctor of Humane Letters (Honorary) - Siena College (1975)
- Ideal Citizen of the Age of Enlightenment Award–All Possibilities, Research and Development Award - American Foundation for the Science of Creative Intelligence (1976)
- Vincent J. Schaefer Award, for scientific and technical discoveries that have constituted a major contribution to the advancement of weather modification - Weather Modification Association (1976)
- Citizen Laureate - University at Albany Foundation (1980)
- Doctor of Science (Honorary) - York University (1983)

My Friendship with Ray Falconer

When I first went to the Mount Washington Observatory I met Ray Falconer, one of the observers at that facility. I recognized him to be considerably different from the rest of the observers, having wider interests and a flair for original research. At the time he was operating his own short wave radio, and had many contacts around the northeast and more distant places. He soon became our contact point at the Observatory and carried out our experimental measurements to perfection.

When I discovered a method to modify supercooled clouds and a variety of new activities developed in the laboratory, I approached the Boss and suggested that we offer a job to Ray in Schenectady. Langmuir, who was also favorably impressed with Ray's abilities and enthusiasm, approved my suggestion; the job was offered and accepted, and Ray was soon a member of the Laboratory staff.

Within a short time Ray had taken over all the activities having a meteorological aspect, and his presence proved to be invaluable. Within a year or so he was deeply involved with our Project Cirrus activities.

When we decided to phase out our Project Cirrus program, Ray had a fine meteorological observatory established in a tower in the new laboratory at the Knolls, and was actively involved in applying his expertise to different projects for Laboratory personnel.

One of the most interesting of these was the anomalous propagation of television signals due to the presence of the jet stream.

Before Project Cirrus ended, the Boss had Ray carry out an elaborate analysis of the seven day periodicity in pressure and precipitation that seemed to be correlated with the output of a silver iodide smoke generator located at Socorro, New Mexico. This was operated for only a single day each week. A widespread periodicity in pressure and rainfall was found by Ray using the Climatological Records of the Weather Bureau. Despite vociferous objections from meteorologists in the Bureau and the universities, Langmuir persisted in his claims, and to this day the anomaly has not been adequately explained.

Ray remained at General Electric for a while after I had left to join the Munitalp Foundation. Within a year, however, I had established a position for him in the Foundation, at which point he left the Company to work with me. During the following five years, his main project was the development of a long-range (30 day) weather forecast method based on the position of high-pressure cells over the Pacific Ocean. It was remarkably accurate for predicting the detailed weather over our region, and seemed to be as good or better than any other experimental methods developed by such persons as Willet, Krick, and Namias.

The continuation of his career after I left the Foundation was based on my subsequent activities, first at Rensselaer Polytechnic Institute, and finally at the State University of New York, where he worked with me until his retirement in 1984.

During this extended period, his ability as a weather announcer became recognized, and over the years he developed a large and very loyal audience which continues at the present time.[43]

My Contact with Dr. Peter Debye

Among the many world famous scientists that visited the Boss when they were in our region, none of them impressed me more favorably than Dr. Peter Debye[44] who was a Professor at Cornell University. He had the same type of enthusiasm that I had learned to treasure about the Boss and Doc Whitney. That characteristic, along with the twinkle in his eyes, set him apart as a delightful person.

The first time he visited us at Schenectady, he showed us the characteristics of a remarkable crystalline material– TRIOXANE. Its structure was such that it was highly symmetrical in one plane, and very asymmetric in the one at right angles. This may have accounted for its remarkable electric properties. When a small chunk of such a crystalline material is placed in a transparent container (such as a glass bulb) and heated locally with a flame, it readily melts. As it solidifies, it sends out long hair-like fibers that are highly charged. If two bits of crystalline trioxane are so melted and permitted to crystallize on opposite sides of the chamber, the long fibers grow rapidly toward each other until they are about one inch apart. As they increase in length, a strange thing happens. They either dive toward each other and merge, or the tops avoid each other and swing past.

Some years after his visit at Schenectady, I visited him at Cornell with eight of my Natural Sciences Institute students. We had a wonderful time as he gave an informal talk to the students, who were equally impressed with his attitude. At that time I asked him if he knew any further things about trioxane. He smiled and shook his head in the negative, but was pleased that it had intrigued me also. I have since made several attempts to find out more about this mysterious electrification effect, but to no avail.

Peter Debye (now deceased) was one of those great scientists who, although they had worldwide eminence and fame, chose to be a teacher, and in the process was a great and kindly man who had the respect and love of everyone.

The Formation of Iron Fibers

On March 7, 1953, my attention was directed towards an interesting problem related to the formation of iron needles that were needed to improve the magnetic properties of various fabricated structures.

Within a couple of weeks I had found a number of mechanical and surface chemistry processes which looked favorable for the formation of tiny needles of the range of 0.02

43 Ray Falconer was nicknamed "Mr. Weather" and was known on local radio stations for his reports. He died in 2002.

44 Debye (1884-1966) was a Dutch physicist and chemist. He won the Nobel Prize in Chemistry in 1936.

microns diameter and 0.2 microns long. This included reactions involving the formation of iron films on the surface of clean mercury, and in double layers with water on top of a pool of mercury held in the bottom of a Langmuir trough. Such films could be formed while under an electric potential and/or a magnetic field.

I also dissolved iron sulphate in a 3% water solution of polyvinyl alcohol and cast it as a thin film on a glass slide. It was then heated in the reducing part of a gas flame and found to produce a very shiny film of iron that was highly magnetic. At the time I proposed that this might be a good way to produce a magnetic coating on a plastic film. It also occurred to me that it would be easy to form filaments of PVA containing iron sulfate that could then be placed in a reducing atmosphere to possibly produce iron needles. If this were to work, the needles could be made of any size desired.

I also found it possible to produce highly reflective films on the surface of a solution of silver nitrate dissolved in a water solution of polyvinyl alcohol. When dried of water, a beautiful reflective coating of silver coated the underlying solid. Successive coatings could be made in this same manner so that films of any desired thickness or combination of reduced metals could be formed. In making such films, I prepared one of silver and gave it to Dr. Frank Studer on January 14, 1954, who determined that it was a very interesting filter for passing ultraviolet light, having a transmission maximum in the 3,250Å area of 55%, ranging from 2,400 to 4,000Å (0.24-0.40 microns).

These were my last research activities at the GE Research Laboratory. I left the Company in the late afternoon of February 26, 1954 to become full-time Director of Research of the Munitalp Foundation.

(From left) Don Fuquay, Vernon Crudge, Jack Dieterich and Vince on Mt. Gisborne in northern Idaho. Courtesy of Kathie Miller.

My Notebooks and Patents

The primary reason for using the GE Research Laboratory Notebooks was to provide back up (if needed) for establishing the background and dates of idea conception that might eventually relate to the formulation of a patent for the General Electric Company.

The Laboratory Notebooks were of two types. What seems to have been the earlier one was a leather bound, page-numbered notebook 5 1/2" x 8 3/4" of 333 pages. There was also a larger one, also leather bound and page-numbered, having the dimension of 9 1/2" x 12" and of 150 pages.

The following list indicates the official number, the type and the period of use.

No.	GE Research Lab Notebook No.	Type	Start	End
1	2474	Large	October 14, 1934	April 13, 1936
2	2642	Large	April 16, 1936	May 13, 1937
3	2750	Large	May 6, 1937	November 15, 1937
4	2836	Large	December 3, 1937	February 8, 1938
5	2891	Large	May 3, 1938	February 20, 1939
6	2991	Large	March 2, 1939	March 14, 1940
7	3132	Large	March 29, 1940	September 7, 1943
8	3211	Small	September 23, 1940	January 10, 1942
9	3398	Small	December 3, 1941	February 9, 1942
10	3652	Small	March 22, 1943	September 17, 1943
11	3771	Large	October 8, 1943	February 3, 1945
12	3986	Large	January 1, 1945	February 24, 1950
13	4651	Small	June 25, 1948	February 10, 1950
14	4652	Small	July 14, 1948	July 18, 1950
15	4653	Large	July 19, 1948	February 26, 1954

It will be noticed that not all of the notebooks are in strict sequence and that some overlapping of time occurs. That is because at several times the Boss needed information recorded in my current notebook and would borrow it, sometimes for extended periods. When this happened, I would obtain another notebook and would enter pertinent information in a sequential fashion. Whenever this was done, the dates of entry indicates my activity during such periods.

I used the smaller notebooks for specific projects. The first three relate to special unclassified defense projects carried out during World War II. The last two of the small size cover activities carried out during Project Cirrus activities.

I much preferred to use the larger notebooks, since that format provided a better opportunity to mount photographs, make sketches, and include graphs and tables without becoming unusually bulky.

Based on information recorded in these notebooks, I was granted eighteen patents which are listed below.

US Patents of Vincent J. Schaefer

Joint Patentee	Patent No.	Title
	2108616	Treatment of Materials
	2264892*	Electrical Devices Containing Free Films and Processes of Forming Such Films
	2352976*	Light Dividing Element
	2374310*	Methods of Producing Solids and Desired Figuration
	2374311*	Cathode Ray Tube
With C. Bachman	2421979	Production of Fluorescent Coating
With I.Langmuir	2439963	Methods and Apparatus for Producing Aerosols
	2462681	Method of Forming Germanium Film
	2492768*	Cloud Moisture Meter
With K. Blodgett	2493745*	Methods of Making Electrical Indicator of Mechanical Expansion
With B. Vonnegut	2527230	Method of Crystal Formation and Precipitation
	2532822*	Electrical Moisture Meter
	2570867	Method of Crystal Precipitation in an Aerosol
With K. Blodgett	2589983	Electrical Indicator of Mechanical Expansion and Method of Making
	2721495*	Method and Apparatus for Detecting Minute Crystal Forming Particles Suspended in a Gaseous Atmosphere
	2726211	Heat Storage Material
	2731713	Method of Making a Focused Multicell
	2924535	Polymer Containing Metallic Compositions

* *Drawings that accompanied these patents granted to Vincent Schaefer appear on the following two pages.—Editor.*

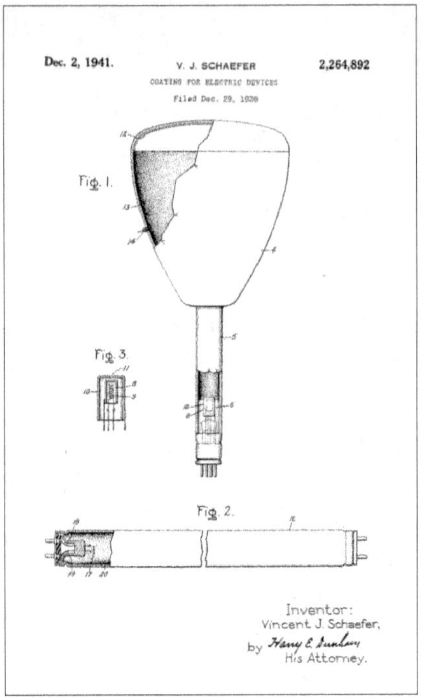
Dec. 2, 1941.
V. J. SCHAEFER
2,264,892
COATING FOR ELECTRIC DEVICES
Filed Dec. 29, 1939
Fig. 1.
Fig. 3.
Fig. 2.
Inventor:
Vincent J. Schaefer,
by Harry E. Dunham
His Attorney.

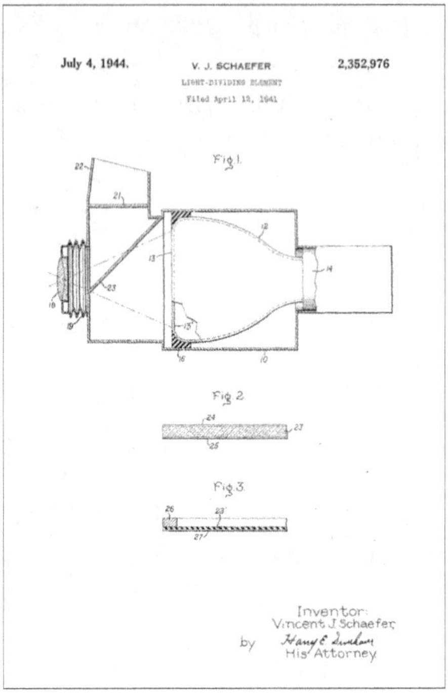
July 4, 1944.
V. J. SCHAEFER
2,352,976
LIGHT-DIVIDING ELEMENT
Filed April 12, 1941
Fig. 1.
Fig. 2.
Fig. 3.
Inventor:
Vincent J. Schaefer,
by Harry E. Dunham
His Attorney.

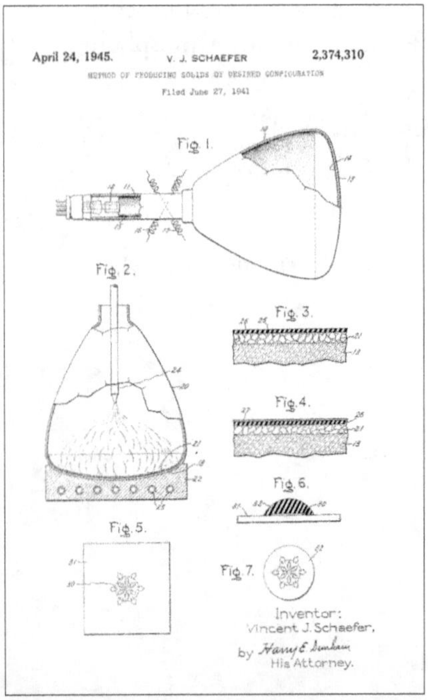
April 24, 1945.
V. J. SCHAEFER
2,374,310
METHOD OF PRODUCING SOLIDS OF DESIRED CONFIGURATION
Filed June 27, 1941
Fig. 1.
Fig. 2.
Fig. 3.
Fig. 4.
Fig. 6.
Fig. 5.
Fig. 7.
Inventor:
Vincent J. Schaefer,
by Harry E. Dunham
His Attorney.

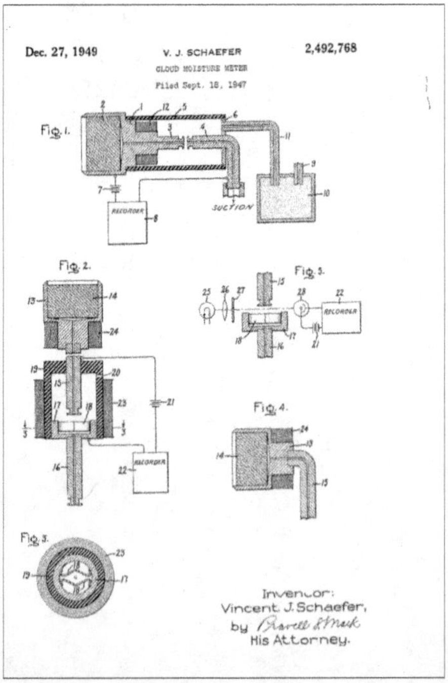
Dec. 27, 1949
V. J. SCHAEFER
2,492,768
CLOUD MOISTURE METER
Filed Sept. 18, 1947
Fig. 1.
RECORDER
SUCTION
Fig. 2.
RECORDER
Fig. 5.
RECORDER
Fig. 4.
Fig. 3.
Inventor:
Vincent J. Schaefer,
His Attorney.

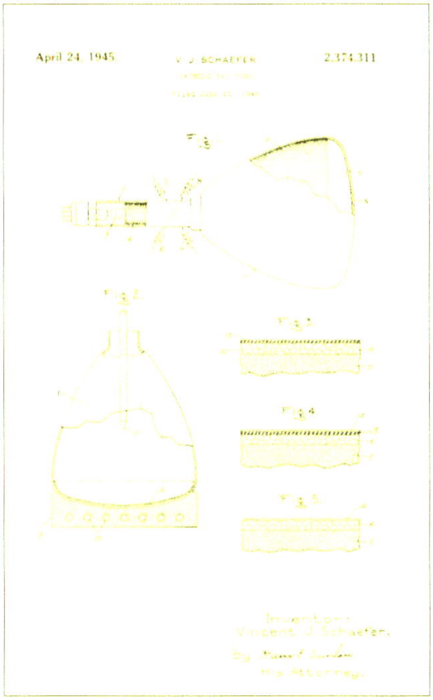
April 24, 1945
V. J. SCHAEFER
2,374,311
Inventor:
Vincent J. Schaefer,
His Attorney.

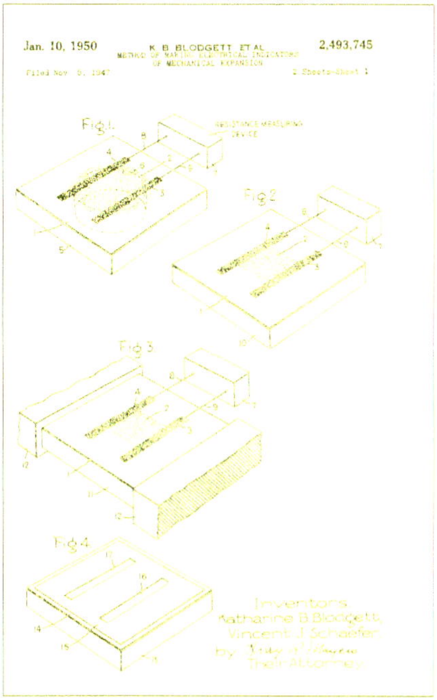
Jan. 10, 1950
K. B. BLODGETT ET AL
2,493,745
Inventors:
Katharine B. Blodgett,
Vincent J. Schaefer,
Their Attorney.

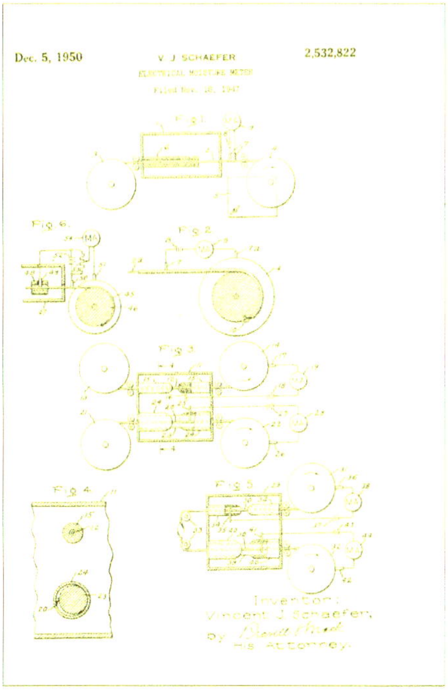
Dec. 5, 1950
V. J. SCHAEFER
2,532,822
Inventor:
Vincent J. Schaefer,
His Attorney.

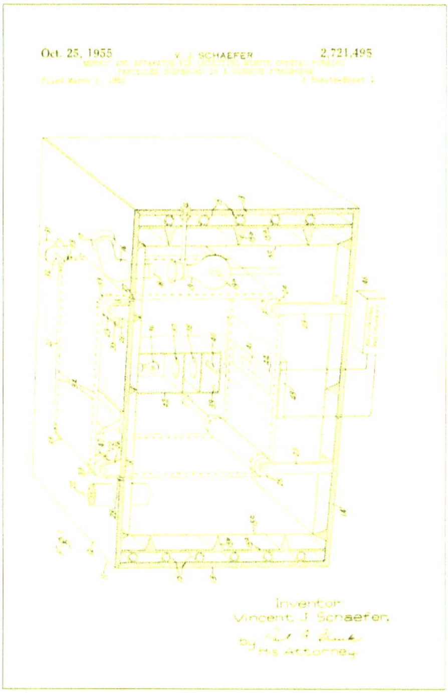
Oct. 25, 1955
V. J. SCHAEFER
2,721,495
Inventor:
Vincent J. Schaefer,
His Attorney.

The End of my University Training

One of the marvelous features of attending "Langmuir University" was the sense of freedom that I enjoyed from the time I climbed upstairs, until the day in 1954 when I left the Laboratory to join the Munitalp Foundation.

Beside this sense of freedom was an implicit understanding that whenever I needed advice, help, or understanding, it was available from the Boss, Katy Blodgett, Doc Whitney or the ever-widening host of friends that I soon developed.

The happy combination of assigned projects and personal interests permitted me to blend them together in such a way that I never lacked something to do. This was particularly valuable, since the Boss was frequently away for considerable periods of time.

I made it a point to work closely with Katy Blodgett, and I believe I was quite helpful to her in many ways. By having identical Langmuir Troughs, which I designed and obtained from the Machine Shop, I frequently carried out complementary experiments in close cooperation with Katy Blodgett. I also learned a great deal from her on many things. We had lots of fun and much satisfaction in making steady progress toward reaching an understanding of the effects of trace metal ions and pH and the purity of water on the properties of monolayers that were forming. I discovered the first evidence of the skeletonization process, and pointed out the importance of a tiny trace of copper dissolved in the water to eliminate the cracking of thick, built-up multilayers.

As I gained confidence, ability, and knowledge under the guidance of the Boss and began to discover new phenomena, I was able to write papers on my own and get them published. This, of course, was preceded by the preparation of the joint papers by Langmuir and me, in that I wrote much of the descriptive detail, while Langmuir provided the analytical (and sometimes) mathematical treatment of the subject.

With the onset of World War II, much of this writing effort was shifted to the preparation of lengthy reports to the sponsors of our defense activities. While these rarely entered the realm of reviewed papers, they had a wider circulation and, in some ways, they were of considerable help to others who where in the field covered by our work, but they weren't included in the permanent literature. Others profited from these reports and occasionally followed up on our findings, and were later credited with having originated the discoveries. Neither Langmuir nor I were concerned with such happenings, though occasionally were disillusioned with the scientific integrity of such individuals.

With the end of the war and my discovery of dry ice seeding techniques, a new era of experimentation and publication was entered. In a sense I had graduated from Langmuir University. After our adventures in New Mexico, Puerto Rico and the Schenectady area, my activities gradually drifted away from close contact with the Boss and Katy Blodgett.

I was in the field more and more, and thus less at the laboratory bench, or when there, was deeply involved in simulations of cloud physics situations.

When possible I kept in touch with Langmuir, who retired in 1950 but continued actively as a consultant to the Research Laboratory until his death on August 16, 1957.

The last of my Langmuir projects was an experimental attempt in the laboratory to understand the circulation mechanism that occurs in the trade wind areas of the Pacific.

At the time (in the mid-fifties) I had greatly expanded the facilities of my home laboratory. When I left GE I was able to buy, for a token amount of money, all of my experimental equipment, ranging from tools to cold chambers to Langmuir Troughs. I converted most of my basement to this purpose and had an ideal setup to continue my laboratory research.

My departure from the Research Laboratory was the result of several converging factors. Some years earlier the Directorship for the Laboratory passed from Dr. William Coolidge to Dr. Guy Suits. Meanwhile, the Laboratory had greatly expanded its size and numbers, having occupied a fancy new building along the Mohawk River east of Schenectady. With increased size, the sense of family slowly changed, and for the first time the staff was divided into teams or groups that were managed by others of the staff, who were designated as Managers.

This gradual drift away from the freedom that we enjoyed during the Whitney and Coolidge eras was disturbing to me, and I found myself watching the clock– something that I had never done in my earlier days.

In 1952 Vernon Crudge approached me on behalf of a small foundation that had been set up by a wealthy mining engineer living in Greenwich, Connecticut. I was told that the foundation was seeking to develop a coordinated program in the atmospheric sciences to support some phase of that field, and that they had approached a number of persons to solicit ideas. This idea intrigued me, and I spent several evenings putting together a plan that would be patterned after the Whitney Laboratory, i.e., searching for outstanding persons and finding a way to assist them in research areas they were enthused about. While there was not much money available ($50,000-$60,000 a year) I felt that if a portion of this (perhaps half of it) could be used to partially support such research, it would fill a real need. I also built the program about my own interests, putting together a dream plan that would embody complete freedom to do as I pleased. To my surprise, the Trustees of the Munitalp Foundation approved my plan in totality! Since I still had a full-time job, I was in a dilemma. I proposed that for the next two years (1952-54), the Foundation pay half of my salary to the Research Laboratory so that I could get things started, and for two years the Foundation Trustees would see whether my program and progress was to their liking. I discussed my plans with Dr. Langmuir and other friends at the Laboratory, including Drs. Hull, Dushman and Blodgett. They all thought it appeared to be an exciting opportunity.

Before embarking upon my original effort for the Foundation, I had a serious discussion with Dr. Suits about my future at the Laboratory. I had been urged by Dr. Malcom Hebb, the designated Manager of our group, including Katy Blodgett and Ray Falconer, to consider going into the field of semi-conductors. Dr. Bernard Vonnegut and Duncan Blanchard had decided to go to Woods Hole Oceanographic Institute.

I told Director Suits that I had been happy during my years with Langmuir, but that I had noticed a gradual shift in philosophy of the Laboratory management toward a more formal, regimented pattern of operation. This was a major departure from the Whitney Laboratory style of management, and I wondered about freedom to operate in the old manner. I was told in no uncertain terms that there was no change, and there would be no change. With this attitude, (which I understood in view of the greatly enlarged staff that was triggered by World War II and aftermath) I decided that it might be well for me to consider leaving the Laboratory as Bernie Vonnegut had done. Thus, when a few months later I was asked to devise a program for the Munitalp Foundation, I wondered whether this might be the opportunity I was looking for.

Once I received the offer of becoming Director of Research of the Munitalp Foundation, I went to Dr. Suits and told him about the opportunity offered me. He raised many objections, stressing in particular that I would be abandoning most of my potential pension income from General Electric. However, when I compared the exciting prospects of the unknown with Munitalp, and a continuation at the Research Lab without Langmuir, with a Manager that I didn't enjoy and activities that didn't involve surface chemistry, cloud physics, fine particles, or thin films, it became quite apparent what I should do. We arranged a program in which I would spend half-time on Laboratory work, and the remainder on Munitalp program development during the 1952-1954 period. As the two-year arrangement was ending, this had gone so well with Munitalp, while the GE Program continued to be less attractive and fun. Thus on February 26, 1954 I ended my formal association with the GE Research Laboratory.

My Last Day at the GE Research Laboratory

On February 26, 1954 the following statement ends my association with GE Research Laboratory:

> This records my last hour at the General Electric Research Laboratory. I shall become Director of Research of the Munitalp Foundation Inc. It is with many regrets that I leave the Laboratory. It took me two years to decide, but finally, in order to have peace of mind, I decided to take a whirl at it! I have enjoyed marvelous treatment from my many friends in the Laboratory and other parts of the Works. I shall miss the close contacts with them, but hope it will be possible to retain my friendship with these many good people.
>
> Vincent J. Schaefer
>
> 3:30 PM. Feb. 26, 1954

Prior to my decision to leave the Laboratory, the Boss had retired at the end of 1949. While he still retained his office next to my desk and laboratory room, which I was now sharing with Katy Blodgett and Bernie Vonnegut, our joint research activities began to taper off after the phase-out of Project Cirrus began. His periodic seeding analysis took up an increasing amount of his time, with Ray Falconer providing him with the meteorological data that was needed in his studies.

I found myself under increasing pressure from the new Laboratory management to shift some of my interests toward semi-conductors, a field quite foreign to my interests and capabilities. Thus I sensed that the remarkable freedom of the "Whitney Lab" was coming into disfavor and the managed, team approach mode could be glimpsed ahead. Thus my move toward the Munitalp Foundation with its promise of continued complete freedom, and an active involvement in field and laboratory experiments in an area full of unknowns and challenges, promised a continuation of the remarkable opportunities I had enjoyed with Langmuir.

This turned out to be a wise move and for the next five years, I had a fascinating time with the Munitalp Foundation.

From time to time during the following 3 1/2 years, my cooperation with the Boss continued. He helped me with the beginning of the Project Skyfire Fire Observers School at Missoula, Montana, and my final cooperative research experiments took place in my basement laboratory on Schermerhorn Road, in our studies of the movement of atmospheric air in trade wind systems.

An Adventure in the Fog Forest of Maui

Toward the end of our Hawaiian expedition– Project Shower, a group of us including Wedell Mordy, Erik Erikkson, Duncan Blanchard, Jack Workman, Pat Squires and I, led by Robert Bruce, manager of the East Maui Irrigation District, were taken to the edge of the great volcano at the edge of the National Park. We were transported by Irrigation District vehicles to a picnic area, where an indistinct trail headed into the ancient lava flows.

The objective of the trek was to visit a picturesque waterfall that plunged for hundreds of feet off the cliffs of the northeast side of the volcano.

Since this waterfall and its surroundings were mostly continually in clouds because of the Northeast Trade Winds that carried moist air against the mountain, and were forced to rise to surmount the obstacle to its flow. Even when there were no clouds over the sea, this region would be in clouds, due to this forced lifting of the moist air causing an orographic effect.

It had been proposed by Jack Workman's group that some of the water from the waterfall be diverted into pipes leading to primitive scaffolding that was erected at the brink of the waterfall. There it would be connected to a series of spray nozzles that would inject water into the oncoming cloud, thus "seeding" it with large water drops, onto which the cloud droplets in the local up-rushing cloudy air would collide, producing precipitation. This would be carried high enough up the mountain so that the falling drizzle could be collected by an irrigation ditch. It would then be carried down the mountain for us as irrigation water for the sugar and pineapple plantations on the lower regions of the mountain. We started our hike in high spirits, but soon found that the trail was a dangerous one at best. After crossing the ditch was nearly dry, the trail worsened and became difficult. At times it actually went over the edge of the cliffs, supported entirely by the tangle of roots from the lush shrubbery that made up the "cloud forest" of the upper slope of the mountain. At other places, the cloud water draining from the leaves of the dense shrubbery collected in mucky pools that were several feet deep. All of us were soon soaked all over with contact of the dripping vegetation, our boots were muddy and our feet soaked in water inside our boots. After struggling for what seemed to be hours, we finally arrived at the test site. It was a crude scaffold made of local saplings and trees found in the region, bound together with rope, and projected over the cliff edge. Spray nozzles had not been secured, so Duncan Blanchard and Erik Erikkson clambered up the makeshift structure and, holding their thumbs over the end of the pipes that had been flowing water, they briefly sprayed water into the dense up-rushing cloudy air.

While it was not possible to see any effects, it served to give me some good movie sequences.

We then headed down for the sea, using ropes that had been placed there earlier by Workman's field workers. While it was not necessary to rappel, the ropes gave us some degree of assistance in passing nearly vertical portions of the volcanic rocks.

As we lost altitude the shrubbery gave way to trees, and the verdant and lush growths obstructed our progress to a considerable degree.

The cliffs finally were left behind, but the trail didn't improve to any extent. The new trouble that we encountered was "rubbery" knees. Our downhill exertions had affected our legs to the degree that it was quite difficult for us to walk easily when more level terrain was encountered. At times, for no apparent reason except for lack of control of our walking equilibrium, we'd suddenly fall down. This was particularly experienced if we tripped over a stone or a projecting root.

It was during this struggle that we learned about some of the logistical difficulties that beset the project.

At the start it had been hoped that the aerial drops could be utilized to bring in nozzles, piping, and other needed equipment to the cliff edge area. After a number of drops missed the target and were lost in the dense forest above and below the cliff edge, it was decided to use burros for transporting the heavy and bulky equipment. It so happened that the most that could be transported was in a single trip for each animal. Nothing could induce an animal to make a second trip! They absolutely refused to budge once they sensed that it involved traveling on either the upper or lower trail! This behavior of the animals was quite understandable to all of us who eventually straggled out of the woods at the edge of Hana Road.

Nowadays I suppose a helicopter could solve the transportation problem, but they were not available in those days.

When we finally arrived at civilization, we went to a local headquarters of the Irrigation District, where Bob Bruce had arranged for all of us to have a steak dinner.

I have never had a bigger or a better steak.

I am stilled kidded by Wendell Mordy about the quick and complete manner in which I made that steak disappear!

While this adventure occurred after I left General Electric, the friendships and connections I had developed while attending Langmuir University permeated the entire operation. Most of the key people in Project Shower were first encountered during the early period of my "attendance," and like the prerogatives that one enjoys after graduating from a regular university, the connections that I made while with Langmuir has only become better over the subsequent years.

What good time I have had and am still having!

The Beginning of My Post-Grad Studies

My association with the Boss continued, and although my contacts were less frequent, my learning experiences continued. My last major trip with him was to Woods Hole. We left my home (Langmuir driving his car) early afternoon of Sunday, September 16, going by Thruway to Catskill, where we crossed the river and continued south to Red Hook. There, after some wandering, we located Tom Henderson, an old friend who was in charge of a cloud seeding operation using 3cm radar, an airplane, and 81 ground generators, in what seemed to be a good operation in hail prevention. Leaving the Red Hook area, we headed east and spent the night at a motel near East Hartford, Connecticut. The next morning, after an early breakfast, we headed to Providence, and finally to Woods Hole, where we arrived in the early afternoon. After meeting the Director, Columbus Iselin, and then Roman Vishniac, a mutual friend, we went to see Drs. Herbert Riehl and Joanne Malkus. Then Joanne gave a lecture on Trade Wind Cumulus, something that Langmuir and I had a great interest in. Joanne posed a number of questions about them and described the atmosphere in which they formed. The Boss made a series of extremely interesting suggestions about their formation and life history. The next day we continued the discussions, with Langmuir making many original suggestions concerning the nature of trade wind clouds. All of his suggestions seemed to be valid and there were no arguments.

That evening Langmuir and I parted company, he visited his nephews at Oak Bluff on Martha's Vineyard, and I took the local train to Boston and then the "Owl," a train to New York City. I slept all the way to New York.

Following his trip, Langmuir visited me in my laboratory, and I showed him a mode of a simulation of the circulation within the polygonal array that the trade wind clouds assume over the ocean. I had obtained some beautiful 3-D stereos of such patterns when approaching the Hawaiian Islands in 1954, and had patterned my model after these observations. He was highly pleased with the results of my experiments that nicely demonstrated the circulation patterns that developed when towering cumulus formed along the edges of the polygonal arrays and moved the saturated air upward, and then horizontally away from the clouds into the weak inversion that existed in the cloud free areas toward the center of the polygons. This was the last experimental activity that the two of us enjoyed. The following June, I, with my family, headed west to Priest River to continue the joint research program of the Munitalp Foundation and the US Forest Service, which was called Project Skyfire. It was a research effort to modify lightning storms by cloud seeding.[45]

Four years previous June 24-26, Langmuir had journeyed to Missoula to participate in our first training school for mountain top fire observers, who were going to help us locate the path of lightning storms and the presence of jet stream clouds over the Rocky Mountain area, stretching from the Canadian Border to the Hells Canyon of the Snake River. Some 25 mountaintops were utilized to provide us with daily observations that would cover an area of about 20,000 square miles.

The Boss gave an excellent talk to the school participants and showed a keen interest in the nature and behavior of the clouds of the Northwest.

On June 25, 1957, I headed back to Montana for the fifth season of Project Skyfire. Many interesting and valuable experiments were conducted during the summer. Dr. Ted Fujita joined us for a two-week period and conducted some extremely useful studies of wind flow patterns in our target area. Dr. Herbert Diehl was also with us for a week or so.

On August 16th, at the end of a very active day, I received a telegram from Schenectady that Dr. Langmuir had died. My first reaction was to drop everything and fly to Schenectady. On second thought, however, I had a strong feeling that the Boss would prefer that I carried on.

Two days later I received more sad news– a call from Vernon conveyed that Dr. Carl Gustaf-Rossby had died. He had been a great friend of ours. As Editor of *Tellus*, the international magazine on meteorology he published my first article on "Clouds of the Jet Stream," the first time color was used for illustrations in a meteorological magazine. He also was the first meteorologist to recognize the importance of Atmospheric Chemistry, and sought a grant from our Foundation to support a graduate student in the field. From that modest start, which we funded with $6,000 in 1954, developed the Swedish studies of that subject, which eventually demonstrated the ravages of acid rain.

While I decided to remain in the Northwest carrying on my research, I prepared a eulogy to Langmuir that I telephoned to my secretary in Schenectady. It was read at the Memorial Services honoring the memory of Irving Langmuir. Later I wrote a comprehensive tribute to him which is published at the beginning of Volume VII of his *Collected Works*, and shorter ones in the *Bulletin of the American Meteorological Society*, the *Journal of Colloid Science*, and in *Science*.

So ends my story of Langmuir University. I doubt if many individuals have been as privileged as I to have known intimately such a remarkable individual, and to have this association continue over a period of a quarter century! It was serendipity of the first order!

---

45 You can listen online to a news story about the project, broadcast on October 21, 1959, by visiting *http://www.cbc.ca/archives/*, and searching for "Project Skyfire."

A Contribution *In Memoriam*

An interest in the atmosphere and all its phenomena held the attention of Irving Langmuir throughout his lifespan. His initial interest was probably aroused while taking hikes and climbs in the Alps of Switzerland when, as a young boy, he spent considerable time exploring those mountains.

He once told me that his parents gave him considerable freedom to travel alone in the mountains, after obtaining his promise to observe several inflexible rules. One was that he learn to read a map and that he always have a map of the area in his pack; another that he was never to devote more than half of the available time for climbing. This assured him adequate time to return. With these simple rules he was often able to climb high into the mountains and to know them better than many older mountaineers. Because of these expeditions his self-reliance developed at a very early age and it was natural that he should become well-versed in the ways of the weather.

As a student in Germany, he continued his interest in atmospheric effects, and while hiking in the Hartz Mountains, he noted and became quite interested in the phenomena related to subcooled clouds. Many years later ,while working with him at Mount Washington on aircraft icing problems, he recalled these experiences in Germany and noted that the nature of the rime deposits and degree of subcooling tended to be similar, although the winds on Mount Washington were much more severe.

Shortly after joining General Electric at Schenectady, Langmuir, in company with his new friend "Appie" (John S. Apperson of Schenectady), made one of the earliest winter ascents of Mount Washington in the White Mountains, and Mount Marcy and Whiteface Mountain in the Adirondacks. It was due to this active interest and knowledge of mountains, gained from first-hand experience, that led him early in 1943 to suggest to the writer that we utilize the high winds of Mount Washington for conducting studies of precipitation static. It was this work which led us into aircraft icing studies and, subsequently, to the much broader investigations of subcooled clouds, growth of cloud particles, and cloud seeding techniques.

In the early 1930s, Langmuir became interested in aviation and purchased the first of several small single-engine airplanes. One of the first experiences I had with him was in his open cockpit Waco monoplane. I had known Dr. Langmuir for several years previously, having been introduced to him by John Apperson, an ardent conservationist with whom I was associated in efforts to preserve the State Constitution's provisions that protected the Adirondack Forest Preserve from persistent developers. I had just organized a winter sports club and was in the process of negotiating with several railroad companies toward operating a snow train from Schenectady. Langmuir offered to take me on an exploratory flight to search for suitable ski areas.

I met him at his home and went to the airport where, with some apprehension, I climbed into the two-seater open cockpit airplane. Incidentally, this exploratory flight was my first experience in an airplane, and one that I shall never forget. In a rather casual way he showed me how to fasten my seat belt, a mechanic spun the propeller, and in a short time, we were airborne. The day of our flight was cold and, of course, colder at the 2,000-5,000 foot level where we flew. Reaching about 5,000 feet, Langmuir showed me the basic maneuvers, the stick and rudder controls, and then told me to take over! After flying until I had the feel of the aircraft, he had me stall and recover, bank, turn, climb, and dive. We then headed for the Catskills and spent the next several hours exploring the area for suitable ski slopes near railroad facilities. During this part of the flight I took my first aerial photographs. The flight lasted about an hour, but with the novelty of a first experience, I didn't notice that I was getting cold. When we landed and parked the plane at the Schenectady County Airport, I suddenly realized how cold I was. My teeth were chattering and I was feeling a bit groggy– the first phases of hypothermia. I didn't get warm for a half hour or more. Marion Langmuir plied me with hot tea and cookies when we reached Langmuir's home, so that before heading home, I was feeling quite normal.

During the flight we found several excellent ski slopes which were among those considered, when we finally selected North Creek and Gore Mountain as the destination of our Snow Train, which first ran on March 4, 1934. Throughout all of this period, Langmuir was most helpful at critical times in our negotiations with railroad officials.

One of Langmuir's joys in flying consisted of "cloud dodging." On a day with the skies well-populated with cumulus or stratus clouds, he would go aloft alone, or with a friend, and made a game of brushing the clouds with the wing or wheels or cockpit of the plane. It was during one such flight he noticed that stratus clouds were often so stable that persistent wheel tracks could be made on their upper surface when touched during precision flying. During these maneuvers, he also became familiar with some of the optical effects produced by sunlight falling on clouds. I learned a great deal from him about coronas, glories, halos, sun pillars, sun dogs, under suns, and related phenomena. He never tired of telling about these effects, photographing them, and getting others to notice them. During his visit to Russia in 1945, one of his major interests on the flight across Siberia was in observing cloud structures and optical effects.

During the period of his active interest in aviation, Langmuir became quite concerned about improvements needed in producing better light signals to increase the safety of aerial navigation. With Westendorp,[46] who had recently arrived from Holland to join the Research Laboratory staff, he conducted a series of simple but very pertinent experiments dealing with the light signals. Such things as the relative visibility of various colored lights, the comparative value of steady versus blinking lights, light intensity, and visual acuity were explored. The results of these studies had a considerable influence on the development of adequate facilities in the then rapidly expanding field of aerial navigation. In these recent years when Doppler navigation, the use of radar, radio beams, and many other electronic aids are in commonplace usage, the importance of the pioneering work of Langmuir toward improving flight facilities during the "kerosene lantern and fencepost" period has all but been forgotten. Langmuir finally decided to abandon private flying. His reasons were

46 William F. Westendorp. Westendorp and Herbert C. Pollock constructed the first synchrotron, invented by Edwin Mattison from the University of California-Berkeley in 1949.

adequate. He became convinced that the uncertain weather of the northeastern United States made the use of a private plane of questionable value for business trips. Perhaps the more important reason for his decision lay in the increasing number of rules and regulations that more and more restricted the flier who flew for the fun of it. The crowning blow, as I remember it, was a ruling that flight log books not only had to be of a certain size, but must have a cover of a certain color. This to Langmuir was too much, and he went out and sold his plane!

In the late thirties, while completing some surface chemistry work on the nature of protein monolayers, Langmuir suggested I build an apparatus for studying the nature of certain types of films which earlier work had shown could, by their presence, reduce the rate of evaporation of a liquid substrate.

One of his early papers dealt with the evaporation rates of small spheres. Subsequently, with the assistance of his nephew David, also employed at the Laboratory, one summer he showed that a single layer of certain types of surface-active molecules, spread over the surface of a liquid having a high vapor pressure, could reduce its rate of evaporation to a measurable degree.

The apparatus I assembled was extremely simple. It consisted of an ordinary Langmuir Trough, equipped with a Chainomatic surface balance, plus an absorption cell for measuring the rate of evaporation of water. This latter device was simply an inverted, shallow tin can, covered with metal screening, on which a layer of anhydrous calcium chloride was spread. The rate of increase in weight of the hygroscopic chemical was used to assess the effectiveness of evaporation reduction properties possessed by various types of films compressed under varying pressure, as compared to a cleaned water surface. The paper describing the results of these studies was published during the midst of World War II, when most research-minded persons had more urgent problems under consideration. Subsequently, the significance of the results of these studies was recognized. In the early fifties, the Bureau of Reclamation of the Department of the Interior inaugurated a series of field studies under the direction of personnel of their Chemical Engineering Laboratories. The fruits of these field studies are now being manifest, and it is now recognized that the highly important bulk quantities of water can be saved by coating large reservoirs, farm ponds and irrigation ditches with monolayers of hexadecanol and similar materials.

Langmuir also studied the interaction of the atmosphere with the surface of water in another context. Again, an early interest in hydrodynamics was evidenced by his study with Mott-Smith of the flow patterns in rotating liquids. Sixteen years (and about a hundred papers later) he summed up a series of observations made on lakes and during ocean voyages concerned with wind streaks, which commonly appear wherever a fetch of wind affects an open expanse of water. The helical vortices he studied and described were only a small portion of his interest in the interaction between air and water. His largest unpublished work is concerned with the heat and energy budget of Lake George. Over a period of more than twenty-five years, one of Langmuir's consuming scientific hobbies was a study of the circulation patterns of the waters of Lake George. Using one of the first bathythermographs invented and constructed by Dr. Athelstan Spilhaus, Langmuir initiated a series of temperature-depth soundings that eventually consisted of more than two thousand such observations, made at various stations on Lake George. Based at his Crown Island camp and using motorboat, ice skates, or skis, depending on the nature of the lake surface, Langmuir would measure the temperature of air and water, and wind velocity and direction and other pertinent atmospheric and weather conditions, during all types of weather, and at many stations on the lake. Since many of his stations were at locations where the water was more than 150 feet deep, he used triangulation with shoreline and mountain top reference points to fix his positions. After World War II, a war surplus lifeboat sextant was a treasured instrument. Shortly before his death, he had assembled all of his notebooks, maps, slides, voluminous graphs, and reduced data, apparently with the plan to write a paper on this labor of love. That was not completed. All of the data is now in the collections of scientific papers assembled by the writer and donated to the Library of Congress by Mrs. Marion Langmuir. It probably still represents the finest collection of such field observations in existence.

Early in 1940, Langmuir was asked to work on improved designs of filters for removing microscopic and submicroscopic aerosols from the air. With characteristic energy, he set about to develop a basic theory of filtration mechanisms and, within a few weeks, had developed a number of ideas that needed testing. Meanwhile, as was his custom, he had asked me to start experimenting with fibers of various types so that time would not be lost in testing out ideas that might arise as a result of the experiment or theory. I concentrated on working with glass wool and asbestos fiber, since they seemed to be in good supply and had the dimensions that appeared desirable.

Test filters made of varying combinations of glass and asbestos fibers showed considerable advantages over cellulose and wool filters, which were then the best available. A small smoke generator was constructed for testing the filter efficiency. As I remember that rather hectic period, the filters made were adequate for testing Langmuir's theories, and a detailed report was prepared by him, which was turned over to the working committee concerned with the development of better filters for a national defense project.

At about the time our filter work was terminating, Langmuir received a request for suggestions on ideas for producing artificial fog. This was an interesting request to us, since our studies of filtration had taught us a great deal about the problems involved in making aerosols, especially in the form of finely-divided particles of liquids and solids. Langmuir suggested that I consider making a larger model of the smoke generator I had built for testing our filters.

When the quantities of smoke needed were established, it became apparent that a radical change in our method of making smoke was needed if we were to meet the desired objectives. Where a hundredth of a gram per second had been used for our smoke filter testing, the generator needed for producing a suitable fog would be required to convert at least 10,000 times more material into smoke.

With this in mind Langmuir, as was his custom, retired to his home study, leaving me to approach the problem experimentally. It occurred to me that a modified version of Langmuir's vacuum pump might work, using oil instead of mercury. When I showed him how effective such a unit could be (I used an oil can as a model), he became very enthusiastic and excited, and encouraged me to build a much larger generator embodying my basic idea.

When I next saw him a week or so later after I had made, abandoned, and redesigned a number of versions of such smoke generators, he announced with quite satisfaction that he had worked out a theoretical design he was confident would work efficiently using the quantity and type of material we had previously agreed on. My more successful experiments had shown that a practical design depended on the use of a flash boiler, operated so as to boil oleic acid or SAE No. 30 auto oil at about ten pounds of pressure, with an escape orifice of 3/16" and a boiling point for the liquid greater than 400°C. He grinned when I gave him these findings as he showed me the results of his theoretical computations. They showed a flash boiler operated at 9.4 pounds per square inch, using an oil having a boiling point of 425°C, with an escape orifice of 0.187 inch was indicated. With his agreement, we pushed ahead with a workable design for a generator, quickly built it, field-tested it locally, and then arranged to show it to engineers of the Esso Laboratories. With a few minor changes, and in thirty days, they had a generator with maximum design capabilities.

During this period Langmuir and I were aided by several assistants recruited to speed up the field development and testing of our smaller generator. A test site was located in the Schoharie Valley near Middleburg, about twenty miles southwest of the Schenectady Laboratory. An observation point was established on a flat, glacier-polished rock summit rising as a sheer cliff nearly 500 feet above the flat, mile-wide flood plain of the Schoharie River. Here Langmuir installed a time-lapse camera and, by taking advantage of the drainage winds and the strong early morning inversions of early spring, we rapidly established the efficiency and field properties of our artificial fogs. By the time the large generator was ready for testing, we were familiar with the diurnal characteristics of the valley.

It was at this point that the versatility of Langmuir was beautifully demonstrated. The so-called "big brass" of the defense department were invited to see the behavioral properties of our generator and the artificial fog it produced. Their aides let it be known that it was unlikely we could get the expedition underway before about nine o'clock in the morning. Since we were anxious that they see the behavior of fog under both stable and unstable conditions, it was essential that everything, including observers, be in position and ready to operate or observe before sunrise.

At a briefing the night before, Langmuir explained the situation, announced our prearranged schedule and described what would be seen hour by hour from sunrise to eleven o'clock. He was such a good salesman that not only was everyone at the summit of Vroman's Nose before sunrise, but everyone had pre-sunrise breakfast before the fog generators even operated! Everything worked to perfection, and to this day I am not sure whether the job was sold on the effectiveness of the fog demonstration or the perfect weather and atmospheric behavior forecast given by Langmuir the night before!

The exigencies of war and the need to get on with other problems never permitted time for an adequate story to be told about the details of our smoke generator activities. A brief description in one of Langmuir's papers is all that appeared in the literature of this interesting wartime development, which assisted materially in saving lives during the Rhine crossings, the war in North Africa, in the protection of troops in the bays and beaches of Italy, and during the kamikaze attacks in the Pacific arena. All of these developments came from Langmuir's small laboratory and the rural reaches of Schenectady, where there were no security classification restrictions to hamper progress, and only rapid developments and close-mouthed workers to explain the high degree of secrecy attained in this development.

As soon as the field demonstration was over and we all had indulged in a couple of days of sleep, we tidied up the loose ends of our field project and entered into a new and entirely unrelated project concerned with the use of binaural sound for the detection of submarines– an interesting sequel to the work done by Coolidge, Langmuir, Hennelly and Ferguson during World War I. This was an interesting activity, but beyond the scope of the present review.

Late in the summer of 1943, Langmuir called me into his office and asked if I would like to do some work on the summit of Mount Washington studying certain properties of snowstorms. This sounded like high adventure, and since it was also a field of mutual interest and apparent importance to the war effort, we were soon involved in intensive instrument design and construction designed to study triboelectric effects produced by the breakage of snow crystals hitting airplanes. Such mechanism led to the electrification of an airplane. When the airplane became charged to its capacity, leakage occurred in the form of St. Elmo's Fire (spark discharges of static electricity) that produced serious levels of noise on aircraft radio communication systems. Shortly after installing equipment for studying precipitation static at the Mount Washington Observatory, it became evident that several serious problems were likely to interfere with carrying out the orderly research plan proposed by Langmuir. The progress of the study was dependent on the development of electrification effects on the electrodes exposed to the blowing snow carried by the hurricane winds at the summit of the mountain. To the consternation of all of us, including Langmuir, the polished aluminum and duralumin electrodes we had intended to represent the aluminum surface of the airplane were immediately coated with ice from the subcooled water droplets clouds that almost continuously swirled over the mountain. Since we were primarily interested in the charging of rates of snow on metal, rather than snow on ice, it appeared that much of the intensive effort to build and install equipment on Mount Washington before the winter storms of 1943-44 blocked the road was likely to be wasted. Intensive work in the laboratory with ice-like powders, and with natural snowstorms on top of the laboratory, provided substitute information. As work progressed with these alternate experiments, and with a flying laboratory B-17, Langmuir was asked to recast the work at Mount Washington into a basic study of aircraft icing. Thus without losing time or effort, we quickly shifted our objectives and were soon actively involved in basic studies of the physical and chemical nature of icing clouds sweeping over Mount Washington. Few places on the world surpass Mount Washington in providing better conditions for such studies. No existing facility in the United States remotely approached the Mount Washington Observatory in providing better research conditions and facilities. Accordingly, for the next two years, intensive studies were carried on with the hearty cooperation of the Observatory research personnel led by Victor Clark.

Langmuir visited the Observatory a number of times in 1944 and 1945. Although over sixty years of age, he climbed the mountain on skis several times with the writer. We would go the short steep route, by way of Tuckerman's Ravine and Lion's Head. His climb was a slow and steady pace interspersed with pauses, during which all manner of natural phenomena would be discussed. The presence or absence of ice nuclei under varying conditions, the properties of the snow, sky and clouds, the appearance of timber line trees, the behavior of ice-boring pebbles, the use of stereo photography to determine the depth of snow in the ravine– these and numerous other aspects of the atmosphere and its interactions with the mountains made such a climb a stimulating adventure in understanding.

Upon arrival at the summit, work would really begin as his searching mind probed all aspects of the many interacting phenomena going on outside the door.

During our visits, which lasted generally for three or four days, many of the crew would literally work around the clock gathering data, trying out new ideas, and devising new and crude but workable equipment to test out theories, which were continuously being made or abandoned as we explored the nature of the subcooled clouds that engulfed us most of the time. Occasionally, when the skies would clear, Langmuir and I would don our skis or crampons and explore the summit cone and its environs. Every such excursion would be an adventure, with nothing ever quenching his enthusiasm for exploring and trying to solve the mysteries of water and ice and their role in the atmosphere. Typical of his observations and studies were his reports issued by the Mount Washington Observatory during this period, including the highly useful monograph, "The Multicylinder Method," used by many research groups since that time, establishing cloud droplet size, size distribution, and liquid water content.

A tangible result of his regard for the mountain as a place to do icing research was the increased use made of the mountain by the Navy and Air Force and aircraft engine manufacturers for testing out anti-icing characteristics of airplane engines, airfoils, flight instruments, and components. This eventually led to the construction of the multimillion-dollar Air Force Icing Research Laboratory near the summit and used for a number of years for aircraft icing protection studies.

With the exciting developments of cloud seeding techniques using dry ice and silver iodide developed by the writer and Vonnegut of Langmuir's group, and that were a direct outgrowth of the basic studies conducted at Mount Washington, Langmuir pointed out the probable importance of coalescence in the development of precipitation, especially in warm clouds. His ideas were supported experimentally by the laboratory studies of Blanchard and Smith-Johannsen; other members of the dedicated group working with Langmuir under the then newly formed Project Cirrus Group at the Laboratory. The growth by coalescence, due to collisions and the turbulence wake-capture of cloud droplets, followed eventually by the breakup of large drops due to instability, explained many aspects of heavy rains observed to form in warm sub-tropical clouds. A visit to Puerto Rico to study trade wind clouds, and later to Central America for studying the causes of blow-down storms in the banana plantations, convinced Langmuir of the importance of coalescence, large salt nuclei, and ice crystals working singly and in combination in the production of the various types of precipitation which occur under natural conditions of the atmosphere. As might be expected, Langmuir returned from these field expeditions with much data based on visual evidence, photographic records, and with intense enthusiasm. He was convinced that Nature tended to be inefficient in the production of precipitation, and in his enthusiasm sometimes exaggerated the situation beyond scientific prudence. This occasionally led to considerable disagreement among the more conservative meteorologists, and during the period of 1948-51, his papers or comments and the active discussion that followed them often enlivened the meetings of the American Meteorological Society.

Nothing in Langmuir's later career produced more controversy than his claims of producing a seven-day periodicity in the weather by seeding with a single silver iodide generator at Socorro, New Mexico. By seeding on a specified day each week, and then with the help of meteorologist Raymond Falconer scanning the climatological rainfall data, he found a remarkable periodicity in rainfall which seemed to fan out from the New Mexico area and, in many instances, dominated the weather pattern during many weekly cycles. The voluminous records which Langmuir produced during this period using various statistical tests, and showing these remarkable periodicities, still represents an unsolved and perplexing problem in meteorology. Langmuir and others have urged the experiment be repeated either in the United States, or in an area less interfered with by the now widespread and fairly routine cloud seeding activities of various industrial meteorologists working for hydroelectric power companies, farmers and others. The suggestion by Langmuir that the widespread periodicities in weather parameters subsequent to 1949, when widespread seeding began, should be compared with all previous weather patterns has not thus far been carried out.

Langmuir's exceptional abilities as a field observer of atmospheric phenomena never failed to impress all who saw him in action. It is one of the tragedies in the progress of science that so much of his effort in his later years was devoted to answering criticism of his bold and unorthodox ideas.

Perhaps this is as it should be. Those of us who had the good fortune to know him and to work with him in the field profited to a degree hard to properly evaluate. The catalysis resulting from his scientific fervor and unequivocal attitude stimulated even those who disagreed with his conclusions to become more active in research and study. His influence and effect on the broad field of the atmospheric sciences will be hard to assess for many years because of the host of indirect contributions in his writings. There are few areas where his theoretical calculations, experimental evidence, and field observations do not play a role. Those of us fortunate to have worked with him know that we were highly privileged, and it is unlikely we will meet his equivalent in the science of today.

One of the major challenges of our present day effort for better education is the formulation of methods for developing more "Langmuirs."

A study of his scientific attitudes and competence, as depicted in the volume of his collected works that deals with atmospheric phenomena, provides us with much pertinent and illuminating evidence bearing on the problem. He had a burning curiosity concerning the entire field of natural sciences. Anything he saw or thought about, and which he did not understand, served as a challenge to his intellect. Nothing would divert his probing mind nor dim his enthusiastic endeavor to solve the problem.

Throughout his life he roved the field. Nothing delighted him more than to find something he did not understand. Whenever this happened, one could be sure that a penetrating analysis would take place, followed by a myriad of ideas and hypotheses, most of which would be ruthlessly torn apart and discarded if they showed any evidence of incompatibility with scientific observation. If a discrepancy was found and established without question, the theory would be abandoned without hesitation, and often with an impatience which one would hardly expect for what, a few hours or days before, was probably a cherished idea.

Such was the Langmuir I knew. Would that we had more like him!

IV

The Adirondack Years

Chapter 10
My Adirondack Connections (Part 1)

My First Memory of the Adirondacks

In 1912 my parents took me by trolley car to Warrensburg, and thence by horse and buggy to Tripp Lake. We stayed for several weeks or more as I remember it, in an old farmhouse on the shore of the lake. The farmhouse was the Wilsie homestead; the owner, an old-timer with a flowing white beard.

We had a wonderful time. My Dad took me fishing to the outlet (or inlet) of the lake where I caught a nice speckled brook trout. I'll never forget the excitement I felt at this first of many such catches experienced over the seventy-five years since that memorable day.

My father commuted from Schenectady for the weekends. I suspect my mother went there in an attempt to escape from hay fever, from which she suffered, that was induced by ragweed pollen. The ragweed had not yet established itself along the roads of the North Country.

On one of my Dad's weekend trips to Tripp Lake he encountered a near disaster. Going north out of Warrensburg, the dirt road slowly climbed into the mountainous terrain, and a short distance beyond the forks in the road where a side road headed for Wevertown and North Creek, the Chestertown Road headed up a steep hill. On the west side of the hill road was a steep cliff rising above an area called the Devil's Kitchen. Just as his buggy became opposite the edge of the precipice, his horse– for some reason– started backing up and continued to do so until the rear wheels reached the cliff's edge! Leaping from the buggy my Dad was able to catch the horse's bridle to lead it back onto the road. This near calamity colored that part of the Adirondacks in our minds for many years.

The scene of this escapade is part of the Charles Lathrop Pack Experimental Forest now owned by the State University College at Syracuse. Someday I should seek to locate the Devil's Kitchen. Local inquiries indicate that this is a foreign place in that part of the Adirondacks.

It is my recollection that we went to Tripp Lake several summers. However, the arrival of more brothers and sisters probably caused such a drain on my Dad's finances that summer trips were impossible. It was during this hiatus that my mother acquired a severe case of tuberculosis and was forced to go to a sanitorium at Ray Brook near Saranac Lake.

After this close brush with death my folks decided that in some manner my mother should get away from Schenectady in the late summer and early fall until the frost killed the ragweed. It was for this reason we journeyed to Bakers Mills and Edwards Hill, where my Uncle Frank went deer hunting. He arranged for us to rent a room in a farmhouse then owned by Georgie Morehouse.

Thus began the saga of our life in the Adirondacks that is still underway. My brothers and sisters all own camps in that region and spend extensive periods there, mostly in the summer and fall.

Dogtown, Scoot, and Moonshine Hills

These names– Dogtown, Scoot and Moonshine Hills, were all terms I have heard given to the little settlement now called Edwards Hill. It had a log schoolhouse but little else of an official nature– rather it was a random settlement of mountain shacks, cabins and well-built houses located along the road running southwest of the little hamlet of Bakers Mills, and dead-ending not far from the Second Pond Trailhead. The abandonment of the sawmill and gristmill at Bakers Mills, and the rupture of the mill dam, marked the beginning of the decline in the general economy of this region. These events occurred within a decade or two of the advent of the 20th century and were developments from which the local economy has not yet recovered.

When the Schaefer family first lived in a room in Georgie Morehouse's home, and then acquired Camp Cragorehol in the early 1920s, a portion of the economy beyond a bare subsistence related to the guiding of fishermen and hunters from the cities seeking trout, deer and bear, and in moonshining. Several families also had members who worked in the Barton Garnet Mine near the summit of Gore Mountain.

During the era of Prohibition, when moonshine commanded a cash value, I used to go after dark to the open field near Camp Cragorehol and watch the flickering lights and hear the clinking of glass jars as the moonshiners tender their stills, or at least I thought that was what was happening.

Our little settlement was essentially self-contained and independent of the life in places like Schenectady or Glens Falls. The natives, though poor in worldly goods, appeared to be self sufficient, proved sincere and independent. Those that we got to know were the kindest, most hospitable individuals one could know. Their sense of humor was marvelous, their dialect distinctive, harkening back apparently to their English, Irish and Scottish forebears. Occasionally there was a descendant of French Canada in the area, though I never knew of any living in Edwards Hill. They were storytellers par excellence and a never-ending source of simple wisdom.

Before the Schaefers arrived at Edwards Hill the road continued upward and eventually reached First Pond, a picturesque body of water at the base of Height of Land Mountain. It then continued southeastward downhill to Ross' Mill Pond, where there was a sawmill at the pond's outlet. The road continued northward to where it joined the North Creek-Sodom Road, now called Peaceful Valley.

In late years the road to First Pond was abandoned and the pond purchased by a group of fishermen from the Capitol District area and renamed Chatiemac, an Indian word which my friend, Dr. Arthur C. Parker, who named it, said could be translated as "beautiful place."

Quite a few farms and cabins were built along this old road, most of which have disappeared, although wild apple trees, clumps of lilac and golden glow and an occasional cellar hole are all that remain.

The drainage of First Pond merges with another stream marked by large beaver dams and meadows that heads on the slope of Black Mountain. Just below the First Pond Outlet is the Pug Hole, a favorite bullhead fishing place. Along side of this tiny pond is a trail that branches to go to Second Pond, although the main trail goes around the westerly portion of the main summit of Gore Mountain. It was this trail that was followed by the Hitchcocks and others of Edwards Hill who worked at the Barton Garnet Mines.

Part way down the Chatiemac Road is a small mountain southeast of this road, called Edwards Hill. This oak-covered summit was purchased by Lois and I some years ago, since I have always wanted to own a mountain– even a little one. It has substantial cliffs on part of its summit, with a fine view of the settlement of Edwards Hill to the northwest, along with the northerly slope of Eleventh Mountain. The origin of the Edwards name is obscure and remains a puzzle.

All of the Edwards Hill area consisted mostly of cleared fields, orchards and pastures. There were some wooded areas, including here and there a sugar bush, but the general appearance was one of old clearings. The small streams and beaver dams had good fishing for speckled trout and wound through hay fields, pastures and occasional marshes. Now the same areas are entirely transformed as volunteer trees have invaded the fields to such an extent that only the occasional stone walls and frequent large clumps of boulders attest to the fact that fifty years ago the fields were open fields and pastures. Trees grow so rapidly that in that length of time they often have diameters of 15 to 20 inches, and the canopy is so dense that park-like vistas are now developing in areas that have not been lumbered for the second or third time.

The Snowy Mountain Climb

In 1924 I met Arthur Burgey, who was also serving an apprenticeship with me at the General Electric Company. We had identical interests and quickly began spending most of our spare time together.

His father owned a camp on Indian Lake, about twenty-five miles across the wilderness from Edwards Hill. We spent weekends, short and long vacations there for several years.

During the summer of 1924, I went with my friend and fellow apprentice Art Burgey to his father's camp on Indian Lake. The Burgey camp was near the north end of the lake and was the starting point of many adventures at about that time.

One morning Art suggested it would be fun to climb Snowy Mountain, a large granite monolith that dominates the mountains about half way down the lake. The day was clear, with unlimited visibility. I had never climbed a mountain previously and his proposal intrigued me. Art's mother prepared a lunch and we took off shortly after breakfast. The trail going in to Snowy was nearly eight miles away but that was no problem. By eleven o'clock or so we came in sight of the mountain, and were soon heading up the long sloping trail that led to the summit and its fire tower. The mountain was an imposing sight; the upper part consisting of a massive monolith, as I remember it. We reached the summit by noon and I was thrilled with the experience. The most noteworthy remembrance of the sights that has remained with me over the years, was the extremely flat bases of the cumulus clouds, that by noon were of considerable size and scattered through the region. My memory still recalls in sharp focus the magnificent view of the lake, but most important, the flat bases of the small cumulus clouds that were scattered across the sky. I marveled at the uniformity of the level bases, and since the dew point was just above the mountain, it was quite obvious that the flat bases were forming at identical levels in the sky, just above our elevation of 3,903 feet. This made a deep impression on my mind and explained to me many features of cumulus clouds that heretofore had puzzled me. The uniform bases of dozens of clouds all at the same elevation greatly impressed me and probably played a role in my later interest in the atmospheric sciences.

While Snowy was my first mountain, it was certainly not my last! Since then I have sensed the wonder of the views and the environment of many mountains in various parts of the world, and feel very much at home in them.

After spending an hour or so on Snowy's summit, we headed down and back to camp. On the return from the mountain climb our route led us down Griffen Hill. On the way I took a picture with my box camera that my dad had given to me some years before. It was of the woods and winding dirt road of Griffen Hill. This picture remains as one of the finest I have ever obtained of the many thousand of scenes I have photographed since those days more than sixty years ago.

Adventures on Indian Lake

The Burgeys had two Lake Pleasant Boats– round bottomed craft that sliced through the water nicely. Art and I were both amateur archeologists, and we spent many hours exploring the beaches, points and islands of Indian Lake. This lake had been enlarged from two smaller lakes whose valley was interspersed with gravelly kames left by the last glacier.

We discovered that the low summits of these kames had been used as campsites by the prehistoric travelers in the valley, and it was primarily these areas that were the objective of our expeditions.

On one memorable day we explored the full length of the lake, a round trip of nearly 25 miles. Since the oar handles on our boats overlapped, the standard practice of traveling in this type of craft was to dip an oar– first right, then left– this staggered type of propulsion becoming quite automatic and easy to do.

On our long trip we elected to travel down the center of the lake, thus encountering most of the gravelly islands. At one particular island, in the wide waters and about a mile east of Squaw Brook and the little settlement of Sabael, we hauled up on a small island and started our search. Art elected to scan the shore while I examined the top of the little kame. Suddenly I saw the outline of a cooking hearth surrounded by stones and showing small fragments of a decorated clay pottery vessel. As I was gathering the fragments I glimpsed the outline of a semi lunar chopper or Eskimo knife. I shouted to Art as I retrieved this beautiful artifact. To my dismay I saw that about a fifth of the blade was missing. However, adjacent to it, I found the missing piece! It is one of the finest I have ever seen and appears to be made of volcanic porphyry of a type that, in later years, I discovered only occurs in a sill that was extruded along the shore of Lake Champlain, just north of Split Rock at a place called Cannon Point. This find set the tone of our expedition, and while we found a number of chipped flints during our travel, it was and remains the high point of our adventures on Indian Lake.

The relic hunting at Indian Lake was restricted to the fall months of late September and October. Since the lake is a water storage reservoir, its beaches and islands do not appear until the water is drawn down to levels approaching the earlier configuration. At the same time many ancient stumps appear. We discovered these held rich troves of plugs and spinners lost by fishermen in their quest for the bass and northern pike for which this lake is noted.

One winter Art invited me to join him over the Washington's Birthday holiday. We planned to go ice fishing, a sport that we both enjoyed, on the Mohawk River near Niskayuna. Driving to the lake in Art's car, we carried our gear, blankets and food to the camp, put everything in order and then, with our ice chisels, skimmers and tip ups, went out to the middle of the lake opposite their camp.

We started cutting holes for our tip ups and discovered a rather revolting development! Since the water from the lake was still being released to augment the flow of the Hudson River, we found there were multiple layers of ice that had formed as the waters lowered. After cutting through more than thirty inches of ice, we realized that it was unlikely that we would catch many pike!

As we worked on the ice, it was in a continual state of adjustment. A crack would occur a mile or so away, with the ice vibrating as the crack moved with great speed toward or away from us producing a disturbing sound which we knew would cause no danger, and yet it was a disquieting occurrence.

A landmark near the north end of the lake was a high cliff, on top of which was located the summer home of Frank Moore, a highly respected statesman from Albany, who in later years, was Lieutenant Governor with Thomas Dewey, and who was a prime mover in the establishment of the State University of New York. His aerie was visible the whole length of the lake and was the target we aimed at as we rowed toward home on our many fall expeditions. Little did I know at the time that more than thirty years later I would seek Frank's advice in forming and operating the Atmospheric Sciences Research Center of the State University. I'm sure that my Indian Lake connection forged a common bond between us and helped me immeasurably in carrying out my University activities.

A Harvest of Elderberries

Not long after we began spending the summer at the Edwards Hill settlement, we made the acquaintance of the Deiseroths. Mr. Deiseroth had been the owner of a very successful bakery in Albany, and upon retirement had chosen to build a fairly large city-type house about half way up the hill. He was a big man, gruff but kindly, with a Germanic tendency toward order, self-sufficiency and planning. His wife was a very small woman and the type who always had a full cookie jar and a whimsical tolerance of her husband's lifestyle.

One day George invited me to go with him on an elderberry picking expedition. He made elderberry wine and had found several areas where they grew in profusion.

He hitched his horse and buggy and we headed downhill to the Mills. Reaching that hamlet, we headed south along the dirt road toward Fox Lair, Griffin and Wells. Reaching the village schoolhouse, we headed down a less traveled lane and were soon in the midst of abandoned fields that had a scattering of elderberry bushes loaded with large racemes of juicy black berries.

We continued down the road, crossed a stream, parked the buggy and horse in the shade, and followed the stream as it wound through a boggy meadow. Although the trout season had ended, I was curious as to whether there were any fish in the stream, since it was a new one to me. Approaching a large pool fed by a small spring-fed rivulet, I cautiously peered into it and almost fell in with excitement at what I saw! There were a half dozen of the largest fish I had ever seen! They were brook trout and were about 20 inches long. As they swam back and forth near the surface of the deep pool I could see every feature of these giant fish. They had apparently moved up the stream from the mill pond to spawn. I was sorely tempted to come back later in the day, but resisted the impulse. However I decided to put this stream at the top of my list for the coming year.

This stream was the North Creek, a creek that originated as a number of tributaries feeding into the mill pond at Bakers Mills. This pond was also probably the ordinary habitat of the huge trout I saw in the pool when I was picking elderberries with my friend George Deiseroth.[47] One of these tributaries of North Creek originated in a swamp on the flattish summit of Eleventh Mountain. In the spring, it appears as a beautiful waterfall that plunges in a series of leaps down the mountainside. Another stream that we call Mossy Glen is fed by a number of springs on the northerly slope of that mountain, and includes two that flow past Camp Cragorehol. A third tributary, Balsam Brook, starts as a huge "boiling" spring not far from the Big Rock, which was called Cold Spring, and which presently feeds into a large beaver pond that has been formed near the base of Edwards Hill.

After glimpsing the big trout and gathering a number of large clumps of elderberries from bushes growing on the edge of the big pool and at other places, Mr. Deiseroth and I walked back to our buggy and returned to his home, having gathered several bushels of

---

47 George Deiseroth is also mentioned in *Adirondack Cabin Country*, published in 1993 by Vince's brother, Paul Schaefer.

juicy berries. There I feasted on some of his wife's cookies and then walked up the hill to our camp.

Over the years since that first expedition I have fished the big pool many times. While I have never caught a trout as large as the ones I originally saw, I never failed to catch several good-sized brookies at this location. I frequently would fish down Balsam Creek, starting in the flats below the Cold Spring, following it down to the boggy meadow that marked the old bed of the Bakers Mills Pond, whose dam failed a year or two after my elderberry trip with Mr. Deiseroth. I would then swing over to reach the other feeder stream that had the big pool, and fish that creek up to the road.

After cleaning the fish and placing them in a bed of ferns, I'd head back to camp, climbing over the spur of Eleventh Mountain. I went that way since I had discovered a berry patch that invariably grew the largest blackberries in the region. They were at least an inch and a half long. The first time I found these berries I had nothing to carry them in so I fabricated a "pail" out of a roll of birch bark, using spruce twigs to fasten the bottom to the circular sides, which also were held together in a similar manner. After more than sixty years, my birch bark pail is still to be seen on the ledge of the fireplace at Camp Cragorehol!

Many visitors are not aware that the village of North Creek, near the popular ski slopes of Gore Mountain is named after the stream that flows northward to enter the Hudson at that village. This stream has always been an excellent source of trout by those who know how to fish a stream that flows through boggy meadows and alder beds. Only one ponded area, called Ross Pond or Windover Lake, presently interrupts its northward flow. Eight streams feed into the main channel along its course, each of which contains trout. As the years pass, the course of North Creek becomes increasingly difficult to fish as balsam, alder and other bushy growth engulf its course.

The cold springs which feed into it assure a high oxygen content of the water and the lack of pollution assures the continuing existence of a wild strain of brook trout in North Creek for years to come.

Mountain Number Eleven

The forested and massive mountain which rises above the little settlement called Edwards Hill, about two miles uphill west of Bakers Mills, is called Eleventh Mountain. Its name reflects its location in Township No. 11 of the Totten and Crossfield Purchase established during colonial days.

It is a rather mysterious mountain, having seven or eight rounded hills projecting above the extensive plateau-like summit. Swampy areas are scattered across the summit area.

Shortly after our first summer at Camp Cragorehol in the mid-twenties, we frequently climbed to the "big rock," which was actually an exposure of mountain bedrock at the summit of one of the lesser summits of the mountain. Climbers to the Big Rock could be seen from Cragorehol, as could the shouts be heard from either camp or mountain. I have never attempted to cover the north-south areas of the mountain and have never heard of anyone who did. One day, after fishing the length of Diamond Brook from its headwaters on the western slope of Eleventh to the East Branch of the Sacandaga where it heads south of the Siamese Ponds Trail, I entertained the thought of returning to camp across the summit. Fortunately, I decided to go out to the Wells-Bakers Mills Highway. This was a dirt road. I was very tired and had at least four miles to walk. To my surprise, a horse and wagon came along the road and offered me a lift back to the village of Bakers Mills.

If I had attempted the "short cut" across the summit, I would surely have spent the night up there!

The summit of Eleventh has a beautiful stand of red spruce, with the floor of the forests covered with mosses and lichens, and low growth of other kinds. There is no evidence that the lumberman has ever invaded the summit area. It was probably too remote for the amount of pulpwood likely to be harvested.

The highest source of North Creek is located as a swampy area on the southeast side of the mountain. In the spring snowmelt period and after heavy rains, this source water plunges over a series of sheer cliffs to reach the meadows at the base of the mountain. It is quite a sight during the periods of high water.

The rich humus on the slopes of the mountain serves as a good site for many mountain ferns and other plants. The Braun's Holly and Maidenhair Ferns are especially large and beautiful.

A couple of years ago a rockslide occurred on the upper gully of the east face of the mountain. This is a problem in the Adirondacks. During the last glacial period, which lasted for about 40,000 or more years, and which ended about 10,000 years ago, the Adirondacks were thoroughly stripped of most of their earlier coverings. The hard rock remaining was polished as the great mass of ice moved southward. Thus, there hasn't been much time for

Above: Vince getting ready to climb the Eleventh on August 8, 1937. Right: Vince standing at the peak. Courtesy of Jim Schaefer.

soil to form on the rocky slopes, and that which does is subject to slides whenever the soil becomes saturated with water.

Eleventh is a beautiful mountain and I never tire of looking at it. In fact it is a major view from our lean-to that we built last year on the Old Doug Morehouse Place.

The Adirondack Trail Camp

In the Spring of 1926 I was approached by the Scout Executive of Schenectady County to see if I would be interested and able to lead a group of older Boy Scouts in experimenting with a new type of summer program which would blend trail construction work and camping in a wilderness setting in the Adirondacks. The program had been proposed by Dr. H. S. Liddle of Schenectady, who for many years was the chairman of the Council's Camping Committee.

"Doc" Liddle had a camp near Piseco Lake, south of Speculator. He had obtained an agreement with the New York State Conservation Department to assist in the construction of a portion of the newly established Northville-Placid hiking trail. Our assignment was to clear brush, construct a corduroy trail across marshy sections of the trail, and to construct a few simple bridges. The agreement stipulated that the State would provide a ranger and pay for the food for about fifteen scouts for a two week period. In return, the Scouts would work for four hours each day on trail work, after which they were free for the rest of the day. After the initial week it was decided to work an eight-hour day, interspersed with a free day. Too much time was being required to get to the work site that, within a week, was four miles away from the base camp. This arrangement worked very well.

The beginning of the trail camp was a near disaster. The site offered us for the base camp was an old lumber clearing which had been cleared of growth but, in the five years that had elapsed, stumps had sprouted and new growth had swallowed the site in a dense mass of brush. To compound the problem, it had rained hard when we arrived at our campgrounds. The site was across a wide grassy meadow and up a considerable wooded mountain slope beyond the road. The heavy water soaked the tents; the iron army surplus camp kitchen, food boxes, blankets and personal gear all had to be hand-carried across the grassy field and up the wood road to the camp site.

Before the tents could be erected, rank growth had to be cut and removed. Meanwhile everyone was soaked to the skin. In fact the situation was so bad it became laughable and before long I had a crew that was almost hysterical with laughter. That attitude saved the day.

I managed to get a big fire going near our proposed cook tent. From time to time, the boys from each tent unit would come and dry out a bit before they would resume their camp establishment.

By evening of the first day, although the rain continued, we had all six tents erected including a huge squad tent of the pyramid type. The cook tent was erected with a large tarpaulin in front of it, and a field stove was in operation, having a savory concoction of hot hunter's stew prepared by Irv Cornell, my assistant and cook. The rain tapered off by evening, with the skies promising a fine day forthcoming.

The next day dawned clear and cool, and I was able, after inspection, to compliment the scouts and declare that the camp was ready for operation.

The first morning was spent on reviewing woods safety, axemanship, the use of cross cut saws, the care of tools, and related activities. The following day everyone headed up the trail, each with a specific assignment. Rapid progress was made and it became clear that we had a fine, hardworking group of willing workers, and the project was likely to be a great success.

After four hours of intensive work, we all headed back to camp, had a hearty lunch, and then headed for Piseco Lake, where there was a fine isolated beach for swimming. This was a fine beach and such excellent swimming area that we frequently used it, sometimes in the evening.

Since our evening swims involved a return to camp after dark, I showed the scouts how to find foxfire,[48] the luminescent fungus that often invades punky wood. We then marked our woodland trail with foxfire to guide us "home" after dark. We had fun!

Shortly after starting out on it, the trail went near a series of beaver dams at Priest Valley. In several instances, we caught enough trout there to have excellent evening meals of them.

At one of them I detected a beautiful pair of blue herons. Watching them, I discovered that they had a large nest in a tall tree nearby. On another day I climbed an adjacent tree with my camera and obtained several excellent close-up pictures of this magnificent bird at its nest.

One evening, shortly after the start of the second week, I sensed that something was "cooking." A group of boys would suddenly become quiet as I approached them. I said nothing but after turning in I kept my ears peeled! Awhile after everyone had turned in and the camp had become silent, I heard a distant scuffling, some hushed words and then silence as the "noise" faded into the distance down the mountain. Knowing the apparent ringleaders were responsible youngsters, I did not worry, but was very curious as to what was underway. The next day I learned the truth of what happened.

One of the boys, lets call him "Rick Hudson," was a spoiled "mamma's boy," as the slang went in those days. He was a shirker and quite adept at doing only what he wanted to do and seemed inherently lazy. I had considered sending him home but finally decided to see what would happen. Apparently, when it did it was quite an experience for him. He was surprised in his tent, gagged, tied up and then led down through the woods to the big grassy field bordering the road to Speculator.

There he was placed on top of a huge stump. After that each boy held forth in succession telling him what they thought of him. After everyone had finished, they untied him,

48 Also sometimes referred to as "fairy fire," it is the bioluminescence created by some species of fungi present in decaying wood. The bluish-green glow is attributed to *luciferase*, an oxidizing agent, which emits light as it reacts with *luciferin*, a light-emitting molecule.

removed the gag and, I believe, shook his hand. They then all returned to camp silently, but not so quietly that I didn't know of their return. The camp then became silent.

The next day the transformation of "Rick Hudson" was something to behold! One would never believe it was the same boy. From then on he would be the first to volunteer for a messy job or chore. Such was boy justice, and an act of responsibility that I have not forgotten.

As we neared the end of our work project, our trail improvement took us all the way to the White House at road end on the West Branch of the Sacandaga. As we finished our work one day, we stopped at the White House and met its occupant. He was an old timer named Richard, I believe, and we found him in the process of putting the finishing touches on a fine pair of snowshoes. He was surprised to see us and was much interested in our project. After an interesting talk we headed back to our base camp, a distance of more than six miles.

Thus ended the Adirondack Trail Camp, one of the pioneering experiments in the development of the Explorers section of the older scout movement, and a continuation of my personal interest in challenging young people to stretch their minds and self-motivation. The transformation in attitude and motivation that I watched develop in two short weeks was very impressive. In fact, I know several of the boys who decided to become professional foresters as a result of their adventure. I knew that "Rick's" folks would not recognize their offspring when he reached home!

The Second Pond Flow

One of the distant places that I heard about when we first went to Edwards Hill was the Second Pond Flow. Johnnie Morehouse and others of the mountain community went there each fall to harvest the coarse swamp grass that grew in great hummocks across the Flow flatlands. This area had either been a former lake or a gigantic beaver pond.

I can still hear the creaking hay wagon, the shouts of the driver, and the talk of the natives who preferred to walk behind the wagon, rather than endure spine-jarring jolts of the wagon as it rolled across the many boulders that dotted the course of the Second Pond Flow trail.

In walking this trail as I did many times, it is hard to realize that a loaded hay wagon could traverse the 4+ miles between the "Fifty Acre Lot" at the Eleventh Mountain trail-head, and the Flow. However, that was a late summer ritual that was carried out for many years prior to 1930. It is quite likely that this annual trip was treated as a picnic by all the participants. I suspect that the wagon trip was preceded by a few days to a week between the time the grass was cut until it was cured and hauled out.

The trail went into the woods at the edge of the clearing and continued as a gentle climb to the top of the watershed. All of the water easterly of its ridge flowed to be collected in Mossy Brook, and then into North Creek, and thus into the Hudson. The waters to the west were gathered into Bog Meadow Brook which drained Bog Meadow Flow and eventually went into Second Pond Brook, and thus into the East Branch of the Sacandaga River. There it went in a southerly direction for many miles before reaching Great Sacandaga Lake, and then easterly into the Hudson River at Hadley, some 40 miles south of North Creek and nearly a hundred river miles from its source.

After reaching the top of the ridge, the trail went westerly to a clearing called Bog Meadow. It was probably originally a beaver pond that became filled with sediment from the glacial till and the humus that covered the area. The swale grass of Bog Meadow was also harvested, although it was in much smaller supply than at the Second Pond Flow area.

Bog Meadow Brook wound through the swale grass and had its quota of 6-8 inch brook trout. Leaving the clearing, the brook continued down the long slope toward the Flow, but with few fishing places before the Falls. At that spot it tumbled down a rocky precipice as a stair-like falls, with a deep pool at the base. This pool was always good for several fine trout.

Just beyond Bog Meadow the trail forked, with its northerly branch heading for Second Pond, the westerly one going to the Flow. This latter trail coursed down a gentle slope for several miles until it reached the great meadow. At that location the grass grew in large tussocks, with a fairly uniform density. It was readily cut by the scythe-swinging natives and quickly yielded a wealth of fine bedding substance. Its wiry nature would seem to be a far cry from the tender meadow grass of the fields at the base of Eleventh Mountain, but I have been told that it was highly nutritious and the main winter sustenance for the native cattle owned by the mountaineers.

I had many adventures in the Bog Meadow, Second Pond, Second Pond Flow, and Mud Pond areas of this region. One time with Bernie Doherty, while camping on the island in Second Pond, we were running short of food and went hunting at Mud Pond. Bernie shot what we thought was a duck. Upon closer scrutiny it turned out to be a teal. When cleaned and defeathered, its substance was so small as to be hardly enough to satisfy one thin individual! Nevertheless, we baked it, but upon tasting it we lost all interest because of its strong fishy smell and taste.

An Adventure in the Siamese Ponds Wilderness

There had been a heavy snow that winter in the Adirondacks. As the first day of trout season had been observed on Becker's Brook and our group of fishing buddies began to look farther afield for new waters to try, Bernie Doherty agreed with me that we should go north. This meant Cragorehol and the Second Pond Flow. This trip was planned, as were so many trips during that period, during the lunch break at Bldg. 101 of General Electric, where I was finishing the fourth and last year of my apprenticeship.

As planned, we would go to our camp at Edward's Hill, hike into the Flow (normally a ninety minute hike back of camp) for a day of fishing, return in the evening, spend the night and head home the next day. Such well-laid plans sometimes go awry.

Thus it was that three of us– Bernie, his friend Leo Durfee, and I– headed north for the long Memorial Day weekend. Bernie's car was loaded with food, fishing gear and blankets.

As we left Johnsburg– about three miles from Bakers Mills and four to Cragorehol– the dirt road became increasingly wet, muddy, and rutted. A mile or so beyond Johnsburg, I began to sense we were in trouble. The mud got deeper, the roads wetter, and then suddenly, we were stuck in the mud so deep that it was apparent that we could go no further. Fortunately, we reached this part of the road near an old farmhouse and found that its driveway was solid. By considerable maneuvering we finally extricated ourselves from the muddy ruts, parked the car on solid ground, and prepared to walk the remaining four miles to Cragorehol. There was no one at the farmhouse but I was quite sure our car would be safe, and unless there was a freeze, the road was temporarily impassable.

Shouldering our packs, we headed for camp– three miles away. We eventually arrived but were so tired that we had no energy to do any fishing that afternoon. We decided to turn in early and get an early start for the Flow in the morning.

The night was cold and the early morning found us with our fishing gear ready to go shortly after daybreak. We headed for the Flow, up past the Fifty Acre Lot where Johnnie Morehouse cut hay among the rocks, and into the woods. A heavy frost had occurred and the surface of the trail was frozen. As soon as we entered the woods we encountered deep snow, but since there was a crust on the snow, the travel was easy and fast. There is essentially no climbing on the way to Bog Meadow, which is about half way to the Flow. The land is a rather flat col between Eleventh and Height of Land Mountains. Reaching Bog Meadow, we were tempted to fish the beaver ponds there, but decided to continue on to the Flow since the fish there should be bigger!

Leaving Bog Meadow, the route trends downhill, since we parallel the valley of Bog Meadow Creek. Reaching the Flow, we assembled our rods and started fishing. After an hour or so with absolutely no strikes or even glimpses of fish, we were forced to conclude that we were too early and our expedition for trout was doomed to failure. After lunch we decided to head back to camp.

As we hit the track we discovered an alarming situation. Where during the morning hours the snow had a hard crust and we were able to walk on top of it, the hot sun had melted this crust so that every step plunged our feet down to the ground, the snow depth being knee deep! The leader took the worst beating, the following two being able to use the deep foot marks a bit easier. In a short distance the leader would become exhausted, at which time the lead would revert to the next in line. Occasionally we would encounter a shaded area where the crust would still bear our weight, but then as soon as the shaded area disappeared, it would soften, and just as the next step was about to be taken, the first would plunge to the ground! If anything, this marginal condition was worse than the areas without crust.

By the time we finally reached Bog Meadow, the three of us were about exhausted. As we reached that area we were confronted with another alarming development. Where in the morning the Bog had been a placid area of quiet water, its flow had become swollen with melt water and posed a serious problem for crossing it. I managed to cross on a slippery log. Leo followed and was halfway across when he slipped and was engulfed in the turbid rapids. Bernie, seeing his trouble, heaved his fishing gear across the stream, waded in and grabbed Leo and pulled him out even though Bernie was a considerably smaller person. Since they were now completely soaked, they managed to wade across the stream to the further side.

The remainder of the trip was an experience I never wish to repeat. We finally reached the clearing and snow free ground, after which we wearily staggered our way down to Cragorehol. Needless to say, we all ate heartily, and went to bed early!

The next day I learned that the Warren County Sheriff was looking for us! It seems that over the weekend the farmhouse where we had parked our car had been burglarized. Since someone had seen our car parked in the driveway and had noted the license plate, we were the prime suspects as burglars. Fortunately, they also contacted Johnnie Morehouse, who attested to our honesty and the circumstances of the situation. Since Johnnie was the Patriarch of the region, his intercession was helpful and the sheriff dropped his pursuit.

It was about sixty years ago that we had this adventure, and its details are still a vivid memory!

Our Log Cabin Near Camp Cragorehol

When my father purchased the mountain farmhouse from Charlie Reese in 1924, his nearest neighbors were John and Georgie Morehouse– brothers– and Charlie Smith and his family of wife and several boys, who were living in the log cabin that had been Johnnie's.

Charlie was a moonshiner. He had a crude still which he operated openly, and which was located in the one and only room of the cabin. Charlie was a drinking man. So far as I knew he personally used most that came out of the condenser and had a tendency to catch the dribbling liquid during a run and consume it without delay. As a result, he was frequently somewhat "worse for wear," as the saying goes.

His youngsters were like wild creatures, extremely shy with "outsiders" as they wandered around the region. One day I encountered two of them carrying a heavy rifle between them. When asked what they were up to, they indicated they hoped to shoot a bear!

When we learned that the Smiths were moving away from their cabin home, my brother Paul and I decided to learn whether or not we could purchase the cabin and the ten acres of fields around it. We managed to buy the property and immediately started to put the cabin in order. It had been abused and we found it necessary to use a hoe to clear the floor before we could scrub it. However, before long we had it the way we wanted it. A photograph I took at the time shows that the cabin was central to a very large open field, with only a couple trees on the west side of the cabin. Huge boulders were scattered in profusion across the ten acres, with only patches of green in between. To harvest such grass took much labor with a hand scythe, a tool the mountain men were expert at using.

At that time an active campaign was underway in the New York State Conservation Department encouraging owners of open land to reforest them with evergreen trees, which they were growing in a large tree nursery near Saratoga. Paul and I decided to obtain 10,000 trees that we would plant on our ten acres. We decided on planting Norway Spruce, Scotch, and Red Pines in equal numbers.

When we got them in the spring, we realized that we had a big job on our hands! After a lot of intensive work with a mattock we finally got a thousand trees planted. Although we could hardly afford it, we asked Johnnie Morehouse if he would continue planting during the coming week.

He seemed reluctant to take on the job on a matter of principle. He mentioned how his forebears had spent years clearing away the forest, and now we were planning to undo their labors. When Johnnie saw the tiny trees gathered in bundles of a hundred he said, "Those trees aren't even born yet!" However, within the next couple of weeks all of the trees were planted and a forest was born! Today, after nearly sixty years, the Spruce and Red Pines comprise a beautiful woods, free of undergrowth, with clear boles that reach toward the sky. The Scotch Pine never amounted to much, with scraggly and crooked trunks.

As the forest emerged from that open field, we eventually realized that we had produced an effect that was not anticipated. Our view of the mountains disappeared, and our cabin– after fifty years– is now completely surrounded by a dense forest!

During the past two years about a hundred of the magnificent Norway Spruce have been cut in the spring, the bark stripped and the resulting logs transported to a cabin site on Coulter Road where our friends, Dan and Noel Johnson, have built a beautiful log cabin. They have also used some of our Red Pine to serve as rafters for the slate roof, which is an "earmark" of all of the buildings that my brother Paul has designed, built, or supervised over the past sixty years. The Johnson cabin is one of his finest!

The Indians of the Adirondacks

I have had a strong interest in the prehistoric dwellers and travelers of our region, for nearly seventy years. My searches for arrowheads and other artifacts at camp and village sites in the Mohawk Valley were quite successful, but when I explored similar choice locations in the Adirondacks, I had little luck.

My first success occurred, as I have mentioned, at Indian Lake, where at low water we found relics along the shores and on gravelly knolls at scattered locations in the lake. The best finds were on the tops of knolls that, in prehistoric times, were probably as bug free as could be found and would serve as a vantage point to survey the surrounding region. Fragments of decorated pottery, spear or javelin points of chipped flint, and the remarkable

Eskimo-type Woman's Knife or semi-lunar chopper, found in a hearth among potsherds, were all included in my finds.

At other places I also found traces of the early inhabitants. On a sandy kame alongside the North Creek-Sodom Road, I found an interesting occupation. It apparently was a cluster of campsites, since the area was strewn with fire cracked stones. Fragments of quartzite chippings were scattered about, with an occasional chipped scraper. I never found an arrow or javelin head, but a nephew found a beautiful chipped woman's knife. One small end of it was missing, the relatively fresh break disclosing a layer of patina (weathered, altered surface) that showed this relic to be very old.

In the State Museum in Albany is a beautiful, perfect Iroquois pottery vessel of large proportion, said to have been found in a rock shelter at the Falls on the West Branch of the Sacandaga River, upstream of the White House. It is one of the finest perfect vessels ever found in New York State.

From a rock shelter near the mouth of the Indian River along the Hudson, another large pottery vessel was found by a lumberjack, Ed King, while "working the river." This was in large fragments and was a classic form and texture of an Algonquin pottery vessel. These fragments were obtained by Art Burgey and reassembled by members of the Van Epps-Hartley Chapter of the New York State Archeological Association that I organized in the early '30s. This vessel was decorated with lines of small depressions, and was similar in design to sherds found at prehistoric village sites found along the shores of the Mohawk and Hudson rivers.

Other artifacts have been found, from time to time, at Lake George, near North Creek, at the edge of Lewey Lake, at the canoe carries in the Saranacs, and similar places such as would have been used by wandering bands of hunters and trappers. To my knowledge, no permanent large village sites have been found among the mountains, since they would have been unsuitable for the establishment of such settlements. A number of substantial sites have been located on the Vermont side of Lake Champlain on the edge of the lake and along the marshy streams that flow into it.

It is likely that additional traces of occupation will be found in the future, though I doubt if they will be more than the sporadic campsites found during the past century.

The Origins of North Creek

North Creek is a trout stream that wanders through great masses of black alder that defy anyone to fish it with any degree of fun. However, in other areas along its 10-mile course from its headwaters, it is a fisherman's dream, with undercut banks, deep pools, log jams, and spring holes that reward the careful fisherman.

There are three noteworthy sources of this stream, the highest of which starts in a wet area near the summit of Eleventh Mountain. Only in the springtime, or after a heavy rain has soaked the summit, does this stream appear in its glory. Plunging over the rocky cliffs of the mountain, it provides a beautiful sight as it rushes downward to the placid meadows at its foot. However, over a considerable part of the year it is of little consequence, and the waters which comprise the most southerly extension of North Creek must rely on the seep springs that are fed by the ground water coming from the slopes of Eleventh Mountain and the lesser hills toward the east.

Two other sources of the North Creek are more reliable and provide a year-round source of cold spring water. The highest of these surfaces, on the old Dalaba homestead, is at the uppermost part of the little settlement known as Edwards Hill. Its runoff appears as a small, rapidly flowing stream that crosses under the road that once ran toward Height of Land Mountain to swing by First Pond ("Chatiemac"). It flows downhill northeast of Edwards Hill Road through abandoned pastures to join Mossy Brook, which originates in the slopes of Eleventh Mountain.

The other spring source has always been known as Cold Spring. It originates as a "boiling" spring along an old road that once serviced the Bateman Place, a rather impressive two-story farmhouse which, sixty years ago, was surrounded by hay meadows, gardens and apple orchards. The flow of the large spring is so vigorous that the sandy bottom appears to boil as the ice cold water emerges from the ground.

The water from this spring flows downhill to form the headwater of Balsam Brook, another trout stream that has several large beaver dams. This stream eventually flows into the boggy region that once comprised the mill pond of Bakers Mills. There a sawmill and gristmill were powered by the waters of that pond.

I have caught many trout from these several spring-fed brooks and know where one could still obtain a satisfying catch of brook trout.

The village of North Creek gets its name from this interesting stream. A few years ago, just above the village, an earthen and timbered dam provided a fine swimming hole for the villagers and others of the vicinity, and a fine fishing place for the village youngsters. I hope it can be restored.

Cross Country Skiing in the Adirondacks

Cross country skiing has been a favorite of mine for more than fifty years. In the early thirties, one of our favorite areas for this sport was the terrain of southern Vermont, in the vicinity of Wilmington. We found that the abandoned pastures, apple orchards, and hay fields consisted of a region of rolling land that provide many nice runs, especially in early spring, when corn snow provided easy maneuvering.

At the same time we found that the Adirondacks were replete with challenging runs, especially where the forest canopy discouraged the growth of witch hopple and similar varieties of underbrush. By climbing to the summit of the rounded mountains so common in the Adirondacks, we found that it was quite possible to go almost anywhere when there was several feet of snow on the ground. Thus, we went to such places as Indian Pass, T-Lake Falls, the summits of Gore and Eleventh Mountains, the Sacandaga River, and other streams. When the latter were ice-locked and covered with snow, their rocky courses provided an ideal thoroughfare for exciting runs. By ascending the stream course, it was feasible to work out a suitable run for later in the day, after having a leisurely campfire lunch.

Some outdoor enthusiasts prefer snowshoes for cross country traveling. I tried both but prefer skis. I have found it possible with skis to go anywhere a snowshoer could maneuver and, in addition, to enjoy long glides where the snowshoe requires the plodding routine.

Our Winter Camp During the Olympics of 1932

Shortly after we formed the Hiking Club, we began to hear that the Winter Olympic Games were to be located at Lake Placid in 1932. The Lake Placid Club was deeply involved in the politics of these arrangements. As the plans began to develop in the next year or two, a number of us decided that it would be fun to attend the games. However, none of us could afford to stay in Lake Placid, or to pay the admissions fees to the competitions.

However– we decided it would be fun to be in the region so we might see and possibly meet some of the athletes. Accordingly, Bill Gluesing and I went to Lake Placid in 1931 and approached Harry Hicks, who we knew was very prominent in the Lake Placid Club, to find out if the Club would permit us to stay in the open face Adirondack log lean-tos clustered along the shore of Lake Clear of Heart, at the Adirondack Loj.[49] This was the trailhead of Marcy and others of the High Peaks of the mountains.

We obtained permission and soon began to construct down and balloon cloth sleeping bags of the type designed and tested by John Apperson and Irving Langmuir. Within a short time we had obtained (through the help of the General Electric Purchasing Department) large bolts of balloon cloth and, from a local furniture maker who was a member of the hiking club, nearly a hundred pounds of goose down. We were advised to put three pounds of down in each of the bags and before long, had between 20 and 30 beautiful sleeping bags constructed. Many of these are still in use after nearly sixty years!

As the period of the games approached, everyone was anxiously watching the weather. As sometimes happens in our region, a warm spell of weather is likely to occur during the latter part of January or early February.

On Saturday, February 6, a group of sixteen members of the club left Schenectady early in the morning, loaded their three cars, and headed for Lake Clear of Heart, southeast of Lake Placid. Arriving at the junction of the road, we found it to be plowed and, as a result, headed south to see how far in we could go. With the small amount of snow (which had everyone concerned with the Olympic Games worried), we had no trouble driving all the way to Adirondack Loj. There we found a substantial amount of snow in the woods and the log lean-tos ready for occupancy. In short order we had our camping gear installed, a large amount of dry wood accumulated, and a cooking fire burning.

---

49 The original Adriondack Loj was built in 1890 and designed by Henry Van Hoevenberg, but burned n 1903. The present one was built in 1927 and is now owned by the Adirondack Mountain Club. It was designed by architect William G. Distin.

The group camped in the lean-tos for eight nights and had temperatures ranging from -28°F to +51°F! Although the Olympic ski jumping, cross country and other ski events had to depend on snow hauled in by railroad and truck, the snow at the winter camp was deep and powdery until the last few days of our stay. Climbs of Mounts Marcy and MacIntyre, and to Indian Pass, were made in deep snow with glare ice on the summits.

Several trips were made to Lake Placid to see the skiers and skaters and to go ski joring on Mirror Lake, pulled by Bill Gluesing's Ford touring car. We met the Swedish Cross Country skiers, who had discovered our excellent snow, and had an interesting talk with them. We also visited the German Bob Sled team, who were hospitalized when their sled leaped out of the course and smashed into a clump of trees. Since several of our club members were from Germany, our communication with them was quite satisfactory.

During the week, our group was augmented by nine other club members, so that we had a total of twenty-five campers before the trip ended.

We had such a good time at Heart Lake that our anticipated attendance at the Winter Games turned out to be a lesser attraction than we had expected.

Apperson of Huddle Bay and Schenectady

John Apperson was a remarkable man. Not large physically, but wiry with a firm set jaw, steely eyes and a dominant passion for the well being of Lake George and the Adirondack wilderness. He was a Virginian, but when he came north as an electrical engineer at General Electric, he adopted Schenectady and the North Country with a fervor that was contagious.

I was one of several dozen young (and older) men of the Schenectady area who joined him on his endeavors. He purchased a choice piece of land on the shore of Huddle Bay on Lake George, just south of Bolton Landing. There he acquired a simple, but extremely well-built camp, with a big fireplace, an ice house, a lakeside dormitory, and a place to store his boats. He had a sleek, well kept Chris-Craft that he loved to run on all sorts of missions, and a number of fine canoes. Before I joined forces with him, he had spent a number of years hauling large rocks that he used to fortify the shorelines of some of the State islands in the lake. A lumber company had for years manipulated the level of the lake to favor its hydro operation– keeping the level of the lake unusually high in the springtime, long after the natural spring freshet would have drained into Lake Champlain. While the lake was high, the windstorms common in April and May would cause erosion damage to the gravelly islands that Appie's rip-rapping sought to stop.

For several years I, and a handful of other young folks from the Mohawk Valley Hiking Club, would spend many weekends at Appie's Camp doing various things to help him in his campaign to protect the lake and the Adirondack wilderness.

Appie was a confirmed bachelor, and as such, had developed routines that rarely varied. We would gather at his house on Teviot Place late Friday afternoon, pack our sleeping gear in his car, and head for the lake. On the way, we invariably stopped at a butcher shop where he had a standing order for lamb chops. His butcher friend saw to it that he got choice cuts!

Reaching camp, we would get in a stack for dry wood, start the fire in the fireplace and, before long, would sit down to a fine meal of mashed potatoes, some vegetable, and those luscious lamb chops. Throughout all of this activity Appie would carry on a rapid fire talk of past adventures, current strategy, and plans for the weekend. It was an exciting time! Sometimes several of us would take off with notebooks and movie cameras to record some sort of activity that would assist us in our campaign to alert the voting public on the facts about some activity that we believed was threatening the integrity of the wilderness.

One of my special jobs was to edit the movie film that was obtained in this manner. Some of our movies turned out to be of great importance in defeating attempts to modify the State Constitution.

While most of our activities at Appie's Camp were carried out in the summertime, we occasionally went there in the wintertime too. I remember one weekend when we were shown how to use a skate sail for ski sailing. We went north from Huddle Bay, through the Narrows, and up toward Rogers Rock. While the speed with skis driven by the wind is not as high as with skates, it is great fun and it eats up the distance quite remarkably. At the end of the day I discovered that my skis that did not have metal edges, and were quite the worse for wear!

While our activities with Apperson were intense for several years, we were followed by others who were more dedicated. Our "apprenticeship" with Appie was never to be a forgotten experience, and in fact after more than fifty-five years, we have never lost the "faith."

A Joint Camping Trip in the Adirondacks

Late in the '20s, Arthur C. Parker of Rochester, his wife Anna, and I began talking about forming groups of hikers into clubs which would promote friendship, outdoor education, land stewardship, and good fellowship in the different river valleys of New York State. While Dr. Parker, who was Archeologist at the New York State Museum in Albany, lived in Albany, I joined him and his wife on hikes with the recently formed Albany Chapter of the Adirondack Mountain Club. When the Parkers moved to Rochester, where he was appointed Director of the Rochester Museum, our plans to form valley hiking clubs matured. He and Nan formed the Genesee Valley Hiking Club, and shortly after I formed the Mohawk Valley Hiking Club.

Within a couple of years we planned joint overnight camping trips. The first of these trips was Chittenango Falls, near Cazenovia. It was so successful and enjoyable that we began to plan a second similar venture, this time to the Adirondacks.

We assembled at our Camp Cragorehol and, after everyone had arrived, shouldered our packs and headed into the woods. We planned to set up our camp near the headwaters of Diamond Brook.

Reaching a suitable campsite near the base of Eleventh Mountain, we built a large fire, gathered an adequate supply of firewood, and then scattered through the woods seeking sites for our individual sleeping bags. Although the sky was filled with stars, most of us erected tents or tarps just "in case."

One of the last of the group to return to the campfire was my brother Paul and his friend, Al Getz. They had found nice level spots in a dip south of the fire, and were boasting about the superiority of their location.

We enjoyed the campfire and small talk until nearly midnight, after which we each would make our way to our sleeping bags. However, when Paul and Al went to turn in, they couldn't find their tent site! They had to search ten minutes or more before they finally found it!

Several hours later, we began to hear distant rumbles and flashes of light warning of a distant storm. In a fairly short time we could anticipate that we were likely to get wet. As the rain began most of us checked our tents and tarps and made sure that we had drainage ditches ready to cope with the expected precipitation.

The storm arrived, accompanied with a gusty wind, so that whether our ditches were adequate or not, all of us began to get wet from the wind driven precipitation. The rain became torrential and continued for quite a time.

Fortunately, our fire was large enough and we had enough dead wood so that we continued to have a big hot fire, despite the rain. In a short time everyone but Paul and Al had arrived before the fire. At that point, we began to wonder whether in fact they had found a superior location. As we were about to conclude that this was the case, they suddenly appeared– not only wetter than we were, but with the news that their camp site was about to be washed away! It then turned out that they had inadvertently located their tents in the bed of an occasional stream! Thus, I had an early introduction to the truism enunciated by my later friend, Walter Garstka, who said that "the channel belongs to the river." Anyone who ignores this hydrological fact does so at his peril!

Now that we were all accounted for, we found that in a short time we were dried out, the thunderstorms passed and we all dozed before the fire. As the sky began to brighten, we took some coals from the big fire and in a short while the aroma of coffee, bacon, eggs, and other breakfast fare began to permeate the woods as we ate and planned our climbing expeditions for the day.

Despite the rain shower, our joint trip continued without further untoward incidence and thus our second joint venture proved to be a most enjoyable experience. The deepening Great Depression greatly limited our mobility and objectives, while other interests and opportunities took precedence over our club schedules and activities, as our conservation activities and adult education programs involved our time and energy.

After the lapse of more than fifty years, both of our clubs are still active, prospering, and having good times.

T-Lake Falls, West of Piseco

One of the highest waterfalls in the Adirondacks is located on the outlet stream of T-Lake, west of Piseco. The first time I saw the falls I was on a trip planned primarily to climb T-Lake Mountain.

The second time I visited was about 60 years ago while cross country skiing. We had a fine trip into T-Lake and decide to have our lunch at the falls. As we approached it, we were surprised to see the high concentration of deer tracks in the vicinity of the falls. We then discovered that there was a very large "deer yard" located on the many ledges of the steep mountainside sloping down toward the West Canada Creek. A rough estimate of the height of the falls indicated it to be in excess of 300 feet.

No part of it is in free fall, rather it tumbles down a rounded slope of bedrock to the bottom where it is lost in a tangle of ancient trees and huge rocks which have accumulated there since the glacier left the mountains.

The last time I stood at the top of the falls looking west was on our Metcalf Lake Expedition when I took two of my scouts on a week's trip of wilderness exploration in August of 1934. That trip merits a special chapter since it vividly demonstrates the fun to be had in bushwhacking across the trail-less forests of the Adirondack Wilderness. We traveled by compass and topographic map from T-Lake Falls into the wild reaches of the East Canada Creek across beaver dams, swales and virgin Spruce forest to the wild shores of Metcalf Lake. There we camped for five days exploring the dozen or more tiny lakes that make the Metcalf Lake Region so fascinating.

From the crest of the falls it is possible to see the slopes and crest of what I called Big Spruce Mountain. This is on the east side of Metcalf Lake. At that time it was (and hopefully it is still) pristine wilderness, with huge virgin Red Spruce covering the crests and the mountain slopes of that beautiful wild area. All of the area is State Land, protected by Article 14 of the State Constitution.

My Trip to Mount Marcy

In 1931, with several members of the Mohawk Valley Hiking Club, I headed for the summit of Mount Marcy for the first time. Leaving home after work, we traveled by private car to the Sharps Bridge Camp Site in the Adirondacks, where we spent the night. The next day, we continued on and reached Lake Clear of Heart by mid-morning. We parked our car and headed for Marcy by way of Avalanche Pass and Avalanche Lake. We traversed the trail referred to as the Hitch-up-Matildas, and then the Flowed Lands. There we followed the Opalescent River to Feldspar Brook, and then to its junction with Uphill Brook. We spent the night at the Adirondack lean-to at this location.

Bright and early the next morning, after a substantial breakfast, we shouldered our packs and headed for Lake Tear of the Clouds and the Marcy summit. The day was a good one for climbing, and we made good time.

While at our nighttime camp, I discovered many fascinating rocks in the streambed showing an opalescent sheen, and deduced that these rocks were responsible for the name of the Opalescent River where they appeared. Later I learned that these rocks belong to the Marcy anorthosite and the particular rocks to a type of opalescent feldspar or labradorite. I have since learned of the importance of these crystals in their relationship to differentiating between the Marcy and the Whiteface types of anorthosite.

When we reached the summit of Marcy for the first time, we were duly impressed with the vast view of the surrounding mountains seen from that vantage point. I was particularly intrigued with the mats of tiny alpine flowers found among the rocks, with a few pebbles of foreign rocks, which showed that this highest point of New York State had once been submerged by the Continental glacier. After climbing Marcy, we continued east and climbed Little and Big Haystack Mountains.

During a later climbing trip by our club, we ascended Haystack from the depths of Panther Gorge after spending the night near the bottom of the Gorge. Climbing northward up the slope of Haystack, we encountered a miniature rainforest, with lush masses and a great variety of lichens.

Since I was impressed with the variety of the lichens, I made a small collection of them and sent them to my friend Raymond Tovey, one of the country's top lichenologists. He was much impressed and told me that a number of new varieties for the Adirondacks were in the collection.

Our initial trip, we also climbed Mount Skylight on our return to Uphill Camp, where we spent a second night. Our return from this first visit to the High Peaks was a retracing of our route by the Flowed Lands and Avalanche Lake.

The last of about ten trips I made to the Marcy Country over the years occurred during World War II.

With Dave Harker and Charlie Bachman of the Research and Engineering Laboratories, we went by bus to St. Huberts and headed in toward the lower Ausable Lakes. There we climbed to the summit of Armstrong Mountain, and then headed for the lean-to at the base of Basin Mountain. We saw no other hikers and had the entire mountain region to ourselves, as a result of the shortage of gasoline. After spending a second day hiking over Marcy and neighboring peaks, we then followed Johns Brook out to Keene Valley and caught our bus transportation back to Schenectady.

Of late years, the hiking pressure in the High Peaks has become excessive to the point where severe erosion is causing damage to the mountain sides. This is a severe problem and needs to be alleviated by focusing attention on the other attractive regions of these remarkable mountains.

A Canoe Trip into the Raquette River Country

The western part of the Adirondacks is a canoeist's paradise. This was recognized by the earliest travelers, when the Iroquois and Algonquin Indians traversed the region. They apparently avoided the High Peak Country, although artifacts of their presence can be found among the lower mountains, in the form of campsites occupied on hunting and trapping expeditions.

One of their trans-mountain trails utilized the water corridor from Raquette Lake northward. They apparently went by foot or snowshoe from the southern Adirondacks, following the Canada Creek Valley from the Mohawk to the Canada Lakes, and continuing overland to the south shore of Raquette Lake. There they went by dugout or canoe the length of Raquette and Forked Lakes, and thence by the Raquette River to Long Lake. They would then continue down the outlet of Long to the St. Lawrence or Lake Ontario region, by way of the Raquette River. When white visitors first entered the area, one of the favorite water routes was from Chateauguay Lake across the northern Adirondacks, westerly to Tupper Lake. The network of small lakes and streams made such an adventure feasible and delightful.

About fifty-five years ago a small group of us from the Hiking Club took a canoe trip from the south shore of Raquette Lake to the first falls of the Raquette River. We enjoyed the myriads of bays and rocky points of Raquette Lake leading to the short portage to Forked Lake. From Forked we followed the Raquette River to Long Lake, canoed the length of Long Lake, and then down the Raquette River to the falls. This was the terminus of our trip. We camped above the falls and then, the next day, reversed our route to end our travel on the south shore of the Raquette Lake. We had a marvelous time, even though it seemed that we had a headwind no matter what direction we went! Years later I discovered that at times the wind behaves this way, pouring down at the center of a lake, and then spreading out in all directions.

One of our observations made at Forked Lake was the remarkable appearance of the forest in all directions along the lakeshore. It was as though someone had walked along the shore with a giant shears and trimmed the under surface of the trees to a remarkably uniform height of about 8 feet. Some time later I learned that this phenomenon is called the "browse line," and is sometimes produced by deer eating the buds and twigs of trees as they walk along the shore under the trees. The height of the line is controlled by the ease with

Vince camping and canoeing with friends. Dates and locations unkown. Courtesy of Jim Schaefer.

which they can reach this type of food. The presence of such a line seems to be restricted mostly to wilderness lakes and is a rather spectacular sight.

The Whip-or-Will Camp on Tongue Mountain

From time to time, the Mohawk Valley Hiking Club, of which I was Director for a number of years, scheduled overnight camping trips.

On one occasion, I obtained permission from John Apperson to use one of his camps on Tongue Mountain as the site of such a trip. To get there, it was necessary to go by canoe, since there were no roads. Consequently, we assembled at Bolton Landing, parked our cars, loaded our canoes and headed across Northwest Bay toward our objective. It was a beautiful, fairly quiet day. Our canoes, though well loaded, skimmed across the bay that, at times, can be very dangerous. Under stormy conditions, very rough waves are produced by the Northwest fetch of wind that commonly blows from that direction.

In this instance we reached the pier of the camp, carried our gear to the building, but decided to sleep out during the night since there was every indication that we would have a clear and cool night.

As dusk approached, several whip-or-wills started calling and we all thought this to be a fitting sound to lull us to sleep in our semi-wild environment. As the night progressed and we all settled down to slumber the birds got reinforcements and, by midnight, the air seemed to be filled with their monotonous, repetitive calls.

As the hours passed, I doubt if any of us got much sleep. Finally, about two o'clock in the morning, Larry Shaw (who was normally one of the least vociferous of our entire group) sat up and shouted, "Will you shut up!" While this did no good, it seemed to relieve the tension that had been building up in all of us, and after that each of us drifted off to sleep the remaining hours of the night.

The next day continued to be a fine one and we swam, talked, relaxed and just had a good time. During the noontime period the wind began to blow, though our area of the mountain was quite protected from the west and northwest winds. Since none of us had experienced the rough weather that gave Northwest Bay its dubious reputation, we packed our gear, lashed everything tight and head down to the point of Tongue.

There we suddenly realized we had a serious problem on our hands. While none of us could be considered expert canoeists, neither were we neophytes. Staying reasonably close together, we headed across the bay in a quartering direction and found, although well loaded, we had enough free board so that with care we could survive the trip. A warning went out when a particularly large wave was seen, and we managed to keep an even keel and climb out of each trough without rolling over. At times the troughs were so deep that the neighboring canoe would momentarily disappear.

As we passed Crown Island where Dr. Langmuir had his camp, we collectively breathed a sigh of relief, since it was a place often encountered by a capsized canoe and its occupants. The rough water steadily decreased and before long, we reached the landing, our parked cars and the safety of land.

The combination of noisy birds, good fellowship and a certain element of danger have placed the memory of Whip-or-Will Camp high in our thoughts even now after more than fifty years!

The Long Path of New York

Shortly after forming the Mohawk Valley Hiking Club in January of 1929, it occurred to me that there should be a hiking trail connecting our area with the eastern Adirondacks. There was a trail across the central mountains from Northville to Lake Placid (the one our local scouts had helped condition between White House and Piseco). I wanted one that went from the Mohawk Valley to Whiteface Mountain.

After thinking about it for a while, I decided it could just as well be more ambitious, so I began thinking of one that also cut across the Catskill Mountains, originating at the Bear Mountain Bridge. With further consideration, we decided on having it start in New York City, at the eastern terminus of the George Washington Bridge. I received encouragement from a few very active hikers who lived in New York City, and eventually agreed that Mrs. Florence Fuller and Mr. W. W. Cady would be relied upon to lay out the trail from its southern terminus, across the Interstate Palisades Park and the Shawangunk Mountains, and then into the Catskills. It would continue across the Catskills to Gilboa in the Schoharie Valley. There we would pick up the trail and scout its potential route down the Schoharie Valley, across the Helderbergs and Helderhills to the Rotterdam Hills, to slant down the slope of the Yantaputchaberg to Lock 9 on the Mohawk River. There it would cross the Touareuna part of the Glenville Hills to Wolf Hollow, and from there head northward past Jersey Hill and Consaulus Vlie, and from there head for Lake Desolation and down the long mountain slope beyond Archer Vlie, where the old Glass House was located. Reaching Great Sacandaga Lake, it would cross the lake on the bridge connecting Batchellerville with Edinburg, and there head up into the Stony Creek drainage. This would take the Path to Round Pond and Crane Mountain. From there it would head for Gore Mountain and cross the Hudson at North Creek. From there, the trail would follow the River Road on the north bank of the Hudson to the valley of Deer Creek, and would thence head for the Boreas River, which would be followed northward to the vicinity of Elk Lake, and from there cross into the High Peaks by way of Panther Gorge or Keene Valley, and from there head for Whiteface summit via Wilmington or the Whiteface Mountain Ski Center.

As I had planned it, the Long Path of New York would be a corridor rather than a blazed, signed trail. It would delineate by topographic map a route that could be followed on trails, wood roads, dirt roads and secondary roads. Thus it could be easily followed by any map-wise hiker, but not by just anyone. It would be a wandering route tying together most of the interesting natural and historic landmarks of a region. Portions could be followed by bicycle, but here and there– especially in the Adirondacks– it would require considerable stretches of bushwhacking.

Shortly after completing the trail layout on the 15-minute USGS topographic sheets, my fiend Al Getz and I decided to follow the Path from Schenectady to North Creek. We took off at the Mohawk River and three days later, reached Camp Cragorehol west of Crane Mountain. It was an adventurous and very pleasant trip. At other times, I explored other portions of the Path until all sections had been established as feasible and readily followed. At the time, I published a complete description of the Long Path from the George Washington Bridge to the northeastern Adirondacks, with the cooperation of my friend Raymond Torrey, hiking editor of the *New York Post*. This appeared in successive Friday issues of his column, "The Long Brown Path."

The idea of the trail and its feasibility then lay dormant for many years, until resurrected by someone in New York City who, in an article in the *New York Times*, credited the concept and initial work on the trail to the late Vincent Schaefer! Some years later a short account of the Long Path was written up in the *New York Walk Book*. Unfortunately, the facts– and most of the information about the trail– were completely wrong. The next published account was carried by the *American Walk Book* by Mrs. Jean George. It was well-written and factually correct. Then it was marked on a series of maps and a Trail Guide issued exclusively, describing the route from New York City to the northwestern Catskills. A Trail Marker was devised and work was coordinated, and the maps and book were published by the New York-New Jersey Trail Conference.

I was then invited to prepare an accurate historical account pertaining to the Long Path of New York that was published in the "Trail Walker," the official newsletter of the Trail Conference.

Consideration is now being given by the Conference to a rejuvenation of the Schoharie Valley to Whiteface Mountain Corridor. I hope it will be possible to develop this section so that it adopts our original plan, and that the only markers will be numbers at Landmark locations, which will correspond with a trail guide that will describe the salient information about the Landmarks, and will show the corridor and those areas where the best route is likely to be found between the numbered designations.[50]

Mosquito-Time at Woodchuck Temple

John Apperson had two pieces of property on Tongue Mountain. One of these was on the side of the Narrows and was the site occupied by our hiking club on a windy weekend in 1933. The other was on the Northwest Bay side and was called by Appie, "Woodchuck Temple."

One weekend, after some exhaustive work at his camp on Huddle Bay, Appie suggested that Al Getz and I deserved some free time, and he proposed that we might like to

---

50 The Long Path is being expanded. In 2009 a man hiked from the George Washington Bridge to the Schaefer Observatory on top of Whiteface Mountain.

spend the night at this Woodchuck Temple camp. Accordingly, we gathered food and our sleeping bags, and he took us up the lake in his speedboat, Article VII, section 7 (the number at that time of the State Constitution article protecting the Adirondacks). In a matter of a half hour or so we had landed at his Tongue Mountain camp and, with a wave after promising to pick us up the next noon, he headed back to Huddle Bay.

Since the weather was fine and bid fair to presage a quiet night, I chose to sleep under the stars. Al decided to sleep inside the cabin, since he was a bit leery about sharing his bed site with the rattlesnakes that populated that part of the Lake George terrain. After fixing supper in the camp, I commented on the number of mosquitoes that seemed to infest the interior. Al said he would solve this problem with some bug spray that he found in a cupboard.

After a good supper we relaxed on the hillside, watched the sunset and, just before dusk, had a relaxing swim in the deep water at the edge of the lake.

As dusk approached, I headed for my sleeping bag that I had spread on a tarp covering a mass of deep grass. Al retired to the camp and I heard him operating the spray that was going to solve his mosquito problem.

As things quieted down and the night noises began, including the lapping of the waves along the shore of the bay, I heard a buzzing sound coming from the camp followed by some loud expletives from Al.

It appeared that the bug spray was not lethal, and all it did was to arouse an incredible number of mosquitoes that, for some reason, were in residence in the camp. Before long I heard the door slam and Al appeared with his sleeping gear to bed down near me.

We spent a quiet restful night and awoke in the morning to another beautiful day. We took an early morning swim, fixed our breakfast, and spent the morning exploring the mountain. By the time Appie appeared we were ready to get back to camp to do some more work for him.

I never did learn how Woodchuck Temple received its name, although I would guess that it was in some manner related to that animal. Appie acquired the camp after it had become notorious during Prohibition days as a refuge for bootleggers running hard liquor down from Canada. I suspect that at times to avoid a revenue officers' ambush along Route 9, the rum runners would use Lake George and high speed boats to make the run from Ticonderoga to the village of Lake George or some other rendezvous along the lake.

Appie was always somewhat nervous about the situation. It may explain his precaution to always have a large loaded handgun either on his hip or within easy reach whenever he was at his camp or on the lake. While I never saw him use it, knowing Appie, I have a hunch he knew how to use it!

A Winter Climb of Marcy

During the winter of 1933, a plan we had discussed during the 1932 Winter Olympics camp at Heart Lake began to shape up. This involved a plan to occupy the rock shelter on the summit of Mount Marcy. While I couldn't participate, Larry Shaw, Chairman of the Hiking Club's camping committee, took the lead in organizing the group of three who planned to go there.

My involvement was to climb the mountain from the other side via the upper Ausable Lakes and Panther Gorge. Our plans shaped up in good order. Charles Florsheim and I planned to go to the Prince Camp on the upper Ausable, spend the night there and head for the summit of Marcy early the next day, expecting to reach the summit about the same time as the group from Heart Lake on the west side of the mountain.

We reached the Upper Ausable about noon, and after lunch decided to scout the trail to Panther Gorge, a route neither of us had ever traveled. All went well and we had little trouble finding the route and breaking trail to the bottom of the gorge. The snow was very deep in the gorge and it was quite obvious that the climb to Four Corners Camp, at the base of the summit cone of Marcy, was likely to be a strenuous activity.

Heading back to Ausable Lake Camp, I took quite a tumble and broke more than two feet off the tail of one of my skis! This posed somewhat of a quandary, since it was not possible to repair it, and to retrace our route to the lower lake was quite impossible.

Early the next morning we rose early and, to our dismay, found that snow had been falling most of the night with over seven inches of new snow covering everything! However, after breakfast, we loaded our packs and headed south and west for Panther Gorge.

Fortunately, the landmarks established the previous afternoon were still recognizable though we had heavier travel than we had anticipated. My short ski seemed to work reasonably well, so I had little to worry about being able to make the trip. As we started ascending the steep slope of Panther Gorge, it became clear that this was going to be a tough job. The snow must have been around twenty feet deep, since it received a great deal of additional snow blown by the wind from the col between Marcy and Skylight Mountains. However, by vigorous tramping with our skis, we found it possible to consolidate the snow so that we could gradually get a purchase on the snow, so as to move upward. After a strenuous effort, we finally reached the top of the gorge to find that an icy wind was blowing from the west and was likely to hinder us some in our climb up the rocky ice cone of the mountain.

My companion, who was an experienced Swiss skier, paused to attach seal skins to the bottom of his skis, while I put ropes on mine, using a technique developed by Langmuir and Apperson. By knotting 3/8-inch diameter ropes in such a way that a series of diamond-shaped crossovers formed on the bottom of the ski, that offered a network of rough fiber to the ice. This device was a marvelous ice-climber and served in good stead when Florsheim's skins kept slipping off the bottom of his skis.

As we reached a spot halfway up to the summit, my companion paid tribute to the ruggedness of the climb that he had previously dismissed as nothing compared to the Alps!

We finally reached the summit of Marcy. Just as we reached the top, a temporary lull in the whirling snow and ice storm that had been buffeting us disclosed the shapes of our friends struggling upward toward the summit. It was good to see them! A group led by

Tony Nerad had accompanied Larry and his friends to assist in carrying the gear they needed for their sojourn at the summit hut. After the snow had been removed and some semblance of order established there, we headed down the mountain towards Heart Lake. The trail to the summit was a narrow one and in skiing down it I found that my broken ski was a distinct advantage! I was able to better control my speed and the turns required than if I had two long skis.

The return to Heart Lake was fairly rapid and we arrived there in good time. When we arrived we found our transportation arrangements were in order and our return to Schenectady was carried out without any problems.

A week later, when the summit skiers returned, we learned that it had been a rugged experience. The weather had continued to be very windy and cold; the firewood problem was a tremendous one since most of the available wood was icy and virtually unburnable. The cracks around the door permitted the wind to penetrate to the inside of the room and the living conditions were far from comfortable. In addition, the open snow slopes that had beckoned to the skiers were crusty and of little use.

In retrospect it appeared that Florsheim and I probably had the most interesting and adventurous experience in this whole adventure.

North Creek in Winter in the Thirties

When I first became acquainted with the village of North Creek in the Twenties and Thirties, it was a small, sleepy town that, after the hunting season ended, seemed to go into hibernation.

I developed a strong attachment to the village as I got to know a few of the townspeople, and especially after my future wife talked a local widow into boarding her for a period in the wintertime, when she skied on the local slopes in a natural bowl at the base of Gore Mountain. The second year my sister Gertrude joined Lois and a tradition began. This was short lived however, since the advent of the 1934 snow train shifted the winter time life pattern of many in the village, resulting in ambitious plans for the future. Since it was not apparent that any help could be obtained from the State Government, the plans for additional trails and similar developments depend on the local populace.

At about that time, Bill Gluesing and I decided to help the planning for the future. I approached the local leaders headed by K. (Judge) Bennett and told them that I thought that Bill Gluesing would be willing to give a popular talk on skiing potential for North Creek. Bill was a super salesman and was in charge of "The House of Magic Show"[51] put on all over that world by the General Electric Company. He agreed to put on his show, along with his magic tricks, at the North Creek high school. At the same time our plan was that he would end the show with his "Ride Up– Slide Down" suggestions.

---

51 The House of Magic was seen by more than 13 million people by 1951 and more than 2.5 million saw it at the New York's World Fair. You can view a segments online by visiting *http://www.youtube.com*, and searching for "House of Magic GE."

Our strategy worked, and a huge crowd of local townspeople and others from the surrounding communities showed up, and were delighted with his presentation. As part of our plan, he proposed that a trail-clearing crew from the Schenectady Wintersports Club would join members of the North Creek Ski Club on the following weekend to establish the Rabbit Pond Run.

This joint work crew assembled on the following Saturday and made great progress in clearing the trail for the new run. During the lunch break, I proposed that a new organization be formed to be called the "Gore Mountain Ski Club," that would have joint membership of North Creek and Schenectady skiers. This was done and a fine cooperative group constituted the initial membership. Unfortunately, the club was such a success that the North Creek Ski Club was virtually swallowed up by the new entity, and in a year ceased to exist. Unfortunately, too, the start of the running of Snow Trains from New York City brought into the community such a strange new group of skiers with different attitudes and principles, that the Schenectady cooperative activities tapered off, and within a year or so, the New York City group dominated the scene.

Then came World War II, and since so many of the Schenectady group were professional scientists and engineers, they were forced to work overtime and to abandon many of their recreational activities. After about 1938 the entire pattern had changed and the snow train and other cooperative activities tapered off and then ceased to exist. After the war the surge of prosperity provided many skiers with their own cars and the need for Snow Trains went into an eclipse and, despite sporadic attempts to revive the concept, such plans never materialized.

The "new look" at the Gore Mountain Ski Center under ORDA (Olympic Regional Development Authority) will change many things at North Creek. Whether for the better or worse, only time will tell!

The Ancient Corner Tree

Many years ago I discovered the Annual Reports of the New York Adirondack Survey. Written by Verplanck Colvin, the land survey superintendent, they are fascinating accounts of the day-by-day activities of the survey crew which Colvin led throughout the Adirondacks. An important feature of his activities was the reestablishment of the location of survey lines established in the 1700's by the Colonial land surveyors. Colvin's books are illustrated with excellent photographs, maps and sketches prepared by him, and which constitute a fascinating record of the condition of the mountains more than a century ago.

One of his most intriguing accounts tells of the finding of the ancient "corner" which marks the junction of Albany, Tryon and Charlotte Counties. These were subsequently divided into a number of smaller units. Albany County, for example, was partitioned to include Saratoga, Schenectady, Rensselaer and Albany.

Colvin tells about directing his chief surveyor, Francisco, to follow Stoney Creek north of Hope Falls to a sharp bend in the stream, and from there head southward casting about for evidence of an ancient blazed line. He tells how Francisco found what appeared to be an ancient blaze, whereupon he "based" the tree, that is, he cut a 90-degree block out of the tree at the blaze location. If the tree had been blazed a hundred years or so earlier, the scar produced by the blaze would show up in the growth pattern. Counting the annual rings from the present to the scar would then tell the year when the line had first been surveyed. When Francisco did this, he discovered that it matched the field notes of the Colonial surveyor and thus established the reality of the blazed line.

A group of my friends from the Hiking Club joined me and we followed Verplanck's description in detail. We headed up the mountain and actually found several of the boxed trees. Taking a compass bearing along this line we encountered a flat swampy area such as he described. It was not a deep swamp, but one that fit the description. I then headed east and cast about until I found evidence of the east-west line that joined the north-south one. Shouting occasionally we then converged and encountered a huge hemlock tree bearing unmistakable large blazes. Surrounding this ancient tree were "witness trees," which were marked to further mark the corner tree.

I then photographed the tree and later had its picture published by the *Schenectady Gazette*, with a descriptive article in my brother Paul's "Woods and Waters" column. Since this ancient corner tree is in a swampy area it is likely that it still remains there and will exist another century or more.

This was only one of a number of items published by Superintendent Colvin that I checked out and found to be exactly as he said it was. This preciseness and sensitivity to natural and cultural values is one of the reasons why Colvin was a key individual to the successful establishment of the New York State Forest Preserve and its constitutional protection.

The Moonlight on Wallface

In some of my earliest reading about the Adirondacks, I found a book by T. Morris Longstreth that waxed eloquent about Wallface Mountain. He said that one of the most sublime views he had ever witnessed during his many world travels was to see the illumination by moonlight of the 1200 foot rocky face of Wallface.

Thus, in the mid thirties I organized a camping trip among my hiking club friends which would take us to Indian Pass so we could see this massive cliff as illuminated by the light of the full moon.

We parked our car near Adirondack Lodge, shouldered our rucksacks and headed for the Pass. We arrived by mid-afternoon, located our campsite at a spot where we had a full view of the precipice and then spent the afternoon exploring the ice caves of the jumble of huge rocks that, over the centuries, have tumbled from the side of Wallface. We found many icy deposits and figured that some of them were semi-permanent.

After having our campfire supper, we cleaned up the campsite and went to our sleeping bags. Our plan was to take advantage of the cloud-free night to sleep under the moon and stars! With dusk and the rising moon, we were treated to a magnificent sight as the light of the moon illuminated the great precipice. The view was indeed one that, once seen, is never forgotten.

As the moon went higher into the sky, we dozed off and then awoke at dawn. Early in the morning after breakfast, two of us decided to climb part way up the cliff, following the many ledges and crevices that marked the vertical face of the mountain. After climbing for some distance, we reached a point where we decided to retrace our steps, only to find that this seemed to be particularly dangerous without a rope. We then decided to continue upward, and within an hour or so, we had surmounted the cliff. We then went south of the cliff face and descended the mountain down a steep wooded slope that was much safer than trying to go down the precipice.

We had a memorable adventure and one that remains clearly in my mind.

Our next visit to Wallface and Indian Pass was during the Winter Olympics of 1932. A group of us who were camped in the open-face Adirondack lean-tos at Heart Lake during the Olympic Games decided to spend one of our days on a ski run to Indian Pass. Despite the general absence of snow around Lake Placid and the games site, we had plenty of snow blanketing the high peaks region. Thus our cross-country run to Indian Pass was most enjoyable. Reaching the Pass I was intrigued with the deep deposit of graupel, a precipitation form also called soft hail. It is produced by the rapid riming of ice crystals, caused either by the turbulence in supercooled convective clouds, or by the crystals being caught in an intense updraft of cold, cloudy air, such as one could expect with an east, northeast or southeast wind at Wallface. This had apparently occurred and the snow in the Pass consisted of more than three feet of pure graupel! What magnificent skiing it produced!

We then headed back to camp and arrived in a short time, since the route was mostly downhill.

I have not visited Indian Pass or Wallface since those two fascinating trips. However, like so many other of my adventures in the Adirondacks, the trips are so much fun that one or two trips are enough to provide memories which persist over many years!

Chapter 11
My Adirondack Connections (Part 2)

The Adirondack Snow Train

Early in the 1930's I read about the Snow Train running on weekends out of Boston. They would take skiers to parts of the White Mountains for the day and then return them to Boston by evening.

I wrote to Fred Grant, Passenger Agent of the Boston and Maine Railroad, to determine whether he would be interested in running a similar train from the Schenectady area into the Green Mountains of Vermont. His reply was to the affirmative, so I formed a Snow Train Committee in the Mohawk Valley Hiking Club and recruited others from the outdoor oriented clubs in Schenectady. We circulated petitions and were able to show Fred Grant that there was a substantial group of winter sports enthusiasts in Schenectady, and it should be worth his effort to arrange for us to journey to Vermont.

We decided to go to Wilmington, Vermont, where a number of us had been cross-country skiing. The branch line of the B&M took off of the main line just beyond the Hoosac tunnel. Since the B&M did not have a line in Schenectady, we planned to board the train in Scotia. Thus, in the winter of 1931, all was in readiness for our first train. The date was set, my committees were ready but– there was no snow! This meant postponing the trip to the next weekend and notifying everyone by postcard of the new plans. You guessed it– there was no snow the second week! This meant another batch of postcards. These were the days before computer labels!

And so it went, week after week, until we abandoned our plans for the winter!

Meanwhile, however, we had developed a substantial following and, as a result, it was decided to form a new club to concentrate on winter sports activities. This was done and in 1932, the Schenectady Wintersports Club was organized. I was elected President, an office I held for several years.

As the winter of 1934 approached, we decided to try to operate snow trains again. This time we planned to utilize the Delaware and Hudson Railroad. The previous fall of 1933, I flew with Dr. Irving Langmuir in his open seat Waco airplane to areas within an hour or so of Schenectady. We explored the hills of the Catskills within range of the railroad and found good-looking terrain around Mount Utsayantha near Schenevus. When the time came to make decisions, Mr. Fred Gelhooley, Passenger Agent for the D&H, proposed that we consider North Creek in the Adirondacks.

The first Snow Train from Schenectady to North Creek on March 4, 1934. Photo spread from the Delaware & Hudson Magazine, *Vol. 16, No. 3, March 1934.*

Since I know the North Creek area very well, I enthusiastically endorsed this plan and made firm plans to go there. I formed many committees, the largest being the First Aid Committee.

Since I had heard of the casualties that accompanied the Boston Snow Trains, I proposed having skiers trained in First Aid techniques, equipped with toboggans, splints, bandages and other similar gear, and a doctor ready for any emergency.

On March 14, 1934, the first New York Snow Train made up at the Schenectady Station and headed for North Creek. At the last minute, a second engine was hooked on since the turnout was much greater than had been anticipated. Upon arrival at the "Crick," the villagers and surrounding mountaineers had a fleet of trucks at the station. As each truck was loaded, it took off and headed for the Barton Mines, located near the top of Gore Mountain. Reaching that spot, everyone clamped on their skis and headed for the several mountain trails that had been established by the North Creek Ski Club Members, with our cooperation. The Pete Gay and Rabbit Pond runs were available at that time. These were old wood roads established many years earlier by loggers. The overgrown trails had been cleared of brush and while rather narrow, had grades that were not unduly steep. As a result, everyone had a fantastic time. Trucks located at the base of the two trails were ready to take the skiers up the mountain for a second, and even third, run during the day. Thus it was that Bill Gluesing's saying, "Ride up– Slide down," became the popular slogan for North Creek.

At the end of the day, members of the First Aid Committee, headed by Miss Lois Perret, a registered nurse, "swept the trails" making sure that there were no injuries or laggards on the trail to delay train departure. As we left for home, I checked for casualties and found

– none! The most serious was a sprained wrist! Thus was born the first Ski Patrol in the United States.

Those witnessing the State-run Gore Mountain Ski Center, with its access road, miles of wide machine-groomed trails, lifts ranging from T-bars to gondola, and a massive snow-making network, and which often accommodates 5,000-7,000 skiers on a weekend, will find it hard to believe the primitive nature of the early days. But I dare say– we had more fun!

The Establishment of Skiland

Shortly after the advent of the first successful Snow Train to North Creek, which ran March 4, 1934, my brother Carl heard about a device developed near Woodstock, Vermont, consisting of an endless belt of rope powered by an automobile engine, which hauled skiers to the top of a ski hill. This permitted skiers to get many more downhill runs than had heretofore been possible. My younger brother has a very inventive and mechanical aptitude toward things, and he soon devised a similar device, which he installed on what was called the Village Slopes, a piece of hilly land ending in a natural bowl which, I believe, was donated to the North Creek inhabitants by Father McMahon of the St. James Catholic Parish of that place. The tow operated for several years. Meanwhile, Carl and his wife Margaret moved to North Creek and secured a parcel of land along the North Creek-Sodom Road, where he also installed a rope tow and developed the North Creek Ski School. Carl's activities thus were responsible for the first ski tow in New York State, and the first Ski School in the state, following the impetus to skiing generated by the 1932 Winter Olympics.

Carl developed a fine terrain setup as part of his school, calling the development "Skiland." His ski activities were progressing very well until a disastrous fire demolished his home near Skiland. This was more than Carl's resources could withstand and, as World War II was declared and travel was severely curtailed, he decided to return to Schenectady and seek employment at the General Electric Company. At the same time, he retained his ski tow at Skiland that became a family and friend activity, and became the scene of many enjoyable winter weekends.

As the plans for the Gore Mountain Ski Center developed, we were dismayed to learn that the access road to the base of the center would cut across the best part of his ski tow. The land acquisition team of the state was quite demanding and uncooperative. They offered him a pittance for the land they wanted, in fact at one time proposed to take all of his property. A compromise was eventually developed where he retained the lower part of the land which included the snow bowl, a strip of land including a brook which carried a stream draining a portion of Gore Mountain, and a strip of land extending out to the County Road and North Creek. Prior to the Ski Center plan being announced, I acquired the strip of land extending to the County Road and the stream, but then following my appointment to Governor Harriman's ski policy committee, I sold the land so that I could not be accused of self-interest in the decisions that I could see would be made.

Carl relocated his tow to a side hill about half as high as the one which he had before the state took his land. This tow provided the young folks and us older ones with lots of fun over the years, and after more than fifty years the Skiland tow continues to operate every winter! The grandchildren are now enjoying the facility as well as their parents and grandparents![52]

Our Metcalf Lake Expedition

I have always been intrigued with maps. In fact, I can easily spend hours studying them and, in my mind, traveling vicariously along the rivers and trails, or bushwhacking through wild country that they depict. So it has been with the early maps of the Adirondacks, especially, the first maps issued by the US Geological Survey. I have been told that some portions of these early maps were drawn in the back room of a county store, with a bunch of old timers advising the cartographers about the drainage systems. In a few cases I am afraid there may be some truth to such rumors, since I have been badly misled in searching for streams that just were not there!

However, in pursuing the Piseco sheet of the survey, I became much intrigued with the wealth of intricate data depicted in the vicinity of Metcalf Lake, west of the Piseco post office.

Accordingly, I made plans to visit this area, accompanied by two of my Scouts. We laid our plans carefully, and on one fine day, drove to the trailhead west of the little village of Piseco, parked the car, shouldered our packs, and headed up the trail to T-Lake Mountain. Arriving there, we continued on a branch trail to T-Lake Falls where the trail ended. The Falls is a ledgy one, having a combined drop of more than 300 feet. Several years earlier I had visited the falls area on skis and discovered a large deer "yard" occupying the ledges below the top of the falls.

From the top of the Falls on this day, we could see the deep valley of Canada Creek and, in the distance, the forest slope of an unnamed mountain. Our plan was to descend the Falls, following the stream to its confluence with Canada Creek, cross the stream and head up to the top of the mountainous ridge which we had seen from the top of the Falls.

Reaching Canada Creek we crossed it, only to find that it was bordered by a massive wetland that was essentially impenetrable. Modifying our route we went downstream for quite a distance before the wetland ended and we could head up the slope of the mountain. Heading by compass, we eventually reached the top of the ridge and there encountered a magnificent virgin forest of giant Red Spruce. Heading down through the park-like woods, I heard an unfamiliar bird sound and saw my first Arctic Three Toed

52 There are various pictures and stories online of Carl's first tow mechanism, which can be found by typing "Carl Schaefer ski tow" into any major internet search engine.

Woodpecker.[53] Pausing here and there, I made core borings of some of the huge spruce trees, and sampling the soil underneath them, I found that the trees were more than 300 years old, and that they were growing in a white, quartzy sand, obviously of glacial origin.

As it was getting late in the afternoon, we then hurried westward and suddenly through the trees, we saw the glistening waters of Metcalf Lake.

What a magnificent lake it was! Edged by low rocky cliffs, it appeared to be quite deep and crystal clear. At the place where we reached the lake we found a campsite and nearby, a fine canoe with paddles, cached in a protected place. Borrowing it and leaving a note for its owner, we embarked upon the lake, and on its western shore found a perfect camp site on a shaded rocky point, about ten feet above the lake's surface.

We quickly erected our tarp tent, built a cooking fire, gathered enough firewood for the next few days, had supper and then, before dark, explored a portion of the lake up to its inlet. There we found several bare rocky islands, each having several sea gull nests. These were ancient nesting sites since the gulls' most recent nests were at the summits of cylindrical piles of mud and sticks, attesting to many years of occupancy. The next morning we decided to explore the chain of small lakes shown on the map, which originally intrigued me with its minute detail. We found the tiny lakes just as depicted, and had a fascinating time exploring each one. On the edge of one of them we found a trapper's cabin. It was one of the filthiest dwellings I have ever encountered. The trapper apparently had no concern whatever for clean living. Dried up garbage was strewn on the floor around the doorway, with cans and bottles strewn about where they were thrown. My young friends had a vivid object lesson in how not to live!

Except for this localized mess, the remainder of the features of the lake and its surrounding waters was pristine and beautiful. The fishermen who camped around the lake were invariably good "housekeepers," and either buried or carted out whatever debris developed during their visit.

I came to the conclusion that the surveyors responsible for preparing the map of the region found it to be such a beautiful place, that they tarried and probably spent an inordinate length of time preparing and completing their surveys.

Although I never returned to Metcalf Lake, its beauty and peacefulness has remained bright in my memory. Fortunately, it is entirely within the Spruce Lake-Canada Lakes Area that has been designated as wilderness and is protected by the Forever Wild concept of the Adirondack State lands, protected by the State Constitution.

White Mullein Near Moxham Mountain

The word we got at Schenectady was that during some heavy rains near North River, a dam holding back the waters of a small lake near the western base of Moxham Mountain had washed out, leaving a wide path of devastation through the forest extending to the Hudson River. Not long afterward, I had a chance to explore the area.

53 A Black-backed Woodpecker (*Picoides arcticus*), also known as the Artic Three-Toed Woodpecker.

Following the river road, I went as far as the road was passable, and then walked along the river to the outlet stream where it meets the river. Where originally it was a narrow stream tumbling down a large mass of crystalline limestone, the flood had stripped all the vegetation off of a wide portion of the rock, leaving a scar which, after more than 60 years, is still quite visible. As the stream courses down the rock face, it has tunneled into it at several places, disappearing and again surfacing as it has dissolved the soluble calcium carbonate rock of that formation.

Traveling upstream along the course of the flood caused by the breaking of the dam was not difficult, since it had cut such a wide channel. Before long I reached the broken dam, and found that it was indeed a scene of desolation.

On the way back to the Hudson, I did some exploring along the edge of the flood channel, looking for flowers and ferns. Not far from the Hudson I encountered some very large specimens of the Great or Common Mullein. One of them attracted my attention since, instead of having yellow flowers, they were all white.

Upon returning home, I learned that while such white flowers are not unknown, they are considered quite rare. I have wanted over the years to revisit the area to see if the white-flowered mullein still grows there. Thus far this has not happened, but one can always hope!

A Rock Shelter at the Mouth of the Rock River

Shortly after Lois and I were married in 1935, we decided to go on a bushwhacking trip in the Adirondack Wilderness.

Parking our car along the Gooley Road that leads from the North River-Indian Lake Road to the wild part of the Hudson River Gorge, we shouldered our packs and headed west by compass. Our destination was the junction of the Rock and Cedar Rivers. I had heard that there was a rock shelter or cave at that location and, with my avid interest in archeology, I felt that this was a lead that merited attention. I was particularly intrigued because our local chapter (Van Epps-Hartley) of the New York State Archeological Association had recently been given an excellent fragmentary, prehistoric pottery vessel that a lumberjack named Ed King had found in a cave on the Indian River not far from our destination.

After traveling through the woods for several hours, we saw the open sky ahead of us and encountered a small, narrow, deep lake. It was on our planned route and we felt good that thus far our bushwhacking plan was working out extremely well. The area surrounding the lake had been badly burned recently, the sun was hot and we decided to go swimming. Shedding our clothes, we were soon enjoying the cool and cold waters of the lake. With this refreshing event, we dressed and were soon on our way to the west.

In several more hours we encountered the Cedar River and, to our pleasure, saw the mouth of a stream entering the river from the southwest that had to be the Rock River.

Doffing our packs, we spent five or ten minutes scouting the local area, and soon located an excellent campsite opposite the mouth of the Rock River, and close to the edge of a beautiful lake. Within a short time we had erected our tarp tent, cut balsam boughs for a soft, springy, fragrant bed, constructed a safe fireplace, and then had lunch.

After relaxing for a bit, we took off for a trip of exploration. At the mouth of the river we did, in fact, find a nice rock shelter, but to our dismay, it had been occupied by a winter trapper, who had left a complete mess of the shelter and its surroundings. Empty cans and all sorts of junk were scattered within throwing distance of the crude door that the occupant had fabricated at the outer edge of the rock canopy that he had occupied.

Exploration of the shelter was not feasible, due to the monumental mess that had been left by the former occupant. How anyone can live under such conditions is beyond my comprehension.

Leaving the rock shelter location (which was in a perfect spot for Indian occupation), we continued exploring the region for several hours, and then finally returned to our idyllic camp site.

As we were fixing our supper, we suddenly were startled by an unearthly sound. As soon as our initial reaction subsided, I realized that it was the call of a loon. An answering call occurred and, for several hours, we were entranced with the sounds that epitomize the spirit of the wilderness of the Adirondacks.

The next morning, after an enjoyable sleep, we packed our gear and headed east toward Gooley Road. This was reached without trouble. We headed for home with a mixed feeling of disappointment about the rock shelter but a profound sense of tranquility that has become the essence of our encounters with the Adirondack Wilderness.

A Visit to OK Slip Falls

One beautiful fall morning, when Lois' father and mother were visiting Cragorehol, I proposed that I guide the male heads of the Perret and Schaefer families to a visit to OK Slip Falls. I assured them that it was not a long, onerous expedition and that we would be home for lunch. With this assurance, the three of us headed for the trailhead on the mountainside part way up the Indian Lake Road, where it climbs Casey Mountain. We arrived there in less than half an hour from Camp, and headed down the trail to the Blue Ledge on the Hudson River.

After traveling for a half mile or so down the trail, I strode through the woods heading for the falls. My associates apparently were a bit disturbed at the wild country we were traversing, but didn't raise any protest. The distance to the gulf appeared to be greater than I remembered, but we kept going, stopping at times to take a breather and to examine natural features nearby. After a considerable time, and at the stage when I was beginning to wonder about the wisdom of our journey, we reached the edge of the deep ravine carved by the falling waters of OK Slip Creek.

When my compatriots questioned the wisdom of descending into the ravine, I assured them that it was the way home. By descending to the creek and following it to the Hudson, we would encounter a fisherman's trail, and thus would have "clear sailing" back to the car.

The ravine was deep and the route rather tricky, but by staying close by, I was able to guide them down to the plunge pool of the falls. Due to the lateness of the season and a lack of much rain, the falls was not the spectacular sight I had promised, but at least was worth seeing. We followed the creek downstream– a rather wild undertaking, but my followers did not protest. We finally reached the river about noontime, but unfortunately I had been over-optimistic about the time required to make the circuit, and had neglected to bring any sustenance, not even chocolate bars or gorp.[54]

Since the nearest food was at Camp, we did not tarry long at the river, but continued down it until we encountered the hill trail. Reaching this, we started back toward the car. This was a steady climb and required many pauses as we headed up the mountainside. We finally reached our cars and I am sure my two elderly friends gave a sigh of relief as they sat down! To their credit– and my relief– neither of them complained at any point of our trip.

Reaching Camp, they climbed out of my car and, before getting to the porch had muscle cramps. Once these were over, they both had a good meal, and then recounted their adventure, that apparently had been enjoyed by both of them to a limited degree!

However– this was the end of hiking trips with my father and father-in-law! While the trip was talked about for many years (both were more than ninety years old before they died), I now appreciate the effort they made and the good sportsmanship they displayed when we visited OK Slip Falls!

GE Vice President J.R. Lovejoy, his chauffeur, Vince and a friend hunting Indian relics at Dunsbach Ferry on the Mohawk River. Courtesy of Michael Sullivan.

54 A slang word for what we would today call "trail mix."

The Caverns of the Adirondacks

From time to time, one hears of caves in the Adirondacks. Upon investigation, most of these features are found to be channels dissolved out of the acid-soluble calcium carbonate crystalline limestone. Ordinary rain, in its fall through the air, becomes mildly acidic as it comes into equilibrium with the gaseous carbon dioxide in the air. This produces a pH of 5.7 to 5.8. During the last few years, the acidity of the rain has increased to a pH of 4.0, and sometimes even lower. This enhanced acidity has been associated with the increase of acid vapors in the atmosphere, caused by pollution emanating from automobiles, power plants and other man-generated activities.

The caves that have developed in the Adirondacks cannot be compared in size or beauty with those of typical limestone country, where the limestone deposits are massive and continuous. The crystalline limestone deposits of the Adirondacks are limited in extent and geographic location, and are extremely variable in their horizontal continuity. They occur mostly in the eastern and northwestern Adirondacks.

The largest such "cave" structure in this formation is the Natural Stone Bridge near Pottersville, New York. This has been formed by the solvent action of the waters of Trout Brook. Being privately owned, it is now a tourist attraction. When I first visited it, the region was in a relatively wild condition and the "bridge" had been forming since the retreat of the great Continental Glacier, some 10,000 years ago. It is well worth a visit.

A very interesting cave has been formed by the presence of a small stream at the base of Crane Mountain, located at the end of the road on the old Eliot Putnam Farm now owned by the Greene family. It is a negotiable cavern that has been traveled for a distance of several hundred feet. Its course is marked by several large sinkholes, at the bottom of which has dissolved the cave channel. Its course can be followed by a crawling procedure. It is unlikely that any cave rooms exist.

Another intriguing cave can be found near the Boreas River, upstream of the bridge carrying the road to the Northwoods Club. It has been formed by the outlet of Hot Water Pond, a fairly shallow body of water, which is located near the eastern boundary of the Northwoods Club. The pond's outlet plunges into a large sinkhole fairly close to the pond, and flows underground for a distance of about a half mile. This stream flows out of a wide cave mouth located not far from its confluence with the Boreas River. The steep slope at this location is covered with ferns, including that rather rare Braun's Holly Fern.

The cave entrance can be entered in a near-standing position but the speciousness soon degenerates into a crawl way. The considerable distance between the mouth of the cave and its origin suggests that this cave might contain some interesting structures. It hasn't been thoroughly explored to my knowledge.

During the period when the great Continental Glacier covered the Adirondacks, it filled the valleys, covered the mountains, and greatly modified these areas with its plowing, grinding, crushing action. As it melted, many new things happened. Kames, eskers, lateral moraines, and localized terminal moraines formed. During this ablation period, crystalline limestone, wherever it was exposed, was modified in some manner. In a number of instances, solution processes produced local modification of the surface rock and, in some instances, shallow caverns, or what became "rock shelters."

Another type of "cave" also formed among the huge tumbled fragments of the mountains that were broken by the inexorable pressure of the slow flowing ice. These resulted in the formation of ice caves and rock shelters where the deep crevices catch and protect the winter snows. Thawing and freezing effects transform the snow into ice and, if of sufficient quantity, this ice persists over the summer.

In the vicinity of such residue, a local microclimate develop,s which causes the formation of deep beds of mosses, lichens and lush ferns. The "tight" valley extending from the depths of Panther Gorge to the col between Mounts Marcy and Haystack is such a place. The moss grows so massive as to hide dangerous crevices and holes among the tumbled rocks.

Another such place exists on top of the cliff along the Ausable River, on its north bank close to Wilmington Notch. On a hot, humid day in summer, a local fog fills the crevices where ice persists into the late summer, and often remains as the late fall snow begins to pile up. A similar place can be found at the base of Wallface Mountain at Indian Pass, and still another on the northwestern flank of Crane Mountain, at the site of what appears to be a massive fault.

Finally, there are a few man-made caves in the Adirondacks, driven into the side of a mountain by a hopeful prospector following a vein of mineral that promised wealth if it got larger. Such abandoned tunnels are fascinating places, especially on hot days when condensed fog drifts out of the mouth of these monuments to human hope. On the Kunjamuk Stream, about halfway between its mouth on the Sacandaga and Elm Lake, such an abandoned tunnel can be found. It is marked on the USGS Topographic Map as Kunjamuk Cave. Similar tunnels exist northwest of Glens Falls, where an iron ore vein was followed into the mountain on a horizontal course. It is an outstanding fog-producer, and I have thought of using it for experiments in particle physics and chemistry. A fine kiln remained near this tunnel that was used for processing the ore. I have not been back there for fifty years, but suspect it is still about, as I saw it then.

A Climb Up Split Rock Mountain with the Kids

One of the great things about the Adirondacks is its accessibility to kids.

The 21 sons and daughters of my two brothers, two sisters, and Lois and me– 7 boys and 14 girls– were often at Camp Cragorehol in force. At any time it was possible to form an expedition to do something, whether it was to haul a flat rock up the hill to form a doorstep, to climb a rocky face of the southeastern peak of Eleventh, to pick huckleberries on the big ridge of Crane Mountain, or to climb some lesser peak in search of adventure.

For some years, while en route to Camp, I had been looking at a rocky knob opposite the mouth of Stewart Creek, on the west side of the East Branch of the Sacandaga River, as a place that merited exploration. It had been denuded by the Great Fire of 1903, and

Vince with his grandchildren (from left) Patrick, Michael and Kathleen. Courtesy of Roger Cheng.

thus showed a deep cleft in the mountain that was probably the site of a rock fault. I proposed a trip of exploration and, in a few minutes, had a dozen or more eager "mountaineers" ready to go. Reaching the vicinity of Stewart Creek, we parked our cars and were soon following the stream down to its confluence with the river.

There we quickly removed our shoes and socks and started to cross the river. In short order we had crossed the stream without anyone falling in, replaced our footgear, and headed up the steep, rocky slope of the mountain.

The going was rough since there was no trail, but by taking things easy and working our way up the cliff following cracks, ridges, chimneys and deer trails, we gradually worked our way to the summit. There we had a fine view of the river and the timbered ridges lining the course of the Sacandaga. We then worked our way down into the fault, and found the rock structures there very interesting and unusual. Continuing downhill, we reached the river, again waded it, and soon were headed back to Cragorehol and a supper such as only a mother and grandmother could make for such a hungry group.

One of the beauties of the Adirondacks is the proximity of everything. The running streams with their fine swimming holes, the brooks having small but beautiful and tasty speckled brook trout (I caught my first one when I was six years old), the winding trails, deep beds of moss and lichens, the "boiling" ice cold springs, the fragrant balsam thickets, the greater variety of colorful rocks in the sandy soils, the remote but easy to reach beaver ponds and lakes– these and many other features are the things that make the Adirondacks so fascinating to youngsters and the older folk who enjoy introducing eager young minds to learn the fascination which attends these mountains. There is little poison ivy or sumac, no chiggers or ticks, very few but highly-isolated poisonous snakes and– except for the period of mid May to early July– few mosquitoes, black flies, deer flies and no-see-ums.[55]

Unlike the most spectacular mountains of the West, it isn't necessary to mount an expedition to obtain a sense of wilderness. The "Forever Wild" provision of the NY State Constitution ensures that a large part of these mountains will remain as they are, of free access, and beautiful to see and to wander in.

As I have often said, "The Adirondacks are for Kids." The memories they acquire will stay with them the rest of their lives. What is more important?

55 No-see-ums are tiny, biting flies or midges (Ceratopogonidae), 1-4 millimeters long, that live near water. They often swarm, forming a "cloud."

My One and Only Trip to Siamese Ponds

One summer's day in the late Twenties, I decided to go to Siamese Ponds. I had heard great stories about this place, its isolation, the twin ponds, and its good hiking trail. I had become familiar with the local trout streams, mountain vistas, and other local landmarks, and decided to broaden my local horizon with a quick trip into the Ponds.

I drove to the trailhead, where the old carriage road to Thirteenth Lake skirted the southerly slope of Eleventh Mountain and headed into the mountains. The trail climbed over a ridge east of Diamond Brook. Once the ridge was surmounted and I descended to the Brook, the trail leveled off and was mostly level for the length of the trail following the Sacandaga River. At a place called Burnt Shanty, the trail forded the river and more or less followed the outlet of Siamese Ponds.

Arriving there, I found the lake to be all it was said to be. Clear water, with a gravelly shoreline, well-used but clean campsites, and a rim of evergreen and deciduous trees. An old timer, whose name I have forgotten, greeted me and told me about the lake. He had a tent platform licensed by the NY State Conservation Department and spent the spring to late fall period there, hunting and fishing. He had several boats to rent for a nominal fee, and served as an unofficial custodian of the Ponds.

I told him I was only on a quick survey trip, but hoped to return sometime when I had more time to spend. After an enjoyable visit, I said goodbye and headed out. As I remember the trip, it took me less than two hours to reach the Ponds from my parked car. Since I had plenty of time, I walked at a leisurely pace and before long, reached the river crossing and then started following the river.

Shortly afterward, I flushed five American Mergansers. They headed downstream, not apparently alarmed, but wary of visitors! As I watched them paddling downstream and feeding as they went, I found a rocky knoll adjacent to the river, where they were bound to pass if they continued downstream. My vantage point had a clear view of the river and a rocky pool below. I hid behind a tree and watched them approach in a leisurely fashion. As they approached my observation post, all five of them went underwater, and I had a fantastic view of them swimming underwater, with a gracefulness and efficiency rarely viewed by anyone. It is a memory still vividly etched in my mind, and one that I will never forget.

It is such events that make the Adirondacks such a marvelous place, where one often encounters the unexpected.

Despite the attractiveness of Siamese Ponds I have never returned, perhaps because it has become more popular, and therefore is visited more frequently by hunters and fishermen. There are so many other wilder places to explore that it didn't have high priority in my mind, and my research activities shifted my interests and activities to the northwestern and southwestern regions of the United States.

Fortunately the Siamese Ponds region have recently been designated as the center of a large wilderness area of the New York State Forest Preserve. Thus, thanks to the "Forever Wild" protection of the State Constitution, and action of the Adirondack Park Agency, it will remain as a pristine place where others can enjoy its clear waters and good fishing in perpetuity. As one gets older, the need to travel and to see and experience first-hand the beauty and fascination of such places recedes in importance, as memories of past expeditions become a vicarious substitute. It is enough to know that others of like mind and younger body continue to have such opportunities to build up their own memories!

My Whiteface Mountain Connection

Despite my familiarity with Mount Marcy, the Range Trail, and Indian Pass, and the Tahawus, Heart Lake, and St. Huberts entrances to the High Peaks area, I had not visited Whiteface Mountain before 1944. At that time, as part of our icing studies for the Air Force, we were asked to cooperate with a group of scientists from Princeton University on a field study of the icing of helicopter rotor blades.

Our joint observations quickly showed that the helicopter rotor was extremely vulnerable to icing from supercooled clouds, and the ice that formed was so tenacious that it resisted all of the available methods for deicing the blades. Thus we were told that a considerable amount of tactical operations against Japan in 1945 were altered as a result of the Whiteface Mountain findings.

This rather brief contact with the mountain sharpened my realization of its importance as a monitoring spot for meteorological studies. However, my activities following the end of the War focused my attention in other directions, and it was not until the late fifties, after I had become deeply engrossed in meteorological matters, that my mind shifted back to Whiteface Mountain.

My activities as director of the Munitalp Foundation caused me to think again about the Adirondack Mountains. I prepared a paper (that was published by the American Meteorological Society) describing my thoughts about developing a Museum of the Atmosphere. This concept I built around the rock-hewn summit building at Whiteface, and was directed toward the thousands of tourists who drove to the parking lot near the summit and traversed the length of the rock tunnel cut into the center of the mountain. There they entered an elevator that transported the visitors in safety up and into the side of the building at the top of the mountain.

In addition, these plans then shifted to the potential use of the old Marble Mountain Ski Lodge that was abandoned as a new ski center began to take form on the opposite side of the mountain.

I envisaged the lodge being converted to a meteorological field station, and I devised a rather elaborate plan for its development, finding it to be perfect as a viable structure that could easily be converted to a highly useful facility.

I originally attempted to get the Rensselaer Polytechnic Institute (RPI) interested in developing a division of Atmospheric Science. While there were a number of faculty interested in such a movement, I found considerable resistance on the part of the RPI administration and, after several futile attempts, I abandoned this effort.

Shortly thereafter, the administration of the State University of New York (SUNY) at Albany, then the State College for Teachers, approached me with a proposition to accept the appointment of Distinguished Lecturer. Since there were things I felt I could do for the students at the College, I accepted the appointment and began to develop a program for high-ability honor students. I was asked to introduce them to the scientific method as I had learned it from Drs. Langmuir, Blodgett and Whitney. I had some excellent students and believed they profited from my program but, after several semesters, I told Dean Oscar Lanford that I thought the program was submitted to the students about six years too late. Their major aim and commitment was to obtain a teacher's certificate and start earning money!

Following this development, Dean Lanford asked me what I would do if I was offered an annual development fund of $50,000 per year. I took my Whiteface Mountain plans, altered a few words and submitted it to him. The warm reception that greeted my proposal was very encouraging! I immediately began to enlarge upon my plans that then included a Research Center at Albany, the creation of an academic department, and the firm establishment of a field station and mountain observatory at Whiteface. I suggested the appointments by President Collins of an Advisory Committee made up of outstanding scientists from the newly-forming State University, and from the Federal and State government. This group came to Albany and enthusiastically endorsed my plans. Thus the Atmospheric Sciences Research Center and the Department of Atmospheric Sciences were endorsed by the Central Office Board of Trustees. Since my good friend Frank Moore, of Albany and Indian Lake, was Chairman of the Board, I had a person in a crucial position at the University who I kept informed of my hopes and plans.

Vince and Ed Halrayd at Whiteface Mountain in September, 1962. Courtesy of Jim Schaefer.

With temporary support from the Research Foundation of the University, and then a budget approved by the Trustees, we were headed toward a permanent role in the University (which is still flourishing in 1987).

Meanwhile, in the Adirondacks our relationship to the mountain continued to develop. Shortly after receiving approval of the Trustees to our program, and the promise of full cooperation from the Conservation Department of the State (headed by Commissioner Harold Wilm), a commercial, politically-strong television station had a law introduced into the legislature which could permit the erection of a commercial television station on the summit of the mountain.

Having had experience at Mount Washington regarding the electrical disturbances imparted on the atmosphere by a radio and television tower, we mounted a strong effort within the national scientific community protesting this action, and pointing out the damage that could accrue to our observational program. The returns to our letter for help were remarkable. Every one of the twenty-five persons contacted replied, many of them with specific examples of the troubles engendered by such a development. Fortunately the law was defeated and our fears were not realized. However, this effort on our part established our presence within the scientific community, and we now enjoy a good reputation for scholarly, innovative work throughout the world.

I asked Ray Falconer, who I hired in 1946 from the Mount Washington Observatory, and who has remained with me until his retirement from the University last year, to assume responsibility for developments at Whiteface. Under his supervision the Observatory and Field Station slowly grew (as funds became available), until it became the finest in America. It is now a major station in a national network for the study of atmospheric chemistry.

Meanwhile at Whiteface, during its early formative stages, many innovative developments occurred. A large array of recording instruments were acquired, which began the accumulation of a host of atmospheric parameters. A summer program of research, conducted by high-ability students, was started in 1962. A series of popular lectures in natural science was inaugurated. Research projects conducted by faculty from other units of the State University system were started. Adirondack Conferences on specialized subjects were conducted for researchers from many parts of the US and abroad, and eventually funds were allocated to construct a summit observatory, now used throughout the year by scientists from many places. It has become an important facility.

The Harriman Ski Committee

Early in 1962, I was asked by Governor Averill Harriman to become a member of his Advisory Committee on Skiing. He wanted a group of citizens and skiers to help him decide on the wisdom and practicality of developing state-owned ski centers in the Adirondack and Catskill mountains.

A constitutional amendment had been passed by the required sessions of two successive legislatures and approved by the voters in a statewide election. The mountains designated were Whiteface and Gore in the Adirondacks and Belleayre in the Catskills. There were seven persons constituting this committee. In addition to me, there was Hal Burton of Long Island, Bill Roden of Trout Lake near Lake George, Ron MacKenzie of Lake Placid, Dean George Earle of Syracuse and, I believe, two others.

After the first meeting was convened, it became quite apparent to me that the main political purpose of the committee was to sanction the development of a new Whiteface Mountain Ski Center.

Following the 1932 Olympics, there was a ski development constructed on the eastern slope of Whiteface on Marble Mountain. This was a disaster! A beautiful, very large log building had been constructed at the terminus of a road on the slope facing Marble Mountain, a T-bar lift had been constructed, running from the edge of White Brook to the top of the Marble Mountain Ridge, and a number of twisting downhill trails had been established, along with some for cross country skiing.

The main problem with the site was that immediately after a heavy storm blanketed the slopes, the wind would start to pick up and it would blow most of the snow from the open slopes into the woods! Thus the Ski Center was a failure in every aspect.

A new site had been located on the other side of the mountain that would extend to the summit of Little Whiteface, with a separate trail going to the summit of the big mountain.

At our meetings it became apparent that little, if any, attention was to be given to a similar development at Gore Mountain or Belleayre. I raised a strong concern about the undue emphasis being placed on Whiteface, and insisted that the other two mountains be given equal consideration. I was particularly concerned about the Gore Mountain development, since I was aware of a great need for both the skiers of our region, as well as the economy of the North Creek area. I announced that I would oppose the Whiteface Mountain development unless some consideration was given to Gore. I was able to get a concession in this respect and was assured that Gore would receive committee attention, although its development would be delayed until the new Whiteface Mountain Center was established.

Meanwhile, several inspection trips were made to Gore by the committee. We went to the Gore summit, and then hiked down the slopes toward the North Creek Valley. All of the committee members became much impressed with the potential for a wide variety of skiable terrain on the slopes of Gore and its satellite peaks, including Burnt and Black Mountains. It was decided that some meteorological data should be gathered especially related to snow depth, wind patterns and snow persistence.

I was asked to erect anemometers at key locations on the mountain in an attempt to determine the wind patterns, and arranged with Ranger Chuck Severance of the North Creek-Sodom Road office to monitor the anemometers. With my son, Jim, I erected three-cup anemometers in areas that appeared to have wind problems to determine, if possible, how serious they were.

Despite our efforts, it became apparent that the Gore Mountain development had been assigned a low priority. It wasn't until the Whiteface Center had been virtually completed that it was possible to revive state attention to Gore Mountain. Part of this delay could be attributed to the onset of World War II, that greatly delayed most of the state's programs of this sort.

My last direct activity with the Harriman Committee took place in the spring, when Ron MacKenzie and a few others went from the summit of Gore over to Burnt Mountain, and thence down to determine snow patterns and the area of Gore best suited for its initial development. We designated the general region that subsequently was first developed, and were again very favorably impressed with the great potential.

In Which We Buy a Mountain

Ever since I knew what a mountain was, I have been fascinated by them. The fact that their roots go all the way toward the center of the earth, the fact that they are still forming as the great tectonic forces crumples the rock and forces some of it upwards, the magnificent views that can be seen from them on a clear day, and the dramatic profile they present as storms develop or the sun rises and sets– these are all features that urged me to acquire one of them. While I realize that such a possession is a very transient thing– others have "owned" it earlier, and others will "own" it later– but a deed in my pocket, which says it belongs to Lois and me, provides a certain sense of satisfaction.

I have never wanted to own a mountain so that I could exclude others from enjoying it. In fact, the actual owners are the plants and animals that use it as their home.

"Our" mountain is Edwards Hill. We bought it from a local mountain man after he finished lumbering it. We eventually found he had "high-graded" it, the forest that is, meaning that he took out the best saw logs easily available, and left the rest.

As we explored the mountain we found the boundary lines, which consisted of fragments of fence embedded in the side of old trees, healed over blazes marking the line, and piles of boulders at the corners of lots. On the east and west sides of Edwards Hill were steep cliffs that presented fine views of the surrounding mountains and valleys. Long wooded slopes were in the other directions that led to the summit.

At the summit we found a massive glacial erratic nearly twenty feet high. It was a fascinating sight to see such a large stone perched on the top of the mountain! The western cliff edge gave a fine view of the settlement of Edwards Hill and Eleventh Mountain, located about a mile or so away. I occasionally walk there from our camp on Chatiemac Road, climbing past the spring from which we get our running water, past the old wood rood, over the ridge sloping towards Balsam Brook, past a big beaver pond, and then along the road that leads to the Dalaba Homestead.

It takes an hour to make the trip, but each time I do it, I find some new feature not seen before. I am still endeavoring to find out the origin of the name of Edwards Hill. Was it one of the famous Jonathan Edwards who were famous ministers of the Gospel, and who were associated with Union College and Princeton University? The folks in the vicinity of Sodom and Bakers Mills were active churchgoers in the old days, and many of them still are so inclined. Or was it some local mountain man lost in history? Some day we may find the answer.

Our Acquisition of Woestyne North

Early in 1969 I learned that some property on the Chatiemac Road was for sale, opposite some that my brother Paul had purchased and built a substantial camp. I met with the owner and came to terms with him, and we acquired the land and a camp.

The camp seemed to be well built as Adirondack mountain camps go, but lacked many features that we felt were essential elements of a Schaefer Camp. Primary among them are natural woodwork and a large fireplace. Paul assured me that he would build an addition to the camp that would rectify this situation. Knowing Paul's abilities and ideals, I had no qualms about the matter and left the design, dimensions and fireplace plans entirely up to his judgment. Just before leaving for my annual sojourn to the Research Center of the Museum of Northern Arizona for the summer, I reviewed his plans and was assured that, upon return, we would find the basic shell of the camp in place and a fireplace under construction. That fall we returned from Flagstaff and found everything as promised. I planned to complete the construction of the siding, as well as the modification of the original camp, according to our needs.

At about that time my brother had made an effort to purchase some very wide white pine planks that we needed for framing the upper part of our fireplace that my brother Carl was completing. He had located a large pile of boards at an old sawmill in the village of Stoney Creek and, upon approaching the owner of the mill, was told that he would not sell a few planks, but was interested in selling the entire pile of boards. The price Paul was quoted was so reasonable that he immediately agreed to the purchase, reported it to me, and on that same day we hauled several truckloads of beautiful, virgin, white pine planks to our camp. That fall, by dint of intensive work on our part, Lois and I not only enclosed the camp with vertical planking, put on a second layer inside the camp, all sanded, chamfered and coated with linseed oil, but also rearranged two small bedrooms by removing walls of plasterboard and converting them into a very attractive bedroom running the width of the original structure. The massive fireplace (as with all of Carl's structures) was very attractive, and highly efficient as a medium for heat production and a place to broil hamburgers and steaks!

During that fall I also cut down a number of medium-to-large, dead red spruce trees to use for fireplace heat and comfort, but also some maple and ash and birch for cooking fires. With the camp purchase we acquired about six acres of forest. This was a mixed wood of

A Christmas Card (year unknown) made by Vince and sent to his friends. Courtesy of Jim Schaefer.

spruce, balsam, hemlock, maple, ash, beech, birch and hop hornbeam. Many of the deciduous trees and hemlocks were quite large and in rather good health, except for the red spruce and some of the beech. A massive beech and another ash were a major reason for acquiring the property.

A dug, boulder-lined well of ancient vintage served as a water supply. However, my explorations of the land had disclosed the presence of a beautiful ice cold spring, located near the base of Edwards Hill that rose nearly a thousand feet above the camp. I soon tapped a portion of this spring with a plastic pipe and led it to a spot near the camp's side door. With time this has become our major source of water and it has never failed us.

Soon after acquiring the property, the question of giving it a name was considered. Since we had named our home on Schermerhorn Road in Rotterdam "Woestyne" (the Dutch word for "wilderness"), we decided to call the original home Woestyne South, and our new acquisition Woestyne North– so it is.

The following year we learned that a local lumberman, Ivan Bateman, had high-graded some 70 acres of land that encompassed the little mountain called Edwards Hill, and was planning to sell it. We met with him and agreed to purchase the mountain.

It is a very attractive place. The summit is somewhat rounded with about half of it consisting of rocky outcrops– some in the form of vertical cliffs– especially on the southwest side, which has a fine view of Camp Cragorehol and Eleventh Mountain. On the eastern side of the summit is a large glacial erratic, probably transported from the higher mountain to the north.

The woods were not devastated by the earlier lumbering and, in fact, except for the occasional stumps, one would hardly know it had been cut over.

Here and there are patches of Trillium, Maidenhair Fern and other shade-tolerant plants with an understory of witch hopple. The northeastern slope has wild apple trees, suggesting that at one time the northern part of Edwards Hill was a mountain farm. To the northwest are the rounded peaks of Height of Land Mountain and Gore. Some years ago I discovered a cabin site where our mountain farmer had lived, along with a cast iron teakettle and a few other bits of houseware.

Woestyne North is a very attractive place for Lois and me, and most free weekends find us heading for it by way of Wolf Hollow, West Galway, Fish House and Batchellerville on Great Sacandaga Lake. We then cross the bridge to Edinburg and go across the sand plains to Northville. There we reach the upper end of the big lake, follow the Sacandaga River to Wells, then its east branch to Fox Lair, and thence over the Eleventh Mountain col to Bakers Mills. It takes us a bit more than an hour and a half to go from Woestyne South to Woestyne North. I occasionally walk from Woestyne North to Camp Cragorehol by a variety of routes, either by way of the summit of Edwards Hill, the Big Rock, or Chatiemac Lake. The route over or around the summit of Edwards Hill is the shortest and the most interesting route, since it passes across the drainage of Balsam Brook, one of the headwaters of North Creek, including a large beaver pond and the place where the Cold Spring serves as the main summer source of Balsam Brook. The Cold Spring is a landmark of the region. It is a "boiling spring," that is, it emerges with such force that the sand bed appears to be boiling as the water emerges from the bed to start its journey to the Hudson at the village of North Creek. It is the highest permanent source of that creek, and during the early settlement of the region, was noted for its crystal cold properties.

Since acquiring Woestyne North and the summit and slopes of Edwards Hill, Lois and I have acquired four other land parcels nearby, so that we now own well over a hundred acres in the vicinity of Chatiemac Road. Its ownership is a source of great satisfaction to us and I believe our stewardship has been in the best standards of land use of mountain property– which is to leave as much as possible to natural forces!

The Old Doug Morehouse Place

When our family first went to spend the summer in the Adirondacks near the base of Eleventh Mountain at Georgie Morehouse's, we were not aware that our association with that area would be of an indefinite period. At the moment, it has lasted more than 65 years.

Initially, all of the inhabitants were natives, that is, their ancestors were the original settlers of the area. These were the Morehouse, Hitchcock, Dalaba, Allen, Rist, Reese, Smith and Farr families. The only outsiders were the Deiseroths, a retired baker from Albany, and the Roightners, who lived across the way.

With our original connection with Georgie Morehouse, it was natural for us to know the Morehouses most intimately. Georgie had two brothers: Douglas, the older, and John, the next in line, with Georgie being the youngest. We got to know Johnnie best, since he was a delightful and warmhearted person. His brother Doug lived down the hill next to the schoolhouse. Doug's wife, Nan, was a wool spinner and for many years supplied us with home spun wool from the sheep that Doug pastured on the mountainside. After Lois and I were married, Lois would obtain all of her wool yarn from Nan that she used for making the marvelous heavy wool socks that we have used for the past fifty years.

Doug had the most substantial house of the three brothers. It was large enough to accommodate the visiting deer hunters that he guided each fall.

This two-story house existed until just a couple years ago when it burned, completely disclosing a remarkable cellar construction. It was made of huge slabs of native stone, quarried from the local mountain rock. Most of these slabs were 2-3 feet wide, 6 feet long and half a foot thick. They were laid vertically, and made a unique foundation.

These stones were obtained by a real estate operator, who purchased the plot of ground that Doug had occupied for many years. Dan Johnson had the stones removed and trucked to a parcel of land he owned, where he was building a large log cabin. My brother, Carl, employed these huge slabs to face the lower part of his cabin, and to serve as a hearth and for other purposes in the construction.

Subsequently, I purchased the land constituting the "Old Doug Morehouse Place," and Lois and I had Leland Morehouse, one of Georgie's sons, construct an open-face lean-to, where we go from time to time to enjoy the magnificent view of Crane and Eleventh Mountains. It is a beautiful location, and is adjacent to the mountain house that was once occupied by Arthur Morehouse, Doug's only son, and the Dieseroth house and lands both which are now owned by my sister Margaret. Thus, fond memories remain as we sit and contemplate the view, or wander across the abandoned lands that are rapidly becoming forested, or climb partway up Eleventh Mountain, where Nan and Doug Morehouse once pastured their sheep, cows, and horses. What rich memories abound!

The Big Rock

Among the interesting landmarks that we discovered, shortly after going to the Edwards Hill settlement in the early twenties, was the Big Rock. This is one of the largest glacial erratics in the Adirondacks. When we first found this rock, it was in an apple orchard and was sometimes used by hunters as a watch rock for the still hunting of deer. In the fall the deer would sometimes come to browse on the fallen apples in the orchard. Since that time, the old road that once passed it has disappeared, and the Rock is now surrounded by forest.

This rock is roughly a square block of syenite that was probably broken off the side of Black Mountain to the north, and carried by the glacier to its present location. It rests on top of the ground and has the rough dimensions of 30ft x 35ft x 20ft, that amounts to about 800 tons.

There are many glacial erratics "downstream" of Gore Mountain. The valley of the stream that originates at Chatiemac Lake (First Pond) is replete with large rocks. A large one, that if intact, would rival the size of the Big Rock, is located in the bed of Chatiemac Creek. However, it is broken into three or more pieces, each one of which is of impressive size.

The northern slope of Black Mountain, directly below the summit of Gore, has many large ones scattered across its north-facing slope. There is a fairly large one on the summit of Edwards Hill, that is one of the wooded mountains west of New York State Route 8, which Lois and I purchased some years ago from the Batemans.

Some of these large rocks have been quarried to produce large slabs, such as were used in constructing the cellar walls of the Doug Morehouse homestead, and the large steps of the Pentecostal Church near the crossroads in Sodom.

Since these rocks were not stratified, the slabs were made by drilling half-inch holes, three to four inches apart in a line. These were then filled with wooden dowels that were then soaked in water. When the wood swelled, it weakened the crystalline rock so that it readily split in extensive slabs.

The Aboriginal Occupation of the Adirondacks

The wild, rugged character of the Adirondacks, and the presence of many lakes and rivers around its edges, which made canoe and barge travel easy, tended to discourage penetration, exploration and settlement among the peaks and valleys of the mountains. Prehistoric occupation of the wilderness was limited to temporary campsites, located principally at the end of portages, since much of the travel in the mountains was by water. Occasionally, however, one finds evidence of prehistoric occupation in isolated spots, probably marking the present of fur trappers and deer hunters.

Such occupation was not limited to one or two cultural manifestations, but covered the gamut of those found at lowland sites, ranging from the so called Paleo-Indian, to the historic period when Iroquois, Mohican, Abenaki and others joined forces with the French, British or Americans.

The evidence of aboriginal presence consists of flint chippings, potsherds, arrow and javelin heads, and occasional spectacular pieces, such as a complete earthen vessel, a semi-lunar chopper, or a stone hatchet head. I have been told of the discovery, some years ago, of a quiver of arrows fletched with blue jay feathers, found in a cave near Piseco Lake. A beautiful Iroquois pottery vessel of prehistoric design was found in a rock shelter near the falls of the West Branch of the Sacandaga River, west of Whitehouse, and another of the earlier Algonquin type was found by a lumberjack in a cave near the junction of the Indian River with the Hudson, above the Blue Ledge. The design of this pottery vessel was similar to the sherds I found in an ancient fireplace on an island in Indian Lake, where I also found a beautiful, carved semi-lunar chopper of the type used by Eskimo women for cutting meat, and which was later made of steel and used by food preparers for chopping meat and vegetables. Before he found it, I had discovered that an ancient camping spot had occupied the sandy ridge and I had postulated that the occupants were likely to be Paleo-Indians, and I had termed the period by the name of Kame, anticipating they were following the retreat of the Continental Glacier.

Some years ago, one of my brother Carl's young sons, while playing on a sandy ridge on the edge of his Skiland property, found a semi-lunar chopper that was made by chipping. This was similar in shape to the carved one I found on the Indian Lake Island, but smaller. A small portion of it had broken, disclosing the fact that a profound chemical change had transformed the surface of the artifact from a black, to a yellowish-white. As knowledge is gained about the leaching rate of such a rock, it may be possible to estimate the age of the tool, possibly as old as 8,000 years.

Johnnie Morehouse

Johnnie Morehouse was one of the patriarchs of the Bakers Mills/Edwards Hill area. He owned a small mountain house that resembled a shack rather than a house. He was very poor in worldly goods, but was a beautiful person.

Soft spoken, with sunken, piercing eyes, he had a massive frame which, in later years, became bent over. His long arms and massive hands made scythes, axes and hay rakes behave as magnificent tools.

On the trail he carried in his shirt pocket a pancake. At lunchtime his pancake– or possibly two– dipped in a stream, served as his lunch.

We all loved Johnnie and wherever possible, employed him to help us. He dug our well, supplied us with eggs and potatoes, planted trees and, from time to time, sold us land which we didn't need, but which he didn't need either as he got older.

Johnnie originally lived in a small log cabin that, I believe, his father had built. Later, when he constructed his shack where it now stands, he moved the cabin log-by-log to a new site close to the small spring-fed stream that originates in a series of boiling springs and seeps that occur on the north side of Eleventh Mountain.

During prohibition, Johnnie had a whiskey still at that location, which produced a small amount of moonshine. I used to enjoy sitting out after dark on one of the massive boulders that dotted the hillsides and watch the nighttime sights and sounds that was the mountaineer's answer to prohibition. The bobbing lanterns, the clink of glass, and the plumes of wood fires that heated the stills, provided a mysterious aura that remains with me as a bright memory.

To my recollection, I don't remember the mountain people being bothered with "revenooers." They had bigger fish to catch along the back road routes to Canada, with the powerful and loaded cars that roared down from the north with their illegal cargoes, heading for Albany, Catskill, or New York City.

Crane Mountain

One of the finest vistas in the Adirondacks is the view of Crane Mountain from Cragorehol and other places along the Edwards Hill Road. It is southeast of that region and with its satellite, Huckleberry, dominates the horizon. Thus, despite the proximity of Eleventh Mountain and its mysterious wilderness summit, Crane continues to be the favorite climbing objective of all the young people who come to Cragorehol. It is also a favorite of the older ones, who have fond memories of past adventures along the trail and at the mountain-top lake and summit rocks. Crane is among my favorites in the mountains. It is said that the Indians called this mountain *Moos-pat-en-wach-o* meaning "Nest of the Thunderer." On some older maps the mountain's name is spelled "Crain," said to recall the name of its first white explorer who was Moses Crain, a local colonial land surveyor. He used the mountain as a key benchmark in establishing the first geodetic survey. Due to its isolation, it is indeed a key spot for locating peaks, ranging from the high peaks to the north, to

Hadley Mountain on the south, Snowy and Blue on the west, and Hoffman and Block to the east.

When we first climbed Crane in the late '20s it harbored great quantities of huckleberries. Over the years the crop has declined, due probably to the absence of fire, so that at the present time there are hardly enough for the bears.

For many years our view of Crane disclosed a sort of dimple on the western slope of its outer ridge. After puzzling about its cause for many years, I finally decided to investigate. Following the outlet of the stream of the summit lake that is the beginning of Paint Red Brook, I traveled part-way down the slope of the mountain and then bushwhacked through the dense growth that blanketed that area. After a while I encountered a jumble of huge rocks that were the result of an intense fault that had occurred in ages past. Many of the rocks forming the dimple were the size of houses, and I have little doubt that their crevices harbor ice caverns.

Continuing past the dimple I reached the base of the mountain, where I then headed in a southerly direction over a host of ridges and valleys, probably the residue of glacial moraines. Eventually I reached the old Putnam Farm where my friend Eliot Putnam lived. There I found my car.

Eliot was the first person to tell me about the Crane Mountain caves. These are located along the mountain trail where a small stream pours into a sinkhole and becomes subterranean. It then reappears some distance below in the valley to become a tributary to Mill Creek. Another side branch crosses under the old road to Johnsburg, not far from the Putnam homestead. On a low ridge along this stream is the "asbestos mine." A deposit of fibrous asbestos an inch long was worked for a short while in the hopes that it would become more extensive. As with most of the crystalline limestone in the Adirondacks, which also constitutes the Crane Mountain Cave, the mineral deposits in them are spotty and of little, if any, commercial value. This doesn't limit their variety however. One of the rarest minerals in the world– Serendibite– occurs in a similar rock that outcrops along the Johnsburg Road a few miles away. It is a boro-silicate, and is named after the island of Sri Lanka, once called Serendip, before it was called Ceylon. Incidentally, it was Horace Walpole who, in recounting a story of the "Three Princes of Serendip," coined the term "serendipity."

The poetess, Jeanne Robert Foster, a close friend of Yeats, one of the literary and artistic greats of Europe, lived for a number of her early years on the hardscrabble Putnam Farm at the base of Crane Mountain. Her previously unpublished poems and other writings have recently been edited by her friend, Noel Reidinger-Johnson, under the title of *Adirondack Portraits: A Piece of Time.* These writings give a vivid picture of the hardships, loyalties and philosophy of the early dwellers of that region.

Our Log Cabin on the Second Pond Trail

A few years after my dad bought Camp Cragorehol and we had established our presence at the Edwards Hill settlement, we became aware of Johnnie's log cabin. This was a small cabin made of square hewn logs, then occupied by Charlie Smith and family. It was located toward Eleventh Mountain from Cragorehol, on the edge of a ten-acre lot, and adjacent to the tote road leading in to the Second Pond Flow.

At that time Johnnie and his neighbors spent a week or so haying at the Second Pond Flow. This was, in our view, a near impossible chore. The Flow Trail was wide enough for a hay wagon, but peppered with boulders of all sizes, some of which made the trail nearly impassable. The sound of the wagon jouncing over the boulders and hollows on the trail could be heard for a mile or more, as it went in empty and came back with a big load of coarse swamp grass. Upon return, the hay was stored in a barn located on the fifty-acre lot, which was at the edge of state land.

While this activity seemed to be superhuman, I had the distinct impression that all of the mountaineer participants treated it as an annual picnic and, despite the hard work, enjoyed the job and looked forward to it each summer.

It was my impression that Charlie Smith wasn't involved in this activity, in fact he seemed to spend most of his time with his moonshine still, which was located in the cabin. It was said that he drunk the output of his still as it came out of the condenser. Whether true or not, he seemed to be in a constant state of drunkenness. His several boys were like wild animals in their shyness and independence. When still less than six years old, they were once accosted on their way to hunt deer. The rifle was so heavy that it took two of them to carry it!

When we learned that the current owner of the cabin had put it up for sale, my brother Paul and I resolved to buy it. Scraping together our meager resources, we managed to raise the purchase price and soon had title to it.

After the Smiths left, along with their still, we had the job of cleanup. We found it necessary to use a hoe to scrape out the organic residue that had accumulated on the cabin floor! Once cleaned and aired, we found it livable. One of our first objectives was to have a fireplace made of fieldstone. This was done by Georgie Morehouse and, after more than 60 years, it is still functional. The second project was the construction of a stone-walled, dug well. This was done for us by Johnnie Morehouse who once lived in the cabin.

It was an ambitious undertaking. He didn't bother to do any dowsing, but agreed to dig it where it would be handy to our log cabin and Cragorehol. It was located adjacent to the Second Pond Trail, between it and the cabin. After digging down about twelve feet, Johnnie encountered a big boulder in the center of the hole, too big to lift out with his tackle. There was nothing to do but get some dynamite to break it up. Johnnie was equal to the task and, while I didn't ask at the time, I presume he knew how to handle and use this explosive from his lumberjack days, when he was the timber boss.

He prepared the charge, and "shot" it. With a loud explosion the rock disintegrated, a large chunk arching out and hitting the side of a favorite balsam tree we had planted near the door of the cabin. The impact tore a considerable chunk of bark from the tree. Since Johnnie knew that we favored the tree, he went over into the pasture nearby, got a big cow flop recently deposited, prepared a poultice covering the wound on the tree, and bound it up with strips of cloth.

Within a year a healthy cambium started spreading across the damaged portion of the tree, and within a few years had healed, leaving a blaze-like scar. The tree grew for many years but eventually became so big we had to remove it.

When we first purchased the cabin property, it was a treeless expanse of ten acres that had been used as a hay field. As with many fields in the area, it was strewn with a large number of glacial boulders, so that most of the harvesting was accomplished by hand scythe.

A few years after acquiring the property, we decided to reforest the land with Norway Spruce, and Red and Scotch Pine. The New York State Conservation Department at the time was actively promoting a campaign to reforest private lands. My brother and I had visions of developing a productive forest and fell for the program. We bought a total of 10,000 trees, that were planted at the rate of 1,000 per acre. We asked Johnnie Morehouse to plant most of them. This was accomplished by sinking a mattock to its full depth, giving it a twist to extract it from the soil and, at the same time, pulling the earth away from the hole. A follower then inserted the seedling in the hole and then, with a heavy boot, kicking the lifted soil so as to fill in the hole with dirt. The seedlings were about three years old. When Johnnie first saw them he exclaimed with wonder in his voice, "Them trees are not even borned yet!" He could not understand why anyone would want to see the land revert back to forest. For many years his ancestors had fought to clear the land and to root out the trees. Here was a force tending to undo the patient work of generations of mountain folk!

We have since learned to rue the day when we planted our forest. In fifty years our action did indeed produce an evergreen forest which eliminated the magnificent view of Eleventh Mountain we once enjoyed, and which threatens the view of Crane Mountain as seen from Cragorehol. The cabin is now surrounded by forest and I am afraid its view is gone forever.

It is true of course that volunteer trees would eventually do the same thing that our planted forest had done. None of our present generation would have the time or patience to harvest the hay which once kept the forest out of the fields.

Within the past year we have harvested some of the very large spruce trees that grew up in the past sixty years. A hundred or more were cut to build a very large log cabin on a hillside west of the Thurman-Johnsburg Road. Butt logs fifteen or more inches in diameter, and more than thirty feet long, were cut in the spring, the bark peeled, and then transported to the cabin site. Roof timbers of red pine from another part of our plantation were cut and used to serve as roof rafters for the same cabin. Thus we have eventually achieved some of the fruits of our earlier dreams.

Ancient Windows of the Earth

When I first explored the Adirondacks, I was greatly intrigued with the wide variety and beauty of the gravel, cobblestones and boulders that were visible in all of our streams.

Many years later, I was able to obtain equipment that permitted me to polish stones by tumbling, and then cutting thin slices from the cobblestones found in streams and gravel banks. These slices disclosed the beauty of the crystalline structures they possessed.

Since, over the years, I had learned quite a bit about the sedimentary, metamorphic and igneous deposits of the state, I began to collect samples from the outcrops and residues of stream and glacial action. Within a few years I had collected rock samples from more than 200 locations in the state, mostly from the Adirondacks.

I also collected from distant places, ranging from the New England states to the southwest. After a few years of collecting I was led to the conclusion that, in the Adirondacks, one can find some of the finest and most interesting rocks in the world. With the wide variety of strata, the vast spread of age, and the complete set of geological forces that formed these mountains, it is indeed a happy hunting ground.

After a few years I was able to develop techniques for cutting rocks so as to obtain intact slices as large as 7" x 12" x 0.040" thick. Many of these rocks are translucent, so they can be used for making decorative objects such as lampshades, plaques, and dingle-dangles. The latter, when suspended and placed in a sunny window, are very attractive as they respond and turn in the slightest air movement.

In 1983, I fabricated a six-foot diameter church window, which I installed above the altar in St. James Church at North Creek in the Adirondacks.

One of the most beautiful of the Adirondack rocks I have sliced is the garnet gneiss of Ruby Mountain found at Barton Garnet Mines. The window in St. James Church contains many of these sliced garnets, since the mines are on top of the mountain directly above the church.

While I have collected from the faces of cliffs and other places where the native rocks are exposed, the finest and greatest variety have come from the stream beds and gravel banks where the rocks have been left by the Continental Glacier. Such cobblestones, gravels and boulders have been subjected for more than 10,000 years to the action of glacial flow and the weathering to which these residues have been exposed since being left by the glacier. Such rocks rarely have cracks or other imperfections, and thus I rarely have any difficulty with their stability when cut.

By studying the color, crystalline nature and general appearance of my slices, I have sought to trace back to the original rock formations the trajectory of the great glacier that transported the rock to where I found it. Occasionally this is possible, especially when the rock possesses very distinctive qualities. For example, the pinkish, yellowish porphyry, extruded as a lava flow from a sill along the western shore of Lake Champlain, is identical in appearance to two boulders I have found recently, one at a gravel bank adjacent to Rotterdam Junction along the Mohawk River above Schenectady, and the other along the upper waters of the Chaugtanoonda Creek at the head of Wolf Hollow in Glenville Hills, north of the Mohawk. These boulders have been carried south-southeast by the lobe of the glacier that buried the Champlain Valley and then spread across the great plain south of Glens Falls, to then be carried by the lobe which ascended the Mohawk Valley. In fact, there may have been two streams of glacial ice responsible for the deposition of these boulders. Whatever the mechanism, it is likely that the boulders originated from this one distinctive location.

This is typical of the fascinating puzzles that emerge as my collecting expeditions continue. They combine to make a life filled with intellectual fun.

St. James Church window created by Vince Schaefer from thin-cut crystals and rocks. Courtesy of Jim Schaefer.

My Adirondack Connection

My connection with the Adirondacks started seventy-five years ago. At the age of six my dad took us into the mountains between Warrensburg and Chestertown, where we stayed at the Wilsey's farmhouse on the shore of Tripp Lake. We went to Warrensburg by trolley car from Schenectady, and finished the trip by horse and buggy. It was on the small, cold stream entering the lake where I caught my first brook trout– an experience still vivid in my mind.

Old Mr. Wilsey was an impressive man with a full white beard and a hard manner. Dad took us by rowboat to various parts of the lake. I remember the dense masses of pond lilies, the dragonflies and the general beauty of the surroundings.

We apparently stayed at Wilsey's for some time as my father commuted on weekends. On one of his weekly trips he had a fractious horse that decided to disobey orders at a particularly dangerous part of the road, and wound up alongside a precipice that marked what was called "the Devils Kitchen." The horse, for some reason, began to back up until the rear wheels were just about to go over the cliff! In some manner, Dad was able to get control of the horse that then went forward and, without further event, took him to his destination.

That summer in 1910 at Tripp Lake represented the first of many excursions into the Adirondacks which I have taken, lasting from a single day to many weeks, during all seasons of the year. Each visit added to the fund of my knowledge about these fascinating wild mountains. Rarely do I go there without learning something new about them.

I have discovered that unlike most mountains in the extensive parts of the world where I have traveled, they are friendly mountains. One doesn't need to mount an expedition to penetrate them, but in less than two hours from my home in Schenectady, I can walk into and be surrounded by trackless wilderness.

It was nearly ten years later before I received the vivid memories about the mountains that still exist in my mind. Meanwhile, my mother suffered each summer from severe attacks of hay fever induced by ragweed pollen. This allergic condition worsened in subsequent years into asthma and then tuberculosis. This latter disease required her to spend an extended period at a sanitorium near Saranac Lake, at Ray Brook.

In 1924, her brother, Frank, suggested that she might avoid her summer troubles by spending the several months when the ragweed pollen was airborne at a place where he went deer hunting, which was free of ragweed. Accordingly, in August 1924 Uncle Frank transported us in his Ford to a mountainside west of Bakers Mills, where he had arranged for us to rent a room in a mountain farmhouse owned by Georgie Morehouse. There she stayed until frost had killed the ragweed plant. It is unclear to me how we managed school attendance. I think my sister and two brothers attended the local one-room log school house, while I was at home with my father.

In 1924, a small mountain house nearby became available for a very modest price and Father bought it. This began our love affair with our mountain camp, subsequently called Camp Cragorehol. Shortly after acquiring the property my father announced at breakfast that he had thought up a name for the camp: CRAGOREHOL. It derived from portions of the names of the four mountains surrounding the camp: CRA for Crane, GOR for Gore, E for Eleventh, and HOL for Height of Land. It was such a perfect name that no one has ever thought of changing it. When father died in 1959 at the age of 91, he willed the camp to mother. When she died in 1961 at the age of 87, she conveyed the property to my sister Gertrude and her husband. They in turn have recently (1987) conveyed it jointly to her six children. Located at an elevation of 1800 feet at the northern base of Eleventh

Mountain, it has a fine view of Crane Mountain to the southeast, though the trees have grown since we first went there into a veritable forest that is so high as to nearly block the view of the mountains.

A small stream runs past the camp and runs into a larger one having brook trout, which skirts the edge of the mountain and bounds the cabin property where my brother Paul and I acquired a log cabin and ten acres of land about 1928. Johnnie Morehouse dug us a well near the cabin not long after we bought it. This has supplied Cragorehol with cold, pure water up to the present time.

Ever since Dad acquired Cragorehol it has served as the gathering point for the Schaefer and Fogarty "clan." On a weekend during much of the summer, family and friends gather for a few hours of pleasant talk, while the children of the large Fogarty (Gertrude and Ed) family spend a week or two enjoying the mountains, its clouds, breezes, storms, sunshine and nighttime skies. Thus its hospitality was legendary when my mother and father were hosts, and so it continues now, and will exist into the generations ahead.

Cragorehol is located close to the edge of the Siamese Ponds Wilderness, a region of mountains, lakes and streams, much of it without trails.

Eleventh Mountain, that slopes up from near the Camp, is one of the most primitive areas in the Adirondacks. With an elevation above camp of nearly 2,000 feet, its summit area covers many square miles. This has never been lumbered. It has eight rounded peaks interspersed by dense balsam and spruce thickets, and with swampy regions in between. Its western slopes drain into Diamond Brook, a wild, almost un-fishable spring-fed stream which empties into the East Branch of the Sacandaga River. The lower slopes of the mountain, on its western side, harbors virgin deciduous trees– birch, elm, ash and maple, along with lesser trees. It is a favorite hunting ground for deer and black bear, and brother Paul has led a hunting party there each year for more than fifty years.

The Bugs of the Adirondacks

Many years ago my brother Paul said, "Some day we should erect a monument to the black fly." Without any serious competition, it has probably done more to discourage the exploiters and developers from running rough-shod over all the beauty spots in the Adirondacks, than any other force. A "city person" has only to experience the plague of the black fly hordes between May and the first of July (dependent on the particular year and its preceding fall and winter; an hour or so spent on a trout stream, near a beaver pond, or on a lake shore on a quiet day during that period, and that individual may never come back! One must really experience the trauma to properly appreciate it!

The last time I had that experience, I went fishing on the stream that runs out of Terrel Pond near Blue Mountain Lake. The day was one with beautiful blue sky, but the air was calm and the time was about Memorial Day, at the end of May. Such a dense cloud of black flies hovered around my head that I could hardly see. They flew into my eyes, my ears, my mouth and all over the rest of my face. Although the trout were rising and everything else was perfect, the flies were so bad that I abandoned the stream and headed out for home. Although I had built up an immunity to an allergic reaction during former years, the extreme annoyance of these creatures was more than I could stand!

The larvae of the black fly lives in highly oxygenated water, often at the lip of waterfalls, where they attach themselves in fantastic numbers, looking much like underwater moss. Fortunately, their life cycle is relatively short and seasonal, extending as it does from mid May to about early July. As they taper off, the mosquitoes, no-see-ums, and deer flies build up, but never to the concentration of the black flies. The no-see-ums (or midges) can be pesky. They frequent beaver dam areas and can make a night at camp miserable, since they are so small they can easily pass through mosquito netting. Their bite is as though a needle has pierced the skin, and their favorite dining area is the scalp and neck of a human. The black fly, on the other hand, operates so that when they bite, one often doesn't know it until a drop or two of blood oozes out of the place where the fly has chiseled out a chunk of flesh! With some persons a serious allergic reaction occurs, and the bitten place swells in a grotesque manner. "Old timers" are rarely affected at all.

The deer flies and mosquitoes are another matter. They seem to have a faculty of zeroing in on the arms and head, and do so in a determined and persistent pattern.

As recounted in an earlier story, I once was camping at Woodchuck Temple, one of John Apperson's isolated camps, located on the wild, western slope of Tongue Mountain on Lake George. Albert Getz was with me. As night approached, Al asked me where I planned to sleep. I told him I expected to sleep under the stars, since it was a beautiful cloudless sky. He warned me about mosquitoes and rattlesnakes, both of which inhabited the region.

Al went into the "Temple," telling me that there was some bug spray there, and he was going to make doubly sure of having a peaceful night by spraying the air before turning in.

As I was peacefully stretched out in my sleeping bag, I heard an audible humming noise from the camp, whereupon Al came bursting out of the camp with word that all the bug spray seemed to do was to wake up thousands of hungry mosquitoes which were inside the building. Consequently, Al changed his plans and spent the night in his bag near me.

The deer fly is, in my opinion, the most vicious of all of our Adirondack voracious insects. It has the apparent ability of biting the instant it lands. It, too, seems to have a tiny, sharp chisel and uses it efficiently! It is rather sluggish in its getaway tactics however, and an alert victim can generally deal with them effectively.

Except for the black fly infestation in late spring, the Adirondack region is relative free of insect problems. We don't have chiggers or ticks or land leeches, and for most of the summer and fall the other creatures are in relatively low numbers, or are completely absent.

Chapter 12
Mountain Tales from Moonshine Hill

A Deer with Fortitude

One of the fascinating things about the natives of the North Country (the Adirondacks) is the droll humor that is part of their heritage. That the art of storytelling is still extant is tested by knowing Leland Morehouse. Leland is the son of George Morehouse who was a brother (I believe) of Doug and John Morehouse. These were grizzled patriarchs when I first knew them as guides.

"Georgie" passed his last years at Doug Morehouse's home, which was located close to the Edwards Hill Road below the second schoolhouse (the first was of logs), and was just above Johnnie's Pond, now filled in with wild growth. (Lois and I have recently purchased the Doug Morehouse property– his old house burned several years ago.)

A typical mountain story told by Arthur (son of Doug) and recounted by Leland tells of the unusual marksmanship of Arthur with his deer rifle and the amazing fortitude of a deer that survived his deadly art.

It seems that the upper part of Doug's land that consisted of the lower one-third of the slope of Eleventh Mountain and which, at that time, was a rough pasture. There was a rock ledge where deer could sometimes be seen. One day Arthur saw a big deer on top of the ledge. It had its head down, feeding with its rear end facing outward. Arthur (who lived then where my sister, Peggy, now has her camp) ran into his house and emerged with his heavy deer rifle. Taking careful aim, he shot at the deer. The deer, according to Leland's account, gave a little jerk but didn't move away. Arthur shot again, again, again, again, again, and again. Except for a momentary jerk following each of the seven shots, the deer remained upright and apparently unscathed. While the distance was great, Arthur prided himself on his marksmanship and was greatly puzzled with the deer's behavior, especially since it showed a reaction with each shot before the sound could have arrived.

Accordingly, he went up to the ledge to see if he could solve the mystery.

When he arrived on top of the ledge he had the answer to the puzzle. He had hit the deer symmetrically with all of his shots! In a neat pile where the deer had spit them out were all seven bullets!

A Remarkable Bear

Ivan Bateman is a native, (a "native" in my definition is a person who is born in the mountains and is the descendant of at least several generations of early settlers, such as the

Morehouses, the Hitchcocks, the Dunkleys, the Batemans, the Harringtons, etc.), a natural carpenter who is also an artist with a chain saw. Earlier it would be an axe or an adze or broad axe.

My brother, Paul, the builder, when constructing the Main Lodge of the Inn on Gore Mountain, employed Bateman in assembling the massive frame of the Inn, which included the timbers of an ancient Dutch barn which Paul found in a tumbled down condition in the Little Nose country along the Mohawk. Ivan Bateman had two helpers, also natives. Ivan decided to have some fun with the two young fellows, so he made up a story about a black bear which had been trapped in the vicinity of Indian Lake (across the mountains from North Creek, near where Paul was putting up the building). According to Ivan's story, as he developed it, this bear showed a great deal of intelligence and in fact, after being taught to dance (as some old gypsy bears did), was able to learn several other tricks and, according to Ivan, was developing the ability to pronounce a few simple words!

The young fellows were much intrigued, found out from Ivan the location of the tavern where the bear held forth, and announced that they would go over to Indian Lake the next weekend, even if they had to walk!

They journeyed to Indian Lake, having considerable trouble getting there. Upon reaching the tavern, they learned to their dismay that there was in fact no bear, and they had been "taken for a ride!" With this disappointing discovery, they resolved to get even.

Monday morning when they reported for work, Ivan asked them if they had seen the bear. They responded enthusiastically to his query, thanking him profusely for the tip and saying they had never seen or heard such an unusual phenomenon. They described the intricate dance put on by the bear, the clever juggling tricks that he performed, and they told him that not only was it now speaking words, but was able to repeat full sentences. But– most remarkable of all– the bear was now singing several songs! Ivan was baffled by these stories and that very evening went over to Indian Lake to see that remarkable bear!

I Wonder If This Will Work?

Paul tells a fascinating story about Doug Morehouse. He had been up hunting or fishing (I can't remember which) using Frank's ancient Wills-Knight. This was a car that never was in the class of the Ford or Chevrolet, so that not many were sold.

Paul and Frank reached Camp Cragorehol without any particular trouble, except for a few flat tires. They had a very successful weekend, staying on longer than expected so that it was late Sunday afternoon when they headed back toward Schenectady. Both of them had jobs and it was essential that they be at work on Monday morning. I should mention that Frank, although not employed as an auto mechanic, had mastered the trade and in fact he had to be good to keep the old Wills operating. They had just started down Edwards Hill Road approaching Doug Morehouse's place, when a fearful grinding noise arose from somewhere underneath the car, and it stopped. Frank looked at Paul, shook his head and said, "I'm afraid we've lost the rear end."

Taking out his large tool kit and, after putting some big stones on the downhill side of the wheels, he proceeded to take off the housing– a messy job at best, since it was bathed in heavy grease.

In a relatively short time he had the housing off and confirmed his diagnosis that the drive gear had virtually disappeared! While he diagnosed the problem, he told Paul that they were in serious trouble. Not only was the chewed up gear a special type, but only a few cars were made using that particular size, since it had been proven to be a poor design and had been replaced with a different unit within a short time.

As the two were discussing their plight, Doug Morehouse came over to find out what the problem was. When shown the damaged gear, his face took on a thoughtful appearance and then he said, "Wait for me." Paul and Frank were puzzled by his words and continued to discuss the serious problem they faced. They couldn't see anything they could do which could get them home sooner than in several days!

After about fifteen minutes or so, Doug came back carrying something in his hand. Reaching the car, he said, "I wonder if this will work?" and handed a gear to Frank. It was a duplicate of the damaged gear and, although rusty, was in excellent condition!

He then told them that some years before, while scything grass on the side of Eleventh Mountain, he saw the gear on the ground– didn't know how it got there– but as is the nature of these mountain folks, he put it on top of a big rock so it could be easily found if needed! With renewed hope Frank accepted the gift and in a relatively short time had replaced the damaged gear assembly and put the cover back on.

Removing the chunking stones, they got the car started and headed home!

The Best Fences

We were talking about fences. These are needed in the North Country so that the cattle they are supposed to confine will not wander into the woods or into the neighbors' gardens, necessitating a widespread search or tense feelings.

One of the standard procedures followed by the early settlers was to use the trees in the neighborhood of the land boundary instead of fence posts. This practice posed later problems when the land was surveyed before transfer of property, especially if the surveyor is a city boy and is not familiar with mountain ways. I have seen more than one survey (and in fact have a couple) which, instead of following the description in the deed such as "North 30 degrees, West 25 chains 10 links," assumes that the line is delineated by the trees used as fence supports. Thus, the surveyed line becomes a strange zig-zag, suggesting that the earlier surveyor had been sampling some of the potent moonshine distributed locally.

As we talked about fences, the talk drifted to the quality and durability of the various types of wood available for making posts when the fence went across a treeless meadow. It was at that point that Arthur Morehouse talked about the merits of black locust posts. Although the black locust is a rare tree in the Adirondacks, Arthur had managed to find some at one time and was familiar with its hardness and other qualities. It was this point that he said the some that he had used had worn out three holes!

Buck Fever

We were deer hunting in the valley of Diamond Creek, including the western slope of Eleventh Mountain. This was in the early days of our hunting activity when we stayed at Camp Cragorehol, before my brother, Paul, formed the Cataract Club, that for many years went into the woods to set up tents for sleeping and cooking, while hunting the slopes of Eleventh and Diamond Mountains, and the Hardwood Ridge.

I had invited several of my shopmates from the Machine Shop of the Research Laboratory, including Art Parr, to join our hunting party.

Art was new to deer hunting, but he had hunted smaller game for some time. I told him before we went that buck fever was a common problem with many hunters, even at times with those having considerable experience with big game hunting. Art was inclined to believe I was "pulling his leg," and assured me that he didn't expect any trouble.

We hunted intensively without any luck, not even seeing a "flag."

I might point out that there is a strange chemistry that assails the hunter out after a deer. After several frustrating days, the mind seems to make the deer an adversary, and the hunter can't wait to see a legal deer in his gun sight.

We had lined up a drive on our final day of fruitless hunting. About half of us were lined up across the col between Eleventh and Hardwood Ridge. My brother, Paul, and several others had similarly placed themselves across the valley about a half-mile away and started "barking."

Nothing happened for a while, and as the barkers began to get near, we on the line felt the adrenalin start flowing as the "moment of truth" approached. We heard a crashing of leaves and twigs as something approached us down the slope, definitely ahead of the drivers. Suddenly, the line erupted in a crescendo of gunfire. I couldn't see anything, but someone between me and Art Parr began shooting and it sounded like the creature had angled down the slope, which would take it close to Art's station. Despite this occurrence, Art was strangely silent, which puzzled me.

When the drivers arrived and we gathered together at the col we found much excitement. A big buck had come off the mountain and was seen by all of the fellows below me, including Art. There was no evidence however, despite the cannonading, that anyone had hit the deer.

When we asked Art why he hadn't shot his gun, he said he had. However, none of us had heard his gun in operation. Since we had gathered close to his station, one of the crowd reached down and picked up a bullet that had not been fired. In short order we found about a half dozen in the same condition. We then were able to understand the lack of noise from his station. Apparently, as each gunshot occurred, Art went through the motion of expelling a spent shell without ever pulling the trigger! He kept doing this until his clip was exhausted! Only after seeing the evidence of his action did he realize that he had experienced a classic example of buck fever!

Sir John Johnson Trail

There are many legends about ancient trails and roads that cross the Adirondacks. Some have been verified by the discovery of abandoned cannon and other gear along backcountry trails, while others are just legends without much evidence of reality.

One of the most amusing was the story of a military road north of Speculator, being built before the War of 1812, and which was continued for a year past the end of the war because the working group was so isolated and insulated from news that they didn't know that the war had ended.

Perhaps the most intriguing of these routes is the one that seems to be unknown to everyone, including historians. It was laid out by Sir John Johnson as the escape route he planned to use after making the devastating raids that he and his regulars, scouts, Tories, and Indians, numbering about a thousand, conducted in 1780, in the vicinity of the Mohawk and Schoharie Valleys.

My friends Dan O'Neill and Austin Hogan had been searching out the great Warriors Trail that passed the prehistoric Stone Heap near Sloanesville, and joined the Schoharie with the Mohawk near Fort Hunter. In searching the historic record, Austin discovered a map among the published papers of Sir William Johnson. It showed a route farther east than any of those that had been accepted as the cross mountain trails used in the French and Indian Wars, the Revolution and the War of 1812.

In studying this map, I was intrigued to see that it went close to Crain (Crane) Mountain, one of my favorite places and one whose countryside I knew fairly well.

It so happened that in 1984 I attended a celebration at North Creek, also attended by most of the leaders of the local community. I approached the President of the Johnsburg Historical Society and queried him about the Sir John Johnson Trail. He had never heard of it!

I then approached my old friend, Chuck Severance, a retired local ranger who knew the countryside probably better than anyone around. He also had never heard of the Trail. However, a few minutes later, he approached me and told me of an interesting occurrence. He said that he remembered that about fifteen years before, when a Town of Johnsburg highway crew was straightening a short portion of the Johnsburg-Thurman Road near Garnet Lake, they encountered an ancient corduroy road that had been overgrown over a wet area. Then he mentioned that as the corduroy logs were being removed, a small cannon had been pulled up along with the logs! This immediately intrigued me, since this location was identical to the route shown on the map of the Sir John Johnson Trail!

I immediately asked Chuck who would know the whereabouts of the cannon. He told me that Sterling Goodspeed, the present Town of Johnsburg Supervisor, should know about it. When I sought to find out how I could find Sterling, he pointed across the room and said, "There he is." I immediately approached Sterling. When asked about the cannon, he said he indeed had it in his office, but WH,[56] who had found the cannon, approached him some years before, had asked for the cannon, and he gave it to him.

It so happened that WH lived for many years near us, a mile or so from our Camp Cragorehol. However, I knew that W died about five years ago. It was likely that his son had inherited the cannon but– he had died within the past year while back in the woods! Thus it was possible that his son, F, might have been given the cannon. When I discovered that F was the one who had just built a new house opposite WH's road, I decided immediately to talk to F.

The following day I stopped at F's home. His wife answered my rap at the door. I asked her if by any chance he had the cannon found along the old road. She answered in the affirmative, but said that F was not home, but would return in a couple hours.

I said I would return. When I did so and met F, he told me he had the cannon, would show it to me and permit me to photograph it. It was a little cannon, eighteen inches long and of the type called a swivel gun. It was small enough to be carried in a pack basket, weighed about thirty-five pounds. The trunions had been demolished– a practice followed by an expeditionary force if they decided to abandon a piece of artillery.

All attempts thus far have failed to learn more about this cannon. I have contacted the Hudson Bay Co., the British Museum, several museums in Canada, Fort Ticonderoga, and our local Watervliet Arsenal Museum. The latter were the most helpful, but the only information I have received is that the cannon was called a swivel gun, it was normally used on small boats, and being cast iron was probably made by a local gunsmith. The cannon is in excellent condition, without a trace of markings on it.

Seeking traces of the Sir John Johnson Trail in the Adirondacks, I uncovered what seems to be an important clue. In studying the Claude Lanthier Map drawn in 1779, I found an area where the trail could have followed the Sacandaga River and crossed it, to ascend the valley of Stoney Creek above Hope Falls. At a river crossing just north of several islands in the river, I found a single small island labeled Canada Island. It is the only one named. Previous exploration had provided me with the information that this was the end of the west road up the river. At this point it was necessary to either retrace the road all the way back to Northville, or ford the river.

I had seen evidence of a fording place during my earlier explorations, and on a copy of the first Geological Survey Map of the Stoney Creek Quadrangle, there was a river ford marked on the map at this location. On the east side of the river, as a continuation of the ford, a small road went back into the Stoney Creek Valley to the tiny settlement of Hope Falls. From there the trail went northward to the headwaters of Stoney Creek and then to the Thurman-Johnsburg Road, which it followed to Johnsburg, then Weavertown and Riparius, where it crossed the Hudson, thence went on the west side of Schroon Lake, and

56 Name withheld on purpose.

eventually reached Bulwagga Bay on Lake Champlain, where boats could be taken to Canada.

Further research is needed, but I think we have made real progress!

A Fishing Expedition to Fish Creek Ponds

About five years ago I was invited by John LaRocca, self-proclaimed Adirondack Elf, who for years had led a trout fishing expedition to Fish Creek Ponds, west of St. Regis Ponds Country, near Saranac. Our group left the Capitol District area after supper, and stopped for the night at a camp near Irish Town, west of Olmsteadville.

Early the next morning, after a hearty breakfast, we headed for Tupper Lake and thence to Clear Pond, not far from St. Regis Mountain and Paul Smiths. There the cars were parked, the canoes unloaded, packs stowed, and we headed up Clear Pond to carry across a swampy area to St. Regis Pond. There we reembarked, and soon went down the Pond to the Long Carry. This was my first taste of packing in about thirty years. I was unsure of my abilities since I had not attempted to carry a heavy pack for a long time. All went well, however, while I didn't have the extra load of a canoe, I had volunteered to carry some of the heavier food needed for our stay of three or four days.

After several hours of sweaty travel, the sparkling waters of Fish Pond shone through the trees and a quick reconnaissance revealed that John's favorite campsite across the lake was unoccupied. In short order we embarked and soon were at our campsite. I elected to take a chance with the weather and selected a bed site about thirty feet away from the Adirondack-type lean-to which made this campsite so attractive in the event of inclement weather.

We soon had rested and took off before supper for our first attempt at the big trout that were said to live in the lake. Our first attempt was without any luck, but that didn't matter since we had many hours of fishing ahead of us.

The evening meal prepared by John measured up to the stories I had been hearing about it and the evidence was quite clear that whether we caught fish or not, we were going to have a good time.

Early the next morning, after another substantial breakfast, we all left for exploratory fishing, each on a different quest. My buddy and I went down to the lake to its outlet where, in a very short distance, the waters of Big Fish Pond flowed into Little Fish Pond. Here in the rapids the fish began to rise and, in a short time, I had hooked and landed several nice brook and rainbow trout.

Toward noon everyone returned to camp– all with fish. We then went to the edge of a rocky cliff and gathered a goodly supply of ice that still remained from the winter's deposits. There was sufficient ice to last for our stay and the trip back home.

On the second day, after fishing successfully in the morning, I decided to take my map and compass and do some exploring. I headed west of camp, climbed a rocky knoll as I worked my way down the lake shore, and then headed cross-country for a small pond marked on my map as Sky Pond. In a relatively short time I encountered this pond and circled it. There were fish rising in the middle, but the many trees which had fallen into the lake from all directions made fly casting quite impossible. A canoe would be needed to fish it successfully, and even then it would be tricky landing a trout.

Incidentally, the spring expedition of 1987 (again led by the Adirondack Elf) was a huge success. John, while fishing by canoe at Sky Pond, caught a very large brook trout (4+ pounds), and other excellent catches were made.

While I am quite sure I could still manage a big pack and the rigors of such a trip, I have had such good times in the past that I am quite content to rely on my memories!

V

Remembering Vince

Chapter 13
Vince's Colleagues Recollect

A Life Inspired by Dr. Schaefer

The first time I saw Dr. Vincent J. Schaefer's name was during my student research in a library on the subject of ice and crystal growth in 1965. I was one of the team members of Dr. Seymour Hess's group at Florida State University. We were in the process of designing a very sensitive hygrometer for Martian exploration for a NASA project.

I wrote to Dr. Schaefer for advice and for suggestions on our task. He wrote back and mentioned a new department of atmospheric sciences was just formed at the State University of New York at Albany, and was a newly emerging scientific field.

I transferred to Albany in 1968 and Dr. Schaefer offered me a job working ten hours a week for $1.25 an hour at his newly formed Atmospheric Sciences Research Center (ASRC). I stayed there for the next thirty-three years!

On one snowy day, I was sitting with Dr. Schaefer in his basement and with a beer in hand I talked with him about the coming new science education arrangement. He pointed to the window and said to me, "Roger, see those tiny ice particles falling from the frosty gutter on the edge of the house? I have been watching this phenomenon for many years. I just could not figure out what is going on there. Why don't you look into the problem?"

Ten months later, I gave him a report showing the way that the ice crystals had been broken up and how they carried negative electric charges. He wrote a paper entitled "The production of ice crystal fragments by sublimation and electrification," and very surprisingly with my name as co-author (Dr. Schaefer had very few co-authors for his publications at ASRC). It was published as the featured cover article in the Journal de *Recherches Atmospheriques* in 1971. This paper was cited later by the *New York Times, London Times, New Scientist,* as well as Russian television and many governmental publications. This research I had conducted was later confirmed in 1991 by scientists at MIT.

This small technical report was the beginning of our ten years of research on the subject of ice crystal formation and related electrical charges generated in a thundercloud. By the order of Schaefer a laboratory to study and assist other ASRC scientists on the subject of atmospheric particulates was created and I was appointed its manager. At his suggestion I began playing with snowflakes under three different types of microscopes and was the author of five technical reports. My photographs of the snow crystals and reports were selected by the editors as the featured cover articles.

During my years with Dr. Schaefer I assisted him with his Natural Sciences Institute, one of his pet projects, and even babysat his children when he was working in the field.

His methods were clearly summed up when he said on August 15, 1960: "As long as we give young minds the privilege and the facilities to roam far and wide, and from time to time to play with impractical things such as snow crystals, they are bound to come up with some exciting new angles on the mysteries of nature."

He always said to his students– and to me as well: "Go and get your dream! Do not give up and let anyone or anything stop you! Please enjoy it and have fun!"

I did.

Roger J. Cheng[57]

Vince and Roger Cheng. Roger directed Vince's laboratory for twenty years. Courtesy of Roger Cheng.

Humor, Science and Everything

Your father's death is a loss many of us have felt deeply. I don't know if you are aware of it, but Ray Falconer isn't the only one of the old Project Cirrus gang that was hired by your father. I was another. In 1947, I was just out of the Navy and college and came to Schenectady to a job with GE. I was a charter member on the so-called physics test program that had people spend only about four months on a project, then move on to another one. This was repeated about four times until you and GE had an idea what kind of work you wanted to do. My first two jobs were routine and not very exciting for me. But the next one on Project Cirrus opened my eyes to a new and fun way of doing science. Your dad hired me on the spot. Of course, at the time, he knew he didn't have to keep me on after four months, so if I turned out to be a real dud, little was lost. (I'm sure it didn't hurt my

57 Roger Cheng has an extensive web site about his research and atmospheric science, which can be found online at *http://www.rogercheng.com.*

chances when a mutual friend told him I was a good skier!) But he could have brought on any number of others in the physics test program. I suspect he gave me a chance because I was a skier and very interested in the out-of-doors and the natural environment. Soon I had met Katy Blodgett, Bernie Vonnegut, Ray Falconer, and Irving Langmuir. At that time, I knew little about how science was done, but I read some of Vince's papers and listened to what he had to say about scientific experiments. Once I began my experiments your father gave me good advice from time to time, but he let me make my own mistakes because he knew that the fun in science was finding out things for yourself. I learned from him the essence of simplicity in designing experiments. Perhaps best of all, I learned about the role of serendipity in scientific discovery. I've always been grateful to him for this introduction to science and the fun that can be had from doing it.

I was intrigued to find that your father worked so well with Irving Langmuir. Here was a high-school drop-out working side by side with a Nobel Prize winner. Through the years, this marvelous symbiotic relationship resulted in one discovery after another. I think it was a few years later, when I was hired by another high-school dropout, Al Woodcock, of the Woods Hole Oceanographic Institution, that I realized that a person's formal education is no real measure of his ability. A natural talent and a self-learning process is the real measure. And in that, your dad took second fiddle to no one. He was unique, and it will be a long time before anyone like him passes our way again.

Not only did he excel in his own scientific work, but through his work with young people, he encouraged hundreds of others to think for themselves, open their eyes and see the fascinating things in the world around them. His can-do attitude always impressed me. At times, when I was having trouble in my own experiments, I'd think of what your father might say: "Your setup didn't work, but have you tried…?"

Whenever I think of Vince Schaefer, I think both of his sense of humor and his interest in just about everything. Often, when talking to him about an experiment, I'd recount the things that didn't work before I told him what did work. With eyes twinkling, he'd always laugh at my tales of woe. But I knew he was not laughing directly at me, but in appreciation of those clever little roadblocks that nature always throws up to protect her secrets. To Vince, a large part of the fun in science was facing nature on her own terms and finding ways around the inevitable roadblocks.

Vince loved to hear about little scientific jokes played on those he knew and he wasn't above getting involved in some himself! One day in 1948, during Project Cirrus, Vince brought a group of visitors to see my demonstration of a large drop of water freely floating in the updraft of a vertical wind tunnel. I had spent several months building the tunnel and was proud of what could be done with it. After a brief introduction, I picked up a dropper full of water, held it carefully in the center of the updraft and began to squeeze out a large drop of water. Now, I had done this successfully a hundred times before, but on this day, the water was torn into a fine mist the very instant it came out of the dropper. I tried again and again. Nothing worked! With a weak smile, I told the visitors the speed of the air stream must be too high, so I reduced it. Same result, a fine mist. At this point, with tiny beads of sweat appearing on my brow, I spotted Vince at the back of the group with a big grin on his face, barely able to contain himself. The truth soon came out. Earlier, he had put a few drops of detergent in the beaker of water he knew I'd use for the demonstration. That so reduced the surface tension that large drops could not remain intact in the blast of air in the wind tunnel. Replacing the water with some from the tap, I got large drops to float beautifully in the air stream. The visitors' faith in science was restored, and Vince had given me a lesson not to take anything for granted.

Many of Vince's numerous accomplishments came out of his interest in just about everything. I have often wondered if these wide-ranging interests would ever have been sustained had he obtained a formal education leading to a PhD? Vince was a high-school dropout; he needed to get work to help his family. His education was obtained on his own, by extensive reading, listening to others whom he respected, observing nature, and doing simple experiments. He saw no formal boundaries, as academia does, between physics, chemistry, and biology. In his mind, all disciplines were seamlessly joined together to make the beautiful fabric of nature. Too many students today are forced to observe these artificial boundaries as they probe ever deeper at the frontiers of knowledge, and end up knowing more and more about less and less. Vince escaped all this, and I think he was the better for it. Certainly the many areas of human endeavor that caught his interest are from surface chemistry to the atmospheric sciences, from science education to the study and preservation of Dutch barns.

Duncan C. Blanchard
Albany, New York

Ray Falconer, Vincent Schaefer, Bernard Vonnegut, and Duncan Blanchard. Courtesy of Roger Cheng.

A Long-Time Friend Remembers

Not long after my arrival in Schenectady in August, 1935, I became aware that outdoor activities, hiking, climbing, canoeing, skiing, etc., were popular in the area and all of these appealed to me. It must have been in the late thirties that Vince and I became acquainted, no doubt through the Wintersports Club. I can't recall ever skiing with Vince but we surely knew each other. I was on the GE Finance Program and left Schenectady briefly, returning early in 1942 when I was assigned to do the war contract accounting for the Research Laboratory. Among my earliest recollections of that job were the "smokes" studies conducted by Langmuir and Schaefer and of their smoke generating experiments at Vroman's Nose. Then came their involvement with aircraft precipitation static problems, and so on to Mount Washington and Project Cirrus. I have just finished re-reading my copy of Barry Havens' 1952 history of the Project. My interest in these activities went well beyond the accounting responsibilities of my job as I had become thoroughly "hooked" on the wonderful and challenging skiing on the Tuckerman Ravine headwall and the Gulf of Slides. Vince gave me, long ago, a copy of the August, 1946 Mount Washington Observatory News Bulletin. I cherish it for several reasons and especially because it contains weather statistics for the summit for every single day beginning 8/1/43 and ending 10/31/45. Fascinating stuff, if you like numbers. I had occasion to use the data quite recently and in a very good cause. My daughter's boyfriend, a very experienced and rugged hiker, was planning to do a solo ascent of Mount Washington last Christmas. I loaned him the Bulletin, he studied the data, and was dissuaded. Thank you, Vince.

Living nearby, I often drive Schermerhorn Road and sometimes would see Vince in his yard. At times, I would scoot on by with just a wave. At other times, I would stop for a chat. One such occasion, I remember vividly, as he invited me in and introduced me to Lois. I noticed and remarked about a colorful and unusual lamp shade, leaded glass, I thought. But we then went to his shop and hobby area, where he showed me that the "glass" was actually stone, cut into very thin slices so as to be translucent and thus reveal their striking colors. Vince was fond of the spiral shape of the chambered nautilus (one of the wonders of nature), and he had made one of his lovely pieces in that configuration. He had combined his artistic sense, his extensive knowledge of geology, and his superb skills with machine tools to create a unique and beautiful artifact.

Neither Vince nor I ever really retired, and we occasionally chatted about "retirement" and how, for some folks, it can be a problem to be faced and somehow dealt with. We didn't, either of us, have such a problem, and agreed that one of the most significant traits for satisfaction over a lifetime is curiosity. No one that I know of was blessed with more, or whose pursuit of the curious was more productive.

I also cherish a copy of Vince's book "A Field Guide to the Atmosphere" which he gave me and in which he inscribed the words: "To my old friend (and present) Ralph Bengtson with deep regards and best wishes."

Ralph W. Bengtson, Schenectady, New York

Slices of Good Memories

I can't remember when I met Vince, but it was over fifty years ago while we both were at GE. Our labs were not far apart and our friendship grew slowly and steadily. After a few years, he was part of a small group of similar souls who met for lunch together in my office. We solved world problems and shared our successes and failures. We were from different fields, but we came to know a little of each other's work. I'll always remember the enthusiasm when one day Vince came in with his sandwich and announced that he had just made it snow by seeding clouds. Vince's sandwiches were a source of interest. One day, they'd be tiny, and the next immense. He took delight in explaining: Lois baked the family bread in large round loaves. On the days when the slices came from the edge of the loaf, the sandwiches were small; from the center, they were very large! He said that with Lois, a slice was a slice regardless of its size!

Vince, of course, was an outdoors person. One of my vivid recollections was one night when we slithered in the cold and slime worming our way through tight passages down in Ball's Cave.

Pleasant are my memories of the cold, still winter nights when we slid on skis on the slope beside the house on Schermerhorn Road. Lots of pleasant memories. We started a drive up to Mount Washington by picking up a peck of apples, which were just in season. By the time we arrived a few hours later, we had eaten the whole peck. I can still taste them!

Vince was interested in the meaning of the little inconsistencies of Nature. We once had a long discussion on a hike in the Adirondacks about an observation that on a certain type of tree, the bark all circled the trunk making a right-handed helix. Never left-handed…

He was no scientific specialist. He ranged far and wide. He was what used to be called a "Naturalist." It was good to have known him.

Charlie Bachman
Moose, Wyoming

Snowflakes of Long Ago

I am a retired GE engineer. I was chatting with my friend, Art Gregg, in early September 1993 when he told me of Vincent's death.

During World War II, I was a field service engineer working out of Building 5 in Schenectady. From December 1943 to June 1944, I was assigned to Dr. Irving Langmuir to conduct radio interference research which required flying in a specially prepared B-17 through snow storms in Alaska, northern Canada and the north Atlantic. In this assignment, I met Vincent who was conducting snowfall experiments and I collected snowflake data for him during our B-17 experiments.

I was very happy to know Vincent Schaefer, and was able to stop and chat with him during my Research Lab assignment and after the war. I always felt welcome; he always had time to see me and his fame never changed him.

I regret that I broke contact with him when I left GE in 1947 and never resumed contact when I returned to GE in 1956 in Philadelphia.

Vince left his mark in this world and I feel proud to have known him.

Albert J. Fiumara
Wayne, Pennsylvania

Consequences—and More Consequences

My respect, admiration and fondness for Vincent Schaefer is complete.

My own contacts with Vince were always a delight and inspiring. One day he invited me in to see his freezer experiment. His breath exhaled into the freezer space immediately formed a fog of tiny water droplets. He then sprinkled it with a bit of dry ice. In the blink of an eye, the entire fog turned to ice crystals. I was amazed that the message traveled so fast throughout the fog. That was a key experiment leading to so much that followed in beneficial consequences.

Then there was that memorable summer when Vince and Irving seeded the sky in Arizona every seven days. That summer, every weekend in New York, it rained, but in Michigan, the weather was beautiful on weekends. That was the summer a Jehovah Witness congregation was building a meeting temple for their use. An older brother of mine was their leader. That fall, he told me that the Lord had controlled the weather so they could complete their Temple. He died before I ever had the opportunity to explain to him just who the Lord was for the occasion: Vincent J. Schaefer and Irving Langmuir. We don't always know how what we do affects other. Vincent's record on that is exceedingly positive and will continue so.

Herb Strong
Schenectady, New York

Rocks and a New Job

I first met Dr. Vincent Schaefer in the 1930s when I joined the Van Epps-Hartley Chapter of the New York State Archeological Association. We met and talked at these meetings and I accompanied Vince on several field trips with the Chapter.

At one of these meetings, I mentioned that I had attempted to secure employment at General Electric several times and was unable to get past the Employment Office. About a month later, I received a phone call from Vince instructing me to visit the Employment Office on a specific day. I did and there was a tremendous difference– I interviewed at three Engineering offices and was hired by Charles Hood, manager of the Equipment Development Lab, as an engineering assistant. This plant was located at the Campbell Avenue Racetrack.

For the next six years, I passed the Schaefer home every day as Schermerhorn Road was the easiest way to the plant. I would stop by occasionally to consult with Vince on chapter business.

Soon after Vince left GE, I was transferred to the Aerospace Division in Utica as a Design Engineer and was thus removed from the Schenectady area and contact with Vince.

The last time I met with Vince was when I asked him to participate in the presentation of the "Gellow" award of the New York Archeological Association to Reverend Thomas Grassman for the publication of his book, *The Mohawks and their Valley.* This was a thirty-five year research effort by Father Grassman, an effort assisted by Dr. Schaefer and me.

With Dr. Vincent Schaefer's passing, the scientific world and the archeological fraternity lost a sincere and knowledgeable contributor and I lost a friend who gave me a tremendous assist when I needed it most. This assist started me on a successful and satisfying career with General Electric. Thank you again, Vince Schaefer!

Henry Wemple
Spring Hill, Florida

Work Introductions Produce Long Friendships

I met Vince in the early 1940s when I was with American Cyanamid Corporation in New York City. I was promoting the Aerosol Wetting Agents and traveled throughout the United States in attempts to promote sales through applications.

Of course, Vince was working with Drs. Irving Langmuir and Katherine Blodgett. He showed me the work they were doing on surface films and he even gave me the horizontal bearings to make my own trough.

He and Dr. Langmuir did experiments on the foaming of Aerosol OT and Vince mentioned this in his historical biography of none sets which he loaned to me recently.

I have kept in touch with your dad from time to time and a few years ago he made a slivered stone trademark T– a photo of which I enclose. He also sent me a print of the "eternity" symbol of the six-foot diameter church window he made of the slivered stones.

He and Lois, Marian and I exchanged Christmas cards for years. Vince was a very fine Christian of high principles and great abilities. He contributed much to our knowledge in many areas. Marian and I lost real friends in Lois and Vince but their souls remain with us in beautiful memories. The Prayer of St. Francis is so fitting.

Clyde A. Sluhan
Master Chemical Corporation
Perrysburg, Ohio

Memories from GE and SUNYA Days

During my 47 years of teaching physics at the State University of New York at Albany, I spent my Wednesdays at the GE Research Laboratory, so I knew Vince in two of his bases. He had bases in many places in the United States (and the world) and kept his associations with each of them.

One of my clear mental snapshots was of an evening at his home with SUNYA students, having a discussion with your father and a demonstration of cloud seeding in the deep freeze in his home.

When a chapter of Sigma Pi Sigma Physics Honorary Society was installed at SUNYA in 1960, the students requested that Vince Schaefer be the keynote speaker. I recall one of his illustrations, where an airplane was following the sun's rays straight at the earth. In the middle of the round shadow of the fuselage was a bright spot which the flyers called a "glory." That bright spot looked so much like the Arago Bright Spot in the middle of the shadow of a penny that I concluded it must be that same diffraction effect. The Arago Bright Spot was the clincher in proving that light is wave motion.

Incidentally, the first president of Sigma Pi Sigma, Mitzi Pringle, worked under Vince's direction in finding projection methods of demonstrating Dr. Katherine Blodgett's step gauge to a classroom. At the Sigma Pi Sigma 30th Reunion in 1990, Vince was invited to speak again.

Vince and Lois live in the thoughts and actions of a host of people, including ourselves.

Kathlyn and Luther Andrews
Orono, Maine

Friends from Switzerland Recall

Our relation to Vincent began in early 1949, when we were in Canada working for the National Research Council. An International Snow Classification was discussed at this time, and it became desirable to bring together scientists who had worked in this field. So a working group was invited by the International Commission on Snow and Ice to gather in Ottawa. Vincent was one of the members besides the Japanese U. Nakaya, the Canadian, G. Klein, and me.

Since then, we have seen Vince many times, conferences, in his Schenectady home and during a stay with a common round trip in Switzerland. We also were received by dear Lois with an unforgettable hospitality (including trips to the Adirondacks). Rita has tried to find out the roots of the family Perret (Lois' father's family). Now all these things are coming back to our memory. We really admired and liked Vince and Lois very much.

Marcel and Rita de Quervain
Davos Dorf, Switzerland

A Bit of Ancient History

My association with Vince Schaefer dates back to the early '50s, when I was teaching at Colorado State University. My wife and I owned several dry-land farms in northeastern Colorado, where we and our renters had been conducting experiments in cooperation with the Soil Conservation Service.

A few years earlier, Vince, laboring in the vineyard of the General Electric laboratories, had discovered that if you introduced silver iodide crystals into the air in a deep-freeze, it would produce a miniature snowstorm. A well-known Cal Tech meteorologist who shall be nameless, had pounced on the discovery and organized a company to exploit it commercially by "seeding" clouds to increase the rainfall in arid regions of the western states. If he succeeded, he would be one of the most powerful men in the world.

In the fall of 1951, his emissary, a highly persuasive gentleman who enjoyed the title of Colonel, turned up in Northeastern Colorado with enthusiastic reports of the company's first two projects, in New Mexico and western Washington. He was irresistibly convincing, with his arsenal of newspaper clippings, data sheets and charts. And, after all, his employer was a Cal Tech PhD with an international reputation. What better credentials could one ask?

It didn't take me long to climb on the bandwagon. If we could increase the average annual rainfall by even two inches, what a boon it would be to the farmers of Northeastern Colorado! Our land should triple in value! Despite the Colonel's air of certainty and the beguiling reports from New Mexico and Washington, I felt that the project was a speculation, but it would cost each landowner only a few dollars to give it a try for a year.

I organized the Northeastern Colorado Weather Improvement Association, covering eight counties, raised enough money to engage Dr. So-and-So's services for a season, and helped three other associations organize and raise money. As I remember it, that made eleven altogether, in Kansas, Nebraska, South Dakota, New Mexico, Colorado, Wyoming and Washington. If they all succeeded, we would revolutionize life in the American West!

But something warned me that things were moving too fast. Was Dr. So-and-So the miracle-worker the Colonel claimed him to be? We needed a way to coordinate the projects and verify the results for ourselves.

I organized the National Weather Improvement Association with a Logan County, Colorado rancher, Harvey Harris, as president, and myself as secretary, and began putting out a newsletter.

Meanwhile, I had written to Vince, expressing my concerns. He answered noncommittally, which I could understand. Shortly after, he turned up in Fort Collins. Very adroitly, without being specific in any way, he distanced himself from Dr. So-and-So's exploitation of his discovery.

"The Great Plains isn't a deep-freeze, you know," he said, with a quizzical little smile. "It's a long leap from one to the other. Some people don't understand that."

I left our meeting with a rather uneasy feeling. I was still more uneasy when I found that Dr. So-and-So's reports of results didn't always jibe with the reports I was getting from our regional associations. It didn't help that he had never really warmed up to the National Association or to me. Quite the contrary.

One day, he summoned me to a lunch date at the Brown Palace Hotel in Denver. He came right to the point.

"Go ahead and have your little association," he said. "It's immaterial to me. But if you ever do anything to hinder my operations, I'll break you in two like a matchstick."

Through the summer, my reports from the regionals established a general pattern of unrealized expectations. Even New Mexico and Washington were falling far short of the first year's remarkable record. Was it just an incredibly lucky natural aberration? But every substantial rain, wherever it occurred, was seized upon by Dr. So-and-So's publicity office as a demonstration of the effectiveness of his silver iodide generators. If there was a bad storm, or hail, the generators had not been operating that day.

I knew what I had to do.

First, I told the whole story to Vince.

Then I composed a letter to the secretaries of all the regional associations, quoting from Dr. So-and-So's uncorroborated reports and his ultimatum to me, resigning as secretary of the National Association, and expressing my profound regret at having become involved in what was obviously a premature and ill-advised promotion by an individual whom I believed to have become seduced by greed and an insatiable appetite for power. I have no doubt that he was confident of success at the start, but…

That was the end of the National Weather Improvement Association. To my knowledge, not one of the regionals reported any discernible benefit from that summer's cloud seeding operations, although the Colonel enthusiastically termed them a great success because the company had made such wonderful improvement in its methods of operation. I don't know of any that renewed their contracts. I understand that the company transferred its endeavors to Africa, Asia and South America, where it harvested new clients for a few years before exhausting the supply and going out of business.

It would appear that cloud seeding isn't practical over large stretches of comparatively level land, at least not in its present stage of development. Another attempt was made in Nebraska a few summers ago, without success. However, I believe there has been at least one highly successful project in operation for many years on the mountain watershed of a California hydro-electric utility. There may be others that I don't know about.

At any rate, my involvement in the Great Plains non-rain-making binge netted me a wonderfully interesting and rewarding friendship that endured for over forty years. It was conducted largely by correspondence, fueled by a broad community of shared interests, perceptions and ideals. Vince's love of life, his concern for the welfare of the planet and its dwellers, his direct, uncluttered thought processes that got right to the heart of things, his generous appreciation for the accomplishments of others, his spontaneous sense of humor, his uncompromising integrity– all those positive personal qualities that made up the man – were pure delight to me, as they were to my son, Steve, who came to know him through their common interest in environmental concerns. His occasional visits to our home on trips to the West were memorable bright spots for Alice and me.

I was truly saddened by the news of your father's passing. He was a remarkable man, an uncommonly perceptive and appreciative friend and a truly gifted scientist. Many scientists become so engrossed in pursuit of their work that they never learn how to live. Vince lived several lifetimes in his 87 years.

Nothing escaped him. Vince was one of those rare individuals who seize every opportunity to enrich their lives. The world was a treasure trove of thrilling secrets to be uncovered, mysteries to be solved, exciting adventures to be experienced, lovely sights to be seen, rewarding friendships to be cultivated and devoted family to be loved. He enriched the lives of all who knew him, while enriching his own. Even the ordinary, routine concerns of everyday living were a rare privilege to be treasured and enjoyed. He left a legacy that will be an inspiration to all of us as long as we draw breath.

Jim Wilson
Wide Skies Farms
Polk, Nebraska

Vincent Schaefer and the ASRC

The American public was profoundly alarmed in the late 1950s when we learned that the Soviet Union had launched Sputnik I. While it was only the size of a grapefruit, we had put nothing into space. As a result, there developed a widespread conviction in the United States that we needed to improve the quality of our educational systems at all levels– especially in the sciences.

At that time, the writer was the chief academic administrator at the College in Albany, which was in the process of becoming the State University of New York at Albany. It seemed to us that we should be making a special effort to challenge and encourage those students who showed extraordinary potential for a career in science. In this connection, Vincent Schaefer's name came immediately to mind. While I did not know him well, I had met him several times at meetings of the American Chemical Society, which had a large and active section in the Capital District, and of course, I knew of his pioneering work in the Research Laboratory of the General Electric Company, especially in cloud seeding.

Thus, in 1960, I wrote to him, explaining in a very general way, what we wanted to do, and asking if he would join me for lunch to give me some advice. To my great pleasure, he responded affirmatively, and we soon got together. The outcome of our meeting was that he agreed to join our faculty on a part-time basis, and among other duties, to offer a seminar for senior undergraduates dealing with special problems in science. He joined the faculty of the college in the spring of 1960.

Almost simultaneously with this, it became apparent nationally, that the United States needed to do much more in atmospheric sciences research and instruction. (The National

Center for Atmospheric Research was then just getting started.) As we got to know him better, it became clear that Vince was uniquely qualified to lead a major university research effort in atmospheric sciences. We soon had convinced him to join the faculty on a full-time basis.

Thus was born the idea that quickly led to the establishment of the Atmospheric Sciences Research Center. Vince and I wrote a brief memorandum to Evan Collins, President of the College, Chancellor (then President) Tom Hamilton, and the University Board of Trustees. All showed considerable enthusiasm for the proposal. Initial limited funding was provided by the State University Research Foundation, and the ASRC came into existence with Vince as Research Director and the writer as Administrative Director.

Our first appointment was Ray Falconer. And so it began. It should be noted that the entire concept was, basically, to develop a major research center around the genius of Vincent Schaefer.

My dictionary defines serendipity as "the facility for making happy discoveries by accident" (coined by Horace Walpole). Many people think of Vince as possessed of this quality to a high degree, and he certainly was, but he was much more, too. Having known his powerful mind and wide-ranging interests, I believe that his accomplishments were much more than accidental. Rather, they were the result of careful planning, broad knowledge and extremely keen observational ability. I'm sure that all who knew him will agree that in these qualities, his equal will be known only very rarely.

All of us who knew him had our lives enriched thereby.

Oscar E. Lanford
Vice Chancellor of the State University of New York (retired)

My Experiences with Dr. Schaefer

In this short piece, I describe how my experiences with Dr. Schaefer influenced me to go into atmospheric sciences as a career.

In 1960, I attended a National Science Foundation-sponsored summer school in atmospheric sciences at the Loomis School in Windsor, Connecticut for soon-to-be high school seniors. The school offered a first-class physics course plus laboratory experiments in atmospheric sciences, which were Dr. Schaefer's favorite. I remember him often sitting in the gardens of his residence that summer writing in his scientist's notebook. I often wondered what kind of new and wonderful ideas were in it.

In 1963-1965, I was a participant in Dr. Schaefer's State University of New York at Albany (SUNYA) Atmospheric Sciences Research Center (ASRC) Natural Sciences Institute (NSI) at the Museum of Northern Arizona in Flagstaff. There, each of us eight or so college students assisted the weather modification organizations working in Flagstaff those summers. I was fortunate to be responsible for wind measurements around the San Francisco Peaks, which gave me independence, a vehicle, and the chance to get out into the very interesting countryside. At the conclusion of my first summer in 1963, Dr. Schaefer gave a slide show of all the various places in the world he had been in carrying out his research. I was inspired by this and decided then and there to go into atmospheric sciences. I also wanted to come back to the NSI in 1964 and asked Dr. Schaefer if that would be possible. I am glad to report that he said, "Yes."

In 1966, I was a part of the ASRC student staff, but this time, with Project Hailswath in Rapid City, South Dakota. There, I took part in chasing hailstorms to gather hailstones for study. Later in the summer I was at the Chalk Mountain meteorological research station of Colorado State University at Climax, Colorado, where I gathered ice nuclei data. In both of these assignments, Dr. Schaefer treated me like a full-fledged graduate student. The enjoyment of the work and the outdoor environs convinced me to change from physics to atmospheric sciences as a career, which is what I did.

Overall, my experiences with Dr. Schaefer over five summer seasons exposed me to the beauty of the atmosphere and its elements, to the possibilities of atmospheric study in the outdoors, and to the mysteries of many atmospheric events. All these factors still inspire me and carry me down the road of atmospheric sciences. For that, I thank Dr. Schaefer.

Alexis B. Long, Associate Research Professor
Atmospheric Sciences Center, Desert Research Institute
Reno, Nevada

Memories from Flagstaff

I will always recall Vince in his plaid shirt and beige twill pants in his house (with Lois) at the Atmospheric Sciences Research Center at the Museum of Northern Arizona Research Center in Flagstaff. Their living room was populated with authentic Navajo rugs that they actually used as rugs. The rest of us had (and still probably have new) them displayed on walls, as possessions. The Schaefers had them spread on the floor as rugs.

There was a big picture window that looked out toward the San Francisco Peaks. Overstuffed couches and easy chairs were arrayed to allow easy view out that big window. I spent many an enjoyable evening watching the clouds and rain gain in intensity as night fell. Some of the most spectacular lightning displays I've ever seen were from that picture window. Vince and Lois made their house a warm and welcome place for anybody from students through project scientists to visiting dignitaries. When in the Schaefer house, everybody was equal and everybody enjoyed that equality. Coffee was endless. Conversation covered all imaginable topics. The Flagstaff cloud seeding projects of the 1960s and '70s most definitely benefited from all Vince offered, from insight through bringing students to the project, to warm friendships. His scientific insight has, and will still be more, highlighted by others. I wish to recognize and express appreciation for his warmth.

Alan Weinstein, Arlington, Virginia

A Trip with Vince: Never a Dull Moment!

Vince Schaefer's ability to see nature, and to understand how it operated, was a gift that few experience today. The Native American must have had this ability– totally uncontaminated by "what should be"– just understanding "what is." On a trip to Flagstaff, Arizona (Vince, Volker and I flew into Phoenix), Vince was driving us northward through Oak Creek Canyon to see an old friend of his. The fact that there were cliffs dropping off several hundred feet into ravines (and no barriers) on this old dirt/rocky road didn't seem to faze our driver as he drove with one hand and pointed out the wonders of nature with the other. At one point, he slammed on the brakes in a cloud of dust, jumped out of the car, and called us out to see a plant sticking out of a crevice in the canyon wall. This was an example of fasciation (*Plateau*, Vol.46, No. 2, Fall 1973) which Vince and Alan Miller had been studying near the Museum of Northern Arizona. (The modern tomato is an example of fasciation– a mutation which increases the size of a plant/fruit.)

Volker and I weren't exactly fascinated with fasciation– but it was Vince's nature to be interested in all aspects of life, and to encourage others to broaden their horizons. Diversity would not be a social issue today if we had only realized that the strength of nature is its diversity .Vince's interests were diverse.

Ronald Stewart
State University of New York at Albany

From Chance Beginnings

Like all meteorologists, I had heard of Vince's Dry Ice Cloud Modification studies, but we had never met. Then came 1962-3, when I was fortunate enough to have been awarded a National Science Foundation Faculty award to study Cloud Physics at Imperial College, London. I first met Vince and Lois in the spring of '63 at an International Cloud Physics conference in Claremont Ferrand, France. Later, in 1968, we attended another conference in Tokyo/Sapporo. After this conference ended, a group of us signed up for a tour of southern Hokkaido, and through that tour, we became better acquainted. I recall getting a touch of Vince's avid curiosity about nature as he rushed from the bus to examine the varieties of flies breeding in the hot water springs of the region.

In 1971, I returned to England on a sabbatical to the Met Office at Bracknell, to do library research on the origins of cloud nomenclature and the early English meteorologist Luke Howard. I had plans of doing an American Cloud Atlas. In the summer of that year, I received a letter from Vince that changed my life– literally. He explained that he had made an oral commitment to Houghton Mifflin Publishers to do a Peterson *Field Guide to the Atmosphere* some ten or twenty years before, but– because of other pressing interests, of which there were many– he had not been able to get around to the writing task. And… would I be interested in being his co-author so he could get the job done? I was surprised and complimented by the offer and said, "Yes". By that response, Vince's and my life became intertwined.

We decided to meet at his cabin at Flagstaff and do intensive work on the book finishing, say, eighty per cent. The remainder we could wrap up through individual assignments. This took place for a couple of months in the springs of 1972 and 1973. Lois was the cook. My young daughter was illustrator. I recall the evenings that we sat and looked north past the San Francisco Peaks and watched the behavior of the contrails laid down by the Denver-to-LA jets and discussed forthcoming weather.

We were informed at the outset by the Houghton Mifflin editors that the project required patience, but had substantial longevity. This proved to be the case. *Atmosphere* was not published until 1981, following the revision of RTP's first book on birds.

Atmosphere and the following *First Guide to Clouds and Weather* have opened many, many doors of opportunity to me. So, for whatever reasons Vince chose to write that 1971 letter of invitation, I am forever grateful!

John A. Day
McMinnville, Oregon

Authors John Day and Vince on the release of their Peterson's A Field Guide to the Atmosphere *in 1981. Courtesy of Jim Schaefer.*

Kliefoth on Schaefer

With a firm handshake and an interesting, mischievous twinkle in his eye, Vincent J. Schaefer welcomed the first group of summer science program students to the Loomis School in Windsor, Connecticut in 1959. Being privileged to be one of that group has always been a humbling experience for me. And to think back over a 17-year association with Dr. Vincent J. Schaefer is an extraordinary delight.[58]

The Natural Sciences Institute, as the outgrowth of our initial program came to be known, had numerous objectives, but most importantly, we were convoked to learn something of the spirit of research. Dr. Schaefer was anxious to show us the "adventures of the mind" and the fascination of discovery for its own sake. He knew that science is fun and wanted us to know also. Affording young people the opportunity to learn in association with great scientists and to discover their own potential has been one of his main goals. Determination that others might experience a promising, exciting time was shown in his actions, words, and writings. The drudgery of his youthful years as a drill press operator was an added incentive to him to help us find a better beginning.

Many adjectives of praise are appropriate in referring to Dr. Schaefer. Recounting our happy and profitable times is a most gratifying remembrance. I was always struck by Dr. Schaefer's incredibly keen sense of observation. He has always been completely attuned to his environment and totally aware– whether it be during a walk in the woods or during the conduct of a scientific experiment. It was he who taught all of us the value of serendipity. He is always open-minded and fair, so that a high school junior's suggestions might seem to carry as much weight with him as a seasoned scientist's. He reveled in fresh observations and a new look at old tenets. But, at the same time, his respect, pride, and adherence to the precepts of his scientific predecessors was always obvious. Langmuir's methods and accomplishments were frequently shared with us.

Innovation through hard work could almost summarize his work ethic. The total number of groups or organizations he founded is unknown to me, but personal experience with NSI and with his Yellowstone expeditions is indicative of the quality of his undertakings. Despite the complexity and significant accomplishments of all of his endeavors, his emphasis was always on simplicity. He demonstrated the use of intellect, observational powers, and careful experimentation, rather than gadgetry and expensive instrumentation.

Whatever the future will regard as his most important scientific accomplishment (and the discovery of cloud seeding must certainly rank high), I suspect that one of his lifelong objectives has been to influence the growth of young people. The legions of students, apprentice scientists, and associates can all attest to his accomplishment of that goal.

The man is more than his work. (But it should be well-known to all who have been associated with Dr. Schaefer, that his work was his life. It was he who taught us the precept that the greatest thrill is to be able to be paid for doing something which one totally en-

58 This was written in the mid-1970s, thus the present tense is used throughout.

joys.) He taught many of us the joy of outdoors. To go for a walk with Dr. Schaefer is an unforgettable experience.

Getting people together is a rare talent that I have seen few people possess as he did: whether it was the students and faculty at NSI, a diverse group of New Yorkers to form the Mohawk Valley Hiking Club, or just an ensemble of families to yearly hike through the snow in the mountains to enjoy a Thanksgiving breakfast together, it was always Dr. Schaefer who was the motivating and instigating force.

Long before the current fad on Indian culture, VJS had explored, catalogued, and enjoyed everything that the Indians, both in New York and in the West, had to offer. His paintings, furnishings, and choice of jewelry (gifts to wife and family), attest to his respect and affinity for their life.

Photography is, perhaps, his greatest avocation. Magnificent photographs from all over the world reside in his collection and are shared with people freely. He has a facility with technique that would make many a professional photographer envious: his knowledge of time-lapse methodology and his complete skill in the darkroom are additional examples of his multi-faceted personality. His dramatic pictures of Yellowstone in the winter, culminated by his publication in *Life* are a couple of his many accomplishments.

I can well remember the hours that Dr. Schaefer would spend in his study composing his papers, letters, and reports. The fact that his prose is most lucid, readable, and correct has never lost its impact on me, especially when I recall that he had only a 10th-grade formal education and spent his early years as a machinist. He taught by precept and example that all of his students should learn the importance of clear, concise scientific writing. In my estimation, Dr. Schaefer is a determined, committed essayist with no serious rivals.

To say that Dr. Schaefer is TOTALLY INTERESTED would be perhaps the best way of encompassing the many facets of his personality and life. The scientific programs that he organized always had input from associated areas, such as geological, historical, ecological, and cultural aspects of society. He is involved in a wide range of activities and in an ever-increasing circle of associates. Not only are scientists, undergraduates, and graduates included in his fellowship, but also ecologists, rangers, and attorneys share his warmth and guidance. The memory of all of the activities which transpired in Yellowstone during the winters can never escape me. One, in particular, the tracking of the hibernating grizzly bear, and tagging in its den, was just one fascinating adjunct to the scientific activities so isolated, and yet so much a part of Wyoming in the middle of winter.

No man stands alone, and Dr. Schaefer's three marvelous children and his devoted, supportive wife have gained immeasurably from him and contributed greatly to him. The fellowship of a happier, warmer family could never be enjoyed more fully than with the Schaefers. I was fortunate to be "adopted" by them and still consider myself very close to them, despite the passage of time and the separation of miles. To be a part of that nucleus is one of the greatest thrills and honors that I could ever have.

While I've studied at several universities and academic centers throughout the United States, I have never experienced the vital awareness, interest, and passion for life as at

"Schaefer University." His students have gone into many fields– architecture, art, social sciences, urban planning, engineering, atmospheric sciences, law, and medicine. The extent of Vincent J. Schaefer's influence has been great. I am currently in the final year of my training in neurological surgery at Washington University in St. Louis. I doubt that Dr. Schaefer ever considered the possibility that one of his charges would end up a brain surgeon! The extent and variety of his precepts has extended like the ripples produced on a still pond by a small stone. To all of those who have had the opportunity to be associated with Dr. Schaefer, we regard him as the modern embodiment of the Renaissance Man in a warm, vital person.

B. Kliefoth, MD
Knoxville, Tennessee

Chapter 14
Vincent Schaefer's Serendipitous World

When Did He Sleep?

I knew Vincent through my involvement with the Long Path. Although I only met him three or four times, we had been corresponding for a few years. In about 1976, I obtained copies of his original descriptions for the route of the Long Path in the New York Public Library with the intention of following the original route from the Catskills to the Adirondacks. As I became more involved with the planning and development of the trail from Gilboa to the Mohawk River, I kept running into what I call Schaefer Coincidences. I think they help to show his all-encompassing curiosity and inquiring mind.

One of the areas the original Long Path traversed is John Boyd Thacher Park. Vincent mentioned a cave known as the Witches Hole in his original description. I inquired at the park office about the location of this and other caves in the area. No one there had heard of the cave but they referred me to a book on Albany County caves. There, I found a description of the Witches Hole and also three or four other caves first explored by Vince Schaefer in the 1930s.

Later on, while looking for old roads or trails along the Helderberg escarpment, I was directed to a farmer whose family had lived in the area for many years. As he was telling me of now abandoned roads, I remarked on the construction of his barn. He told me that it was a Dutch barn and told how an expert on this type of barns was there a few months ago. His obvious pride in his barn prompted me to learn more about these structures so that I would have a common ground with some of the local farmers. A few months later, I was reading an article about Dutch barns, which included a reference to an expert on Dutch barns, Vincent Schaefer.

While reading the descriptions in the *New York Post* from the 1930s, it became apparent to me that there were some places which were special to Vince. I was determined to visit these places above all else. One of them was Vroman's Nose. Here I met Wally Van Houten, who told me of the Vroman's Nose Preservation Corporation. One of the founders was, of course, Vincent Schaefer. I also learned of his work with the smoke generators there.

During my travels in this area, I came across a roadside historical marker which mentioned the Sloanesville Indian rock pile. I stopped to talk with one of the landowners who told me that the rock pile was really on the other side of the road. He showed me the remains and mentioned that a local historian was there recently to measure the pile. Sure enough, it was Vincent Schaefer!

When the NY-NJ Trail Conference decided to extend the trail northward, we received assistance from the National Park Service Rivers and Trails Assistance Program. One of the persons assigned to help us was Karl Beard, who had previously worked with Vince at the Mohonk Preserve. He told me of the time the two of them went collecting rocks which Vince was going to slice with a rock saw. Karl remembered it well, since he was the one who ended up carrying the rocks!

Not long after, I was watching a National Geographic show on TV about wild rivers. It featured the Craighead family. I had read the description of the Long Path in Jean Craighead George's *American Walk Book* but I didn't make the connection until they mentioned taking snow samples from their travels to their friend in Schenectady, Vince Schaefer.

Less than a month later, while flipping through channels on TV, I happened upon a show about a time traveler from the future who assumed identities of people from the past. On this show, he was a rain maker in the mid-west. There was a passing reference to Vince Schaefer's work on cloud seeding. It seemed that no matter where I turned, his influence was felt. As one of my friends said, "When did he sleep?"

Of course, Vince's concept of the Long Path was quite different from that of the Trail Conference. He wanted an unmarked route to be utilized by the woods-wise hiker, similar to an orienteering course. However, the concept of an unmarked "trail" seemed to be too subtle for the hiking community to grasp. When I asked, he said that, as far as he knew, only he and one other hiker had followed his route in the past 60 years (I hope to be the third). Although he frequently spoke in favor of an unmarked route, I always felt that he was glad to see that his original idea of a foot path traversing the length of New York wasn't forgotten.

Ed Walsh
West Haverstraw, New York

Early Letters from Vince to an Indian Chief

I was in Northern Ireland when Vince died. As I was flying back home, and before I knew of his passing, I began planning a research project that was to involve him. One of my first tasks was to call him; unfortunately, one of the first things I saw upon my arrival was his obituary…I will have to find another way.

While doing research at the State Museum about 1991, I found letters that Vince had sent to Arthur Parker when he was quite young. Vince was a would-be archeologist and Parker was both a Seneca Indian and the State Archeologist at the time. His first letter of March 20, 1922 is particularly endearing. Vince was on the threshold of manhood, and he was quite proud of it. Near the close of the letter, he said, "Well, Mr. Parker, undoubtedly you are getting weary of this talk, so I'll close with a short description of myself. I am 15 years old weight 150 lbs. and am 6 ft. tall. I am now more interested in my relics from an archeological point of view and love to collect and study relics better than to eat." I told

Vince about this and other letters at lunch not long ago, and I enjoyed the way in which he could appreciate that fifteen-year-old with the objectivity that a space of seven decades provides.

Dean R. Snow,
Albany, New York

Rare Ferns

It was some 70 years ago that we found common interest in Mohawk Valley Iroquoian archeology. In succeeding years, Vince introduced me to Dutch barns for which his exuberant appreciation was contagious for anyone such as me, formerly uninformed. We shared concern for protection of Adirondack wilderness areas. In writing or in conversation on that subject, he always presented convincing arguments based on both his scientific knowledge and practical experience. Vincent and I had long advocated a special status, possibly as a natural public park, for the Big Nose Mountain area near Yosts (New York). When he alerted Tom Porter to the opportunity that the auction of the Montgomery Manor (County Home) offered as a home for Porter's Akwesasne (St. Regis) Mohawks, I was thrilled at Vincent's innovative idea and this very practical solution to acquiring and preserving this choice area. I think that Vincent's leadership role, his finesse in molding public opinion, his desire to help people, backed by his very extensive scientific knowledge formed an approach to matters of public interest which proved highly successful time and again. People listened to him because they realized that he knew what he was talking about.

I would hasten to say that his in-depth study of every field of interest he explored at once set him apart from those who could only present a casual opinion.

Speaking of Big Nose Mountain reminds me of the delightful volume which he presented me. I refer to *Vrooman's Nose*. Again, his aesthetic sense, appreciation of nature, coupled with an appeal to the public to guard well such a select spot mark, his expertise in the subjects of ecology, conservation and human history.

It was on Big Nose Mountain, fifty years ago, that the extremely rare Harts Tongue ferns Vincent had propagated from spores were sent by him to Orra Phelps, MD, to plant in some safe, wild place. After Orra and I consulted on the matter, we agreed on the limestone soil of the deep Valley of Knauderack Creek on Big Nose. It was a very distinct pleasure for me to invite Vince to accompany me and a small group of highly interested people to inspect this colony of Harts Tongue. We found them thriving and the very evident pleasure it gave Vincent was in itself as rewarding as the sight of the ferns. This was about nine years ago.

One day Vincent brought Professor Carl George of Union University to Fort Plain and we three enjoyed a very happy get-together discussing my collections of fossils, minerals and Mohawk artifacts. Prof. George was especially intrigued by an exceptionally large trilobite I had found and he photographed it.

This past winter, at Vincent's request, I wrote a number of personal reminiscences of the old Erie Canal for a volume he was then compiling, "Landmarks of the Old Erie Canal."

Amongst the many topics upon which Vincent and I exchanged information were discussions of local water falls and specifically the Tekaharawa (Little Cascades) of Judd's Falls, on Canajoharie Creek headwaters about 1.5 miles north of Cherry Valley.

Beside me here, is a treasured volume, inscribed on the fly leaf, by Vincent, for me: *A Field Guide to the Atmosphere.* It is a rare trait in a scientist to find explicit technical knowledge combined with such a deep appreciation of natural phenomena and presented to the average reader with such clear, understandable language.

Vincent and I shared a very real interest in the conical hill in Schoharie County, known since colonial times as Barracks Ouray because of the fancied resemblance of the hill to a type of stock forage housing typical of the early days and called by the Dutch, a barracks.

Vincent's optimism, happy attitude in the face of knotty problems, perseverance in pursuing a subject to a complete mastery– will always stand out in my memory. But over and beyond everything else, the warmth and sincerity of his friendship will be with me whenever his name is mentioned.

Douglas (Dug) Ayres, Jr.
Fort Plain, New York

An Adirondack Outing

Ten or so years ago, Vince joined me and eight kindred souls on what had become for us a semi-annual date with brook trout in the Adirondack's St. Regis Canoe Area. We called ourselves AE (Adirondack Elf) Tours and we paddled and hiked back to Big Fish Pond each spring and fall, trying always to base our operations at the Blagden Lean-to on the west side of the pond. We rarely traveled "light" and Vince shouldered more than his share going in.

Shortly after arriving and setting up camp, we were all off in separate canoes to fish for the waning hours of the day. Vince and his canoe mate set off for what we called "the rip," the outlet of Big Fish/inlet of Little Fish Ponds, and Vince promptly caught two nice Brookies on worms. And that was all he wanted. For the next three days, he prowled the woods sampling the pH of ponds both big and small, observing plants and animals, and examining rocks. And each time we'd reassemble at camp, for meals or rest, he would join the conversation– always contributing, but never dominating (which he could easily have done).

About 9:00 pm on our first evening in camp, as our fire became just a warm glow, we all headed for bed– most of us on the lean-to floor, a few in tents. Vince grabbed his sleeping roll and headed for a grassy spot 40 yards or so behind the lean-to. We all protested– there was plenty of room in the lean-to. "It's better this way," Vince replied, "you'll understand." And in a few minutes we did! Even at 40 yards away in the woods, none of us ever heard anyone snore so loudly or so regularly!!

Our trip that year ended, as always, with a slow paddle and long portage out to Little Clear Pond. Going out we did "go light" having consumed our heavy burdens of food and supplies– at least most of us "went light." Vince struggled with a pack heavier than when we went in– a pack full of rocks destined for his special saw and a beautiful lamp shade or window somewhere!

We do not go as often to Blagden as we once did, but when we do, we always talk about the big fellow with the pack full of rocks.

John A. La Rocca
The Adirondack Elf
Rensselaerville, New York

A Bear in the Cabin, Seeds in the Clouds

I would like to share two especially bright memories of Vince, both involve my kids.

Like John La Rocca, I can attest to Vince's late evening repertoire. Vince asked me once to give a talk to a group at Whiteface Mountain in the Adirondacks. Afterwards, we retired for the night to a cabin where Vince slept on one side and my young son, Benjamin, perched above me on a set of bunk beds on the other. I was awakened by a sudden impact when Ben jumped down from above and landed on my chest. "Dad, there's a bear in this room." I listened. It was Vince snoring, of course!

A few years after that, my son Jeremy was watching television one night and raced in to tell me that Vince Schaefer was on his favorite program. It proved the case. The show was about time travel back to historic events, and the hero had landed in a drought in the 1800s. They called on their hidden computer to solve the problem and out came the answer: cloud seeding as invented by one Vincent J. Schaefer.

Peterson's *Guide to the Atmosphere*...acid rain studies...Dutch Barns...serendipity in science. None of these achievements would rate more than a nod from Jeremy. But honorable mention on the time travel show– now that was something! I called Vince to tell him this story. Characteristically, he was delighted with the humor and impatient with the recognition.

Hal Williams

Farewell to an Adirondack Giant

Another Adirondack giant passed away this year.[59]

He combined his Adirondack interests with other interests and became known throughout the state and beyond. His final project, just before his untimely death, was to get the Mohawk Indians established on their original grounds at the abandoned Montgomery Manor in Yosts.

Vince Schaefer, certified meteorologist and research scientist, was a man of many interests.

The last visit I had with Vince was at Milt Schaber's Deli on East Fulton Street in Gloversville. We went to lunch where we could get some real "double-breaded" Adirondack sandwiches and discuss the old Indian trails and Tory escape routes through the Adirondacks. Vince had located a cannon under a corduroy road near Garnet Lake on the Fort Hunter-Crown Point Trail and was trying to establish some documentation on the existence of the early routes. Other cannons had been located, including some that once stood at Johnson Hall, and Vince had heard that I had some old maps. I did.

J. Yates Van Antwerp prepared a map of Sir John Johnson's escape from Johnstown in 1776. It pointed out where one was recovered in 1937. The old Grass River route, which passed from Johnstown to Fish House and then up to Raquette Lake through Lake Pleasant, was marked on this map. Two cannons had been found on this old military road, moved back to Johnson Hall, and later transferred to the Peebles Island facility. Lumbermen and hunters have reported seeing a cannon in the woods near Route 10 in Arietta.

Another map prepared in 1780, showing General Burgoyne's 1777 campaign, shows a trail from the Mohawk Valley traveling west of the Sacandaga River and over Lake George. Another, printed with a legislative document before 1780, portrays a route from Johnstown to Crown Point named the Sir Johnson Trail. Other maps show "Indian Paths" running through the Adirondacks. After studying all of the old maps, Vince surmised that the route he was looking for forded the Sacandaga River at "Canada Island" in Hope, ran easterly to Hope Falls, northward to Garnet Lake and then followed Mill Creek to Johnsburg and Weavertown before heading north.

Vince's inquiring mind led him into other fields of research.

Some have seen the summit observatory on the top of Whiteface Mountain. The four story, 24-foot diameter field station is the "highest" building in New York with running water. It is there because of Vince Schaefer.

Vince, serving as a General Electric Research Scientist, saw the need for a year-round, long-range, recording of air chemistry, pollution, and clouds over New York and decided the top of the Adirondacks, at 4,867 feet, would provide the purest collection. He made the proposal to officials at the State University of New York at Albany and it was built in 1978. Today, it is one of the world's pioneer mountain observatories for studying air chemistry.

---

59 This article was originally written in 1993.

Vince made his final contribution through his interest in returning the Mohawks to their original "hunting grounds."

When Montgomery County put the former Jelles Fonda (Montgomery Manor) property up for sale, Vince wrote to Tom Porter, former tribal chief and religious leader of the Mohawks at St. Regis, and encouraged him to return to his native land.

Vince had a serious heart attack and could not attend the auction. Tom did make the auction and bought the 300-acre property for $210,000. Some of the money came from an anonymous benefactor.

We may never know who it was, but then again, maybe it was another great Adirondacker who made his mark in history.

Don Williams
The Sunday Recorder, *January 16, 1994*
Amsterdam, New York

Yellowstone Food

I was the chef for Dr. Schaefer and other scientists for fifteen years at Yellowstone National Park, where they did research for several weeks each winter.

Dr. Schaefer didn't fit the stereotype of scientists as people out of touch with reality. He was a man in tune with everything and everybody. He was compassionate, a good listener, and had tolerance for those who were set in their ways. He recognized the good in individuals, and accepted and treated others as equals.

I wonder if he developed his ideas about cloud seeding as he and other scientists gathered about Old Faithful as she blew sky-high on those below-freezing winter nights. They stood around the geyser with spatulas and glass slides, catching the falling ice crystals that glistened like diamonds in the lights.

He balanced kindness and fairness, not an easy thing to do even for a great scientist. I remember the time a self-important ranger told me I couldn't use the telephone to call into town for food supplies. I told the ranger if I couldn't use the phone, he would have to find someplace else to eat. He reported me to Dr. Schaefer, who asked him, "Did you tell Harold he couldn't use the phone?" When the ranger said, "Yes," Dr. Schaefer told him, "Then I guess you will have to find another place to eat." Later, the ranger came to me and told me he'd gotten permission for me to use the phone.

Vince was a man you could count on to do what was fair, what was right. His accomplishments, his scientific discoveries and contributions will help mankind for generations to come, but I will always remember him most for the kind of man he was.

Harold Gooding
West Yellowstone, Montana

From Montana

I was not as closely associated with Vince as was my brother, Frank, nevertheless, I count him among my closest friends and most admired professional associates. He had a warmth of spirit and depth of intelligence seldom encountered. His greatness was evident not only in his work and accomplishments, but also in his understanding of human nature. He treated all as equals and this humbleness endeared him to colleagues and all who crossed his path. No one equaled him in conducting a seminar. I fondly recall those at Old Faithful in Yellowstone, where I admired his keen perception, contagious enthusiasm, mild humor and always fatherly encouragement to the lecturer whether deserved or not. It was a great privilege to attend these and feel a master at work. One departed a better man for the experience. The world has lost a friend.

John J. Craighead
Craighead Wildlife-Wildlands Institute
Missoula, Montana

Vince at Yellowstone in mid-winter. Courtesy of Roger Cheng.

A Glimpse at Yellowstone

On a cold, sparkling Yellowstone day, I was walking with Vince near Old Faithful Geyser. Suddenly, he knelt down and began to closely examine the steaming stream of water which flowed from the geyser into the Firehole River. I was a Park Ranger assigned to winter duty at Old Faithful in 1962 and 1963, and found myself frequently accompanying Vince on his saunters through the geyser basins. Each walk was a delight of discovery and good conversation and most were a test of endurance in temperatures which ranged to fifty degrees below zero; a test for me, not for Vince. He seemed not to particularly notice that ice would accumulate on his eyelashes or that his cheeks would take on the kind of reddish glow that presages severe frostbite.

On this particular day, while he stared into the geyser outflow, he was making a natural history discovery which had evaded all other naturalists throughout the history of the park. Nestled in the water, at just the right temperature, was an orangish color which Vince, with his hand lens, quickly identified as an egg mass. Further examination and a bit of research confirmed that Vince had discovered tropical Brine Flies living and laying eggs in a two-foot section of a geyser stream, surrounded by some of the coldest temperatures imaginable. Somehow, we all took pleasure in the knowledge that little insects were living in a tropical paradise at frigid Old Faithful; it made our numb extremities feel better.

I will always honor Vince as a mentor, a leader, an eloquent communicator, a keen observer of nature, and a skilled scientist who easily shared his captivating enthusiasm and intellect with others.

Robert O. Binnewies
Bear Mountain, New York

Another Yellowstone Winter Tour

In January 1987, Anne and I joined a group tour of Yellowstone, whose leaders were not only skiers but were also geologists, wildlife biologists, etc., so there were a lot of interpretive features to the program. After diner, there were evening sessions, and tour members were also encouraged to participate.

I knew that Vince had Yellowstone experience, and sought him out to learn how to prepare for this trip. He told me that winter conditions at Yellowstone were not too different from the colder parts of the Adirondacks. He also gave me some details for the seminars he organized there to study the huge volumes of super-cooled clouds generated by the geysers.

The last night of the trip, I gave a talk on the history of cross-country skiing and worked in the story of how cloud seeding began and Vince's contributions to winter outdoor recreation. When I got back, I wrote it up and let Vince edit it so as to tell it as it really was.

I am pleased to present a copy to the Schaefer family, because Vince meant a lot to me in my growing up days. I was a leader in a Scout troop at that time and Vince came to a meeting with his photos of winter at North Creek and got us going. And I recall that some of the leaders stayed at the Schaefer camp near North Creek one night. I slept out on the porch testing my new Mohawk Valley Hiking Club sleeping bag that my mother had sewn up for me and my Dad and I had filled with duck down with a vacuum cleaner. Vince took us out the next morning to some steep open slopes that were gullied, and taught us how to ride into the gullies from one side and make a stem turn at the crest of the other side, so as to swoop back down. There was a light fall of new snow on a firm base. This recollection is most vivid in my mind.

Vince was such an outstanding person in so many ways, I can't say enough about him.

Almy D. Coggeshall
Schenectady, New York

A Visit at Schaefer's after World War II

One of the most memorable evenings of my life was spent at the home of Vincent and Lois Schaefer a few years after World War II. They were hosting the famous arctic explorer Vilhjalmur Stefansson, and on this evening, a few "explorer-type" friends were invited in for discussions and general scientific conversation. Stefansson was retired and quite old in years (born 1879), but his mind was very bright and alert, and when triggered with an appropriate question, he would refer back to his vast experience and hold our rapt attention for unrecognized periods of time. The evening was gone before we realized it. The accounts of arctic experiences and safe practices were added to by Vince's accounts of his winter camp experiences on Mount Washington, New Hampshire, and at Yellowstone Park where arctic conditions usually prevailed. So, we "lay persons" mostly sat back and just listened to the talk of these two outstanding scientific outdoor men tell of the problems, their experiences, and the scientific logic of how to deal with the problems safely.

One of the discussions was about proper and safe clothing for continuous outdoor trekking, day after day, without refuge in heated dry buildings. The problems and dangers come from body moisture condensing and freezing in the clothing. Stefansson had found that wool was not safe or practical under arctic conditions. The clothing system of the Eskimos was far better, using fur liners and exteriors with body ventilation to get rid of the body moisture without it condensing and freezing at some depth in the clothing. Also, for day after day operation, it was necessary not to sleep in one's work day clothing. One needed to strip naked and get into one's sleeping bag within an igloo constructed on the site.

One of the greatest dangers was polar bears killing and eating the sled dogs. During a bivouac, one person at a time would have to stay awake (in sleeping bag with gun within reach) to listen for the steps of approaching polar bears. When one was heard, the guard would have to immediately go out naked with gun and shoot the bear. Taking time to put on clothing would make it too late, for the polar bears could kill a tethered sled dog very quickly, in less time than the guard could dress.

Vince Schaefer was always one of my "heroes" from the time I first knew him in the late '40s. His background and life development always seemed to me to nearly parallel those of Michael Faraday, who, with very humble beginnings and practically no formal schooling or training, rose to become one of the best experimental scientists of the world. This came about because of his native intelligence, curiosity, energy, and commitment to science. Vince Schaefer was like that too, and in a very personable way. We shall always remember him as a vigorous, friendly, observant, exploring, scientific person. He lived a wonderful, long, productive life, which we all admire and hold up as an example to our youth.

Francis P. Bundy
Lebanon, Ohio

Vince's and My Elk

During the Tenth Field Research Expedition to Yellowstone, organized as always before by Dr. Vincent J. Schaefer (Vince), we had a wonderful time and a successful mission while investigating the origin of ice crystals in water supersaturated environment close to the Old Faithful Geyser (on January 6, 1970) and in the area of Norris Geyser Basin (on January 7, 1970). In the morning hours of the first day, I was sampling the ice crystals with Vince at temperatures around -30° C (-22° F). Vince used a swinging technique and I used a simple sedimentation for collecting tiny columnar ice crystals on glass slides covered by a thin layer of Formvar.

The following day the weather was sunny and a little warmer, and I went alone far from the research group experimenting with the modification of the microstructure of clouds around the warm pools. While standing in a shallow warm brook, I felt quite comfortable during my ice crystal sampling. Suddenly, I heard a slight noise and deep breathing behind my back. I thought that Vince or someone else was coming to assist with the ice crystal sampling and I started to explain– without turning my head– that I had to finish the sample I was working on first. After a while, surprised that there was no response to my comments, I looked back and was speechless as I saw the huge head of a bull elk very close behind me. I stood silent as I looked into the great dark eyes with long eyebrows and the enormous antlers above the head. I dared not move for what seemed to me an eternity, while the beast was quietly staring at me. Finally, he turned back into the adjacent forest. I hastily wrapped up my sampling tools and marched in a deep snow to join our research group. To my greatest surprise, the elk suddenly appeared and was moving up the slope 100 feet behind me.

Telling Vince about my adventure and showing him the elk peacefully behaving at the warm brook not far from our group, Vince immediately decided to take a photograph of the magnificent solitaire animal. We approached the elk cautiously and the huge bull elk posed like a movie star when Vince shot the photographs from a very close distance. The splendid photograph of Vince's and my elk was later published in the Interim Report on the Tenth Field Research Expedition to Yellowstone (ASRC-SUNYA, Publ. No. 140).

Today, I was looking at the ice crystal samples and photographs from the 10th Yellowstone Expedition and I tried to recall back the wonderful time we all had there– but both Vince and the elk are gone.

Josef Podzimek

Memories of Journeys with Vince

I feel grateful to be included in those friends who can contribute reminiscences to this memorial book. The friendship between the Kahan family and the Schaefer family began in 1952 in Fargo, ND. We have a cherished file of Christmas photographs that enabled us to watch Sue, Kathie and Jim grow up. My final telephone conversation with Vince occurred about a year before his death, after I had belatedly learned, from Tommy Henderson, that Lois had passed away. My contacts with Vince ranged in duration from a few minutes to many days. They were sources of inspiration to me. I learned a lot about what it means to be a scientist, a friend, a husband, and a father from Vince. I regret that I never got around to telling him just how important he was in my life. Maybe he knew without my telling him.

Picking an anecdote to submit has not been an easy task. We had many memorable moments together. We laughed together in Tokyo in 1965 and traveled over deep snow in Yellowstone Park some years later. I have chosen the reminiscence that follows because it reveals a side of Vince's character few may have witnessed.

In the Spring of 1965, Vince, Gene Bollay, and I set out together on an expedition aimed at determining what concentrations of silver could be found in the high level snows that fell on the mountains of the western states of Colorado, New Mexico, Arizona, California, Oregon, Washington, Montana, Idaho, Utah, and Wyoming. Because the melting season was well underway, we drove ourselves hard to cover the entire region before the snow we intended to sample disappeared. We put in long days that lasted well after sunset, taking turns driving, spending the short nights at whatever accommodations we came upon. It was a taxing routine, but Vince appeared to have a boundless store of energy and good spirits.

One dark night, as we were nearing the end of our trip, Vince was taking his turn at driving. Suddenly, out of nowhere, a deer bounded onto the road immediately in front of the car. Vince slammed on the brakes, but it was impossible to avoid hitting the deer. Vince was devastated. He unfairly blamed himself for what happened. Both Gene and I tried assuring him that neither one of us could have done any better, but he would have none of it. I can't be sure, but I think some tears were shed. It was an unforgettable experience to see how sensitive this giant of a man could be to the suffering of a dying wild animal. My respect for him was already quite high, but it increased markedly that dark night.

Archie M. Kahan
Lakewood, Colorado

Wading Through Water and Mud

Quite a number of years ago, Fulton-Montgomery Community College had an evening program on environmental and pollution problems. Vince Schaefer was a speaker.

Vince told of how he and a group of archeologists studying Indian sites wanted to visit one on the west bank of the Cayadutta Creek between Fonda and Sammonsville (New York). There are no bridges from County Route 334 crossing the creek, so the group decided to wade the knee-deep stream. Being fearful that the creek bed might contain trash that would cut their feet, they wore their shoes.

As soon as Vince reached mid-stream, he said he felt his canvas, rubber-soled sneakers loosen. He thought that it was because they were wet. When he stepped out of the water on the west bank of the stream, he saw that the canvas uppers were disintegrating. As time went by that day, the cloth parts deteriorated and fell away, till all he was wearing were the rubber soles.

He concluded that tannery acid from the Johnstown-Gloversville industries had been disposed of in the creek. "It ate my shoes," he said.

On the recommendation of Vince Schaefer, Carl Touhey purchased the Wemp Dutch Barn on Queen Anne Street in Fort Hunter, as a replacement for the original Dutch barn on Touhey's Onesquathaw Road property.

In early summer of 1990, Richard Babcock's crew had removed the upper structure of the Wemp barn. Vince visited the site alone. As he was walking one of the floor sills that recent rain had made slippery, he fell off and dropped three or four feet to the ground under the barn.

The barn had covered this dirt for over 200 years. Now that it had been exposed to several days of heavy rain, it had turned to deep soft mud.

Fortunately, Vince wasn't hurt, but said he sank above his knees into the mud. While falling, he dropped and damaged his Linhof camera. He climbed back out and continued his study of the barn sills and foundation.

Clark Blair
Fonda, New York

A Tribute to Vincent J. Schaefer, Delivered at the 10th Anniversary Annual Meeting of the Vroman's Nose Preservation Corporation, August 1993

Dr. Vincent Schaefer, 87, passed away this year. He has always been a friend of Vroman's Nose and the Vroman's Nose Preservation Corporation, so we would like to take a few minutes to remember and celebrate his life and his contributions.

Dr. Schaefer was an early hiking enthusiast. He organized the Mohawk Valley Hiking Club as a young man in 1929. However, he was a visionary who saw beyond the trails that he followed. In 1931, he conceived of the Long Path as an outlet of the Appalachian Trail. The Long Path would go from the George Washington Bridge in New York City to the Whiteface Mountains in the Adirondacks. Vroman's Nose was a primary point along the trail.

This project has been revived by the New York-New Jersey Trail Conference. Vroman's Nose is an important hiking location along the newest section of the trail to the Mohawk River.

Dr. Schaefer's career led him to work with Dr. Irving Langmuir, a Nobel Prize-winning industrial scientist. They invented the high-efficiency artificial fog smoke screen generator widely used during World War II. The generator's mission was to produce "large quantities of smoke with the hope of using it to screen cities, beachheads, ships, bridges and the like from air surveillance and bombing." Fifty thousand of these generators were used during the war. The technology had first been tested by Dr. Schaefer and the military on Vroman's Nose.

Dr. Schaefer had a long love affair with Vroman's Nose. Fifty years ago, he was interested in buying and preserving Vroman's Nose. He was disturbed in the '70s to hear of plans to build a restaurant on the nose. When the Vroman Family and the Schoharie County Historical Society worked together in the early '80s to purchase Vroman's Nose, Dr. Schaefer was a central figure. He helped involve the public by speaking before the Middleburgh Rotary Clubs. The Vroman's Nose Preservation Corporation was indebted to Dr. Schaefer's long commitment and voted him Director Emeritus.

Dr. Schaefer's love of Vroman's Nose lives on in his book.[60] He has collected the scientific, historical, legendary and aesthetic qualities of Vroman's Nose. Many will learn a great deal for years to come about Vroman's Nose through his book.

But we can also learn about the author as we read his book. His personal experiences are plentiful, thus enlivening the text. We learn of his own exploration of Gebhard's or Clark Cave. He paints a vivid picture of one of his hiking experiences on Vroman's Nose when he saw twenty turkey vultures perched along the edge of Vroman's Nose "taking turns at leaping off the cliff edge into the thermals and soaring upward until they were no longer visible, after which, in long swooping dives, they return to their cliff-edge perch."

---

60 Vrooman's Nose: Sky Island of the Schoharie Valley: A Study. Purple Mountain Press, 1992.

But my favorite personal experience of Dr. Schaefer's comes after his retelling of the famous tales of Timothy Murphy and his inviting the reader to stand on the summit ledge of Vroman's Nose and look carefully for Timothy Murphy's footprints. How appropriate that Vincent Schaefer was chosen to play in a film as Timothy Murphy dressed in leather fringed jacket, breeches, a coonskin cap and carrying a flintlock musket.

Dr. Vincent Schaefer was a man of science and history, but he was also a man of great personal warmth. His smile, his lively eyes and especially his enthusiasm will be missed by us all. It is an honor for the Vroman's Nose Preservation Corporation to pay him tribute. We are thankful for his life.

Susan Vroman Walker, President
Vroman's Nose Preservation Corporation

A Visit to Ball's Cave

Spring of 1941...Party of Three...Vince Schaefer...Bill Smith...Art Gregg

This limestone cavern, near Schoharie, NY, had been explored and documented at some period around the Civil War. Not as large, by any means, as Howe Caverns, but it has some large rooms and an underground waterway. The exposed stream is deep and wide enough and has head room for about 800 feet. The original explorers built a log raft and one or two individuals floated down the stream, illuminating the passageway with lighted pine knots. They found a series of tufa dams (consolidated volcanic ash). Over the years, others have constructed rafts or used inner-tubes to duplicate the feat.

We drove to the cave site, on a working farm, got the farmer's permission to enter and found the pit entrance at approximately 8:00 pm. It was Vince's idea to arrive in the evening. "Why waste daylight to visit a cave; after all, it's dark down there, anyway."

We spent two hours crawling and walking the main corridor, passing by the entrance to the river channel not far from the pit entrance. Upon our return, we crawled into the waterway entrance to look around. Vince, without hesitation, announced he would swim to the first tufa dam, some 100 feet downstream. He disrobed in this chilly constant 42° F temperature. We tied the loop end of our safety rope around his chest and he plunged in. We payed out the rope, keeping a light beam ahead of him, as he dog-paddled to the first tufa dam. He stood up on the dam, surveyed the situation, satisfied he could observe another dam farther along, and carefully swam back to safety.

Art Gregg
Intercourse, Pennsylvania

Ever Heard of Horton Falls?

The enclosed article has so much in it that is so typical of Vince Schaefer. Of all the articles I have written about him for the *Times Union*, this is my favorite. It shows how he worked, always exploring, tying up loose ends, connecting his scientific and historical interests in projects that contributed to our knowledge and understanding of our world. In 23 years of writing about people in upstate New York, no one has ever approached Vincent Schaefer in depth and extent of his contributions. That would have been sufficient, but his gentle, unassuming and cheerful spirit made him even more special. None who knew him will ever be the same– we can only be the better for knowing him.

Ever Heard of Horton Falls?

Scientist's diligence gives waterfall a name

Voorheesville– There's not much in the natural world that escapes Vince Schaefer, who has spent most of his 84 years studying earth, sky and water. One waterfall did slip by him, however. A thorough man, Schaefer set out to correct the oversight.

That's why two years ago the scientist, a man of considerable size and height, was hauling himself hand-over-hand up a rope from the bottom of an 80-foot ravine.

He had just visited the falls, a generous free-falling 80-foot plume of water on Vly Creek as it channels through a rocky section of New Scotland between Krumkill and Normanskill roads.

"I envisioned a waterfall of about 20 feet high," he said during a pre-Christmas return visit this year. "I was very surprised, because I thought I knew every waterfall within 50 miles of here."

Declaring it a beautiful waterfall, Schaefer was quite pleased with his find, which is located on private property and thus known mostly as a swimming and fishing spot by the locals.

Access to the foot of the falls is down an old horse trail on Guido Mazzeo's property on Normanskill Road.

Mazzeo, who graciously allows strangers to park in his driveway and walk down the hill, as long as they stop at his house and ask, says the waterfall turns "beautiful, all silver in winter" with ice formations.

Most people would have been satisfied with such a find. Not so, Schaefer. Studying the US Geological Survey topographic map of the ravine, he noticed that the falls had no name. He decided it should be called Horton Falls, after the eminent hydrologist who lived and worked at the falls.

Robert E. Horton died in 1945 after a lifetime of accomplishment, including the design of the Albany city water supply. His breakthrough work in rainfall runoff synthesized geology, hydrology and meteorology, all subjects close to the heart of Schaefer…

…Well, Horton was a very remarkable fellow," Schaefer said, explaining why he felt the honor was due. "He was a scientist, an engineer, a pioneer in hydrogeology. He wrote several novels. He was known all over the world, but not here."

Both the American Geographical Society and the American Geophysical Union give lectureships, medals and fellowships in Horton's memory. "These are rather prestigious awards," Schaefer noted.

In August, he submitted a packet of maps, photographs and biographical material on Horton to the Geological Survey.

The survey's Board on Geographic Names decides on the official names of natural features such as lakes, streams, mountains and waterfalls.

The board approved his application November. 8, establishing the official entry as follows:

"Horton Falls: falls, 24m (80 feet), on Vly Creek 2.7 km (1.7 mi.) E of Voorheesville; named for Robert E. Horton, early 1900s founder of a hydrological laboratory at the falls; Albany County, N.Y.; 42°39'33"N, 73°54'24"W; USGS map - Voorheesville 1"24,000."

The next time the Voorheesville topo map is updated, the falls will be marked with its new name, according to Ernest Berringer of the board's staff. He could not say when that might be.

Horton's house was right on the falls, as was his laboratory, Schaefer said. The structure has since been converted to a large house owned by Dr. John Waldman of Albany, whose property includes the waterfall.

Waldman and Schaefer have not crossed paths. On learning that Schaefer had named his falls, Waldman said, "Well, I think it's neat."

Horton ran a vertical and horizontal iron water wheel to spin a turbine that generated electricity, which he used for heat, lights and power in his house and laboratory.

The water power of the Vly Creek was put to use as early as the 1700s, when the LaGrange Grist Mill ran on hydropower generated by falling water from a millpond created by a log dam.

Although most of Horton's work was done in the early part of this century, Schaefer noted that several of his basic formulas are still quoted by the scientific community.

It took Schaefer about a year to complete the research for his application. When asked why he went to all that trouble when most people would not have bothered, Schaefer looked puzzled.

"I'm not quite that kind of person," he explained patiently. "I do things because I have a lot of fun doing them. You know, there are puzzles, and you like to solve puzzles."

As for scaling an 80-foot ravine at age 82, he admitted it was a tricky traverse, but asked if he had been frightened, he was much amused at the thought.

"Oh no, why should I be?" he chortled. "I've been down steep places before."

The project also closed a circle that opened more than 40 years ago, when he was active in the American Geophysical Union and heard talk of Horton's having lived and worked in Voorheesville. "I've thought about this for a long time," he said. "It was very definitely unfinished business."

Judy Shepard Wolk
The Times Union, *December 30, 1990*
Albany, New York

A Friend of the Gunks

In 1974, Dr. Vincent J. Schaefer became a Research Associate of the Mohonk Trust. Vince had recently retired as active director of the Atmospheric Sciences Research Center at SUNY-Albany, which he founded in 1960. During the year, he made six three-day trips to Mohonk and Sky Top Tower to study the quality of air in the Shawangunks. Thus began a close friendship with the late Daniel Smiley until Dan's death in 1989, and with the Trust and its successor organization, the Mohonk Preserve, that lasted until Vince's death on 25 July 1993.

Most people connected with the Trust and Preserve over the years knew of Vince, his international prominence as an atmospheric scientist, and his connection with Dan and local research. However, few people knew of the breadth of contributions Vince made over the nearly twenty years of service to this organization. For example, in 1976, Vince joined the Research Committee. With Harvey Flad, Chairman of the Committee, he worked with Dan to strengthen the role of research. Not only to strongly embrace long-term monitoring and maintenance of the database started by Dan over fifty years before, but also to summarize the data in the form of research reports and natural science notes which could be used in stewardship and management of the land.

It was Vince who recognized early on the tremendous value of the unique records and collections that Dan had amassed– not only to the Preserve and Shawangunks, but also as a legacy to the world. In the late 1970s, Vince and noted biologist Frank Craighead recommended that these materials be housed in a fire proof, atmospherically controlled archive– a suggestion that Dan carried out in 1980 at his own expense. The present Daniel Smiley Research Center is the result.

When I became a Research Associate focusing on flora and the vegetational history of the Northern Shawangunks in 1974, I can well remember a very special field trip with Dan and Vince to define the 90-square-mile Northern Shawangunk Study Area, a map still in use today as our research baseline.

As an author, Vince donated signed copies of his books to the Research Center Library – among them *A Field Guide to the Atmosphere* (1981) and *Clouds and Weather* (1991), co-authored with John A. Day as a part of the Peterson Field Guide Series, and, most recently, *Vrooman's Nose* (1992).

Vince was interested in nearly everything– such as the archeological dig then being conducted by Research Associate Leonard Eisenberg at the Mohonk Rock shelter. In the early 1980s, he became interested in the glacial stones being unearthed and what structure pattern they might reveal inside if cut into thin slices, a process that he had developed over the years. This is a good example of his keen ability to take seemingly unrelated events or opportunities, which, by their convergence in time, led him into a new line of discovery.

The lasting legacy to the Preserve is the Schaefer Summer Internship in Research endowed by Vince and his wife, Lois, in 1991. The first student intern, Anne G. Rhoads of New Paltz, worked with us at the Research Center this past summer. Fortunately, Ann had the opportunity to get to know Vince before his death. From now on, each year new interns at the Preserve will benefit from the Schaefer's concern for generations yet to come. As Vince stated: "Our hope is that one or two summers as an intern at the Daniel Smiley Research Center will give…a student the crucial experience and background that will develop the…qualities of leadership we sorely need to solve our planetary problems for survival in the next century."

Our "thank you", Vince.

Paul C. Huth
Mohonk Preserve Research Center
New Paltz, New York

Serendipity and Andy Rooney

Vince, a long-time trustee of The Rensselaerville Institute, would join us faithfully for the semi-annual meetings of the board. It was at one of these that Vince and I cooked up a unique workshop for our "Minds On" program, "Serendipity in Science." Vince wanted to share his joy in approaching ideas and learning with the gifted students that the school districts in this region send to TRI for these full-day programs. When the workshop took place, Vince had a grand time first talking about how the "prepared mind" is ready for serendipity and then sending forty junior high schoolers out in the woods and fields on our campus with the simple– and for them mind boggling– mandate to use their eyes and creativity to let science happen.

It was also at a board meeting that an Institute neighbor met Vince, CBS' Andy Rooney. A personal friend, Andy is chary of public appearances during his Rensselaerville weekends and always made it plain to me that walking over for cocktails with trustees in Stonecrop was a not overly-welcome neighborly duty. That is, until he met Vince. After that, whenever I passed on the invitation to Andy and his wife, Marge, to drop over for drinks with the trustees, Andy would respond: "Is that Schaefer guy going to be there? Now he's an interesting person."

Mary-Ann Ronconi
Rensselaerville, New York

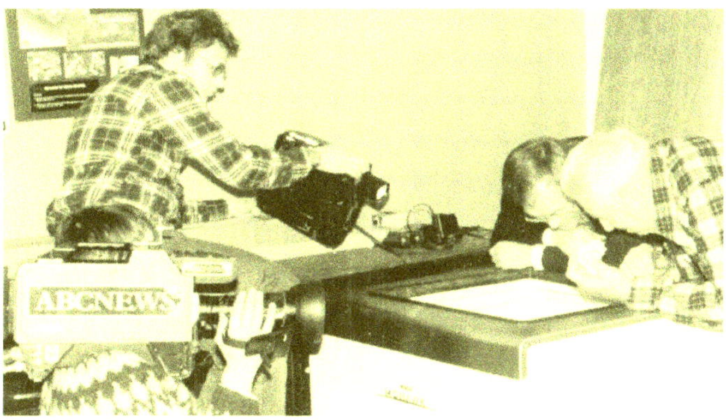

ABC News filming Vince demonstrating his cloud seeing experiment. Courtesy of Roger Cheng.

Chapter 15
Treasured Memories from Friends and Family

A Letter to the Editor

Dear Sir,

We lost a quiet man a while ago, when Dr. Vincent Schaefer left us.

In his 87 years, he achieved an incredible record of accomplishments. A book is needed but here are a few highlights: As a teenager, he produced and edited a newspaper which was distributed in England as well as in the US. As an adult he brought recreation to many in the dark days of the Depression by starting hiking clubs. He brought science to the layman by directing the excavation of Indian sites with professional safeguards and expertise. He was the first promoter of the great cross-country trails which are now being finally achieved. Being a Journeyman Toolmaker was of great help in his many inventions and patents. In World War II with Dr. Langmuir, Vincent developed a superior Smoke Screen to protect our troops. Experiments on this Smoke Screen were carried on near Vroman's Nose in Schoharie. Later, Dr. Schaefer led the movement to preserve Vroman's Nose.

Dr. Schaefer's book, co-authored with John Day,[61] is the definitive work on the subject and was a continuation of his pioneering work on snowflakes. In my search for ancient Indian Trails, Vincent offered not only the benefit of his vast knowledge of archeology, but was an active searcher in the hills at the age of 80! Perhaps one project of his will illustrate the capability of the man. He bought a 1700s Dutch Barn to use, but found it could not be repaired. He took the huge barn down entirely by himself, a feat far beyond the effort of most people. Being Vincent Schaefer, he photographed, measured, and made notes on each mortise. Instead of a pile of beams, he left a detailed study of one of our vanishing treasures.

The epitaph of Sir Christopher Wren might well apply to Vincent Schaefer, "If you would seek a monument, look about you."

Dan O'Neill
Delanson New York
Schenectady Gazette, *August 9, 1993*

---

61 *A Field Guide to the Atmosphere* (Peterson Field Guides).

Tales from Shared Boyhoods

My old hero of my childhood, Vince, lived near us when we lived at 30 Arthur Street in Bellevue/Schenectady. My Dad built the house my brother, Dan, and I lived in then. He went on to become athletic director of Exeter Academy; I finished my studies at Pratt Institute in Brooklyn, worked in newspaper and advertising art studios.

I last saw Vince and Lois about 1988 on Schermerhorn Road...even had a chance to stick our heads in his freezer mist experiment box. YEA! Amazing...great!

Vince started us off as "Lone Scouts" and Indian relic hunters. We probably walked over every foot of land along the "Old Erie Canal," Mohawk River valley and Meyer's Farm Creek, too! Later at his suggestion, we donated our collection– including 100 perfect arrows and spearheads– to the Schenectady Museum. I saw Carl a bit more than Vince, (1915-1922), when Gertrude was still a cute wee ankle biter. The old valley haunts me and calls me to return.

I'd like to find that fountain of youth and drink a gallon or two...take another swim in ol' Lake George, then roll-up under one of those 200 year-old Whites on top of one of those mountains... 'Twas ever thus?

Marty Fowler
St. Louis Park, Minnesota

Neighbors

My first recollection of Vince Schaefer dates from my childhood. I remember my great-aunts, Mary and Alice Schermerhorn, telling me that a young couple had purchased property across the road from their home. They also bought an old Dutch barn from their cousin, Clarence Schermerhorn, and a barn built in the mid-1880s for my great-grandfather. Both of these buildings were used in the construction of Vince and Lois' Early American brick home, which was located adjacent to the Dutch barn.

I vividly recall my aunts telling me of the arrival of the three Schaefer babies and of their delight in having Lois wheel them over to visit.

Years later, Sue and Kathie were occasional baby-sitters for my children, and Jim was a member of the Boy Scout troop of which my husband, Norm Fox, was scoutmaster.

We miss Lois and Vince as neighbors and friends, but are pleased to learn that Jim and his wife plan to occupy the home which was built as a true labor of love by his parents.

Helen (Marlette) Fox
Rotterdam, New York

A Friend from Bakers Mills Recalls

When my mother, Hester Rise Dalaba-Capwill, was in her late years, perhaps ninety, she wrote about the "real Edwards Hill lot," and Vincent Schaefer wrote to say he and Lois owned the lot, explaining its location and sending a photocopy of the land area. We appreciated his thoughtfulness to take time to write and explain.

Schaefers and Dalabas (John and Hester) were neighbors in Bakers Mills. Hester corresponded with Mrs. Peter Schaefer (Vincent's mother). I remember visiting her with Mama at home here. Once, Vince brought me an electric typewriter, lugging the heavy machine up our poor steps. Why to me, I thought, for a patient man such as him, to take the time to bring this into our humble home.

At an earlier time, after he published his book on the clouds and atmosphere, he brought us a beautiful copy of it. I was touched that he included us in his special friendship. We for years felt close to the Schaefer family and often were included in Vince and Lois' friendship circle with all the others of the family– even their 50th Anniversary picnic, when Lois took time to search me out of the circle to visit with me.

My sister, Violet (Dalaba) Heath and Lois were special friends. They are now gone; we have lost friends in the death of Vince and Lois and others of the Schaefer family. But I hope the life of those gone have left behind a rare life of beauty.

Daisy and Earl Allen
Bakers Mills, New York

Friendships Begin in Interesting Settings

My husband, Walter Orr Roberts, who was the founding director of the National Center for Atmospheric Research, was an astronomer turned astrophysicist turned atmospheric scientist, as his career progressed. I remember hearing about Vincent Schaefer's prodigious accomplishments in cloud seeding for weather modification almost from the beginning of our life in Boulder. But I don't think I ever met Vince until the summer of 1956.

In that year, we had finally made the momentous decision to double the size of our small house (by tearing down the outside walls of the living room and expanding into the space in the front yard) in an effort to give our four children a little more breathing room. We had also decided to try to live in the house– all six of us!– during the disruptive work of expansion and remodeling. It was not an easy summer.

Midway in the expansion project, Walter telephoned and casually informed me that he was bringing Vincent Schaefer and his wife, Lois, by for a drink before dinner. There was no way I could make the house presentable, so I simply got out some crackers and cheese, and when Vince and Lois came, we all sat down on whatever was handy (I remember Vince sitting happily on a pile of bricks that were waiting to be used in the construction of the new fireplace!) and had a wonderful, relaxed visit.

Indeed, the visit was so congenial that we exchanged Christmas greetings with the Schaefers every year from then on, until Lois' death just a year before Vince's. I shall miss them very much.

One comforting note: just this last May, on a brief visit to Santa Fe with my youngest son and his wife, I browsed the bookshelves at a garden center while they selected plants for their xeriscape garden. One slim little paperback jumped out at me, and I quickly picked it up and bought it. It was a new Peterson Guide -- this one to clouds and weather, authored by John A. Day and Vincent J. Schaefer.

I shall cherish this fascinating little book. Vince, I only wish you were here to autograph it for me!

Janet S. Roberts
Boulder, Colorado

Vince working with students. Courtesy of Roger Cheng.

A Younger Cousin Writes

After hearing of Vince's death, I found it hard to believe, because he seemed like he would be forever.

My first real memories of Vince were during the construction of the home he built on Schermerhorn Road. I helped my Dad, Joe Holtslag, Sr., in the installation of all the wiring, plumbing and heating. I first learned the trades there.

Vince had me make a sign for his brother, Paul, with the homes of both brothers in outline on the top, painted in color as the homes actually were. It was through this sign that I got to know Paul. So you could say that Vince started me toward the building trades.

More memories might have been possible, but the difference in our ages (14 years) does not seem like much now, but in my early teens, I must have been just a kid to him.

He was a great man.

Joe Holtslag
Utica, New York

Grandpa Memories

I remember when I was little, Grandpa and Grandma were visiting us at our home in the mountains outside Boulder, Colorado. Grandpa was outside testing the air with a big plastic bag with what looked like a stick or something like that inside, and an air pumper. I asked him what that stick in that bag was, and he took it out to let me smell it. After taking a deep breath and smelling it (it was methane), I kind of wished I hadn't asked!! I didn't like that smell at ALL!

The summer of 1972, the family went for a hike together up at Grandpa and Grandma's camp in the Adirondacks, and at one point, we stopped to catch our breath. Grandpa had a beer he was drinking to cool off. I was standing nearby, and, being only about four years old, was just about the height of the bottom of his hand. Grandpa said, "Oh, look, a beer stand," and he put his beer on top of my head! He was fun to take a hike with!

Andrea Miller

A Treasured Relationship

My earliest recollection of Grandpa Schaefer is from the summer of 1967 or 1968, down in Flagstaff, Arizona. The whole family (Millers, Sullivans, etc.) would go to Flagstaff where Grandpa was working for the summer.

Grandpa Schaefer was always measuring the atmosphere and doing other "weird" (to a four or five year old!) things to analyze the air. We grandkids were inquisitive, but didn't really understand it.

The family has always been very close. The best representation of this is depicted in a family picture taken by Grandpa in front of their Adirondack camp, Woestyne North, in 1972. Through the years, we have had various get-togethers, but the most memorable to me was the occasion to celebrate Grandma and Grandpa's 50th wedding anniversary the summer of 1985. After spending part of our time at their home, Woestyne South, in Schenectady, we all went up north to camp to relax and enjoy the mountains. Before we left after a fun day, we assembled all the kids, mothers, fathers, etc., for a family picture, positioning ourselves exactly as we'd been in the family picture taken back in 1972. (It came out GREAT!! What fun it is to compare both!) Though we'd all grown up– in both size and age– we looked not much worse for wear. It was really a lot of fun and made for a great weekend. After that weekend, Grandpa and Grandma and I started to converse more often, mostly because I was in the Navy and they knew that it was important for us to receive mail once in a while when we were deployed. Grandpa was always curious about my job in the Navy and in the submarines I rode. In looking back to my years of living in Groton, Connecticut, I realize that I should have had them down to take a tour of the submarine, because they imagined me aboard a submarine similar to the World War II boats, which wasn't the case at all.

As time went on, I started to write more and more often, especially after Grandma died. In the months after that, I wrote to Grandpa about once a month. He was always very supportive of my position in the military as well as the outside ventures that I became involved in. He played an instrumental part in my decision to exit the Navy, allowing my "wings" to expand and take to flight.

I received a letter from Grandpa about the 7th of July 1993, which was in reply to my letter describing my business mentor with whom I worked after leaving the Navy in Hawaii. Grandpa said it was good for me to have someone like Bob as a mentor, and to pay very close attention to all the different things that he does and to be attentive when he speaks, as he may say something very worthwhile for my future growth.

After I learned that Grandpa had died, I sang a song called Grandpa at a Karaoke bar in honor of him, since I was unable to attend the funeral. Someday, I hope to get to Schenectady to visit both Grandma's and Grandpa's final resting place.

(James) Peter Miller

From the Pen of a Granddaughter

Dear Grandpa,

I want to let you know how much I love you and respect you. I don't know why it is that I waited until you are very sick to let you know, but I know I don't want to leave this unsaid either. You and Grandma have had a very powerful influence on my life. As a child, I remember that you helped develop my conscience. If I was thinking of doing something questionable, I would ask myself if I would want you or Grandma to see me do it.

I have always admired your faithfulness to God and your steadfast love and dedication to your wife and family. You also seem to keep a nice balance between work and play and enjoy working. Your interests ranged far and wide and you are involved in a variety of groups and endeavors. I remember when I was a freshman in high school, when sitting in an Earth Science class, I suddenly saw you and your cloud seeding experiment in the film we were viewing. I whispered to a friend of mine, "That's my Grandpa," and she proceeded to announce that to the whole class!

You were and are also willing to make time for friends and relatives. I remember when Todd and I were in Burlington, Vermont, that you and Grandma drove all the way to see us and have lunch with us, then drove back. You have made for me many wonderful memo-ries, from birthday celebrations together, to hikes, to rides on the back of your tractor, that have helped to form me to be the person I am today. I thank God for you and for having had the chance to know you.

Love,
Kathy

[Kathleen Sullivan, daughter of John and Sue Schaefer Sullivan]
Bangor, Maine
July 20, 1993

Lady Slippers

I found out late that Uncle Vince liked wildflowers, though I should have known, as he was interested in everything in the natural world. The next spring after he told me he had never seen showy lady slippers, I let him know when they were at their peak in the cedar swamp near Brant Lake. This orchid is rare because it needs acidic sphagnum on top but calcium enrichment down below at root depth. There are a few places in the Adirondacks where Grenville marble adds to the numbers of species that are able to grow, but this "rich fen" is the only known site for the three-foot high, pink and white lady slippers. And there are hundreds of them, if the deer don't eat them!

So, on July 4, 1991, we met at the Northway and I guided Uncle Vince and Aunt Lois to the site. As Aunt Lois was already quite frail, I took them in the short, flat way, but that meant stepping over a narrow stream. With Uncle Vince's help, she made it and was then near a group of the eye-boggling flowers. She was content to stay there and admire them, while Uncle Vince went farther in, taking pictures of even more amazing clumps of the beauties. The exclamations of delight were quite gratifying!

Mom had set me on the lady slippers' trail (a well-kept secret as the locals in the know are very protective of them) after seeing Mary Reynolds'[62] journal about them. It had taken me a couple of years to find them, too late for Mom, unfortunately, so I was especially glad that Uncle Vince and Aunt Lois did get to see them. He wrote shortly afterwards saying that they were one of the best birthday presents he had ever had– I hadn't known it was his birthday!

PS: I'm also on the trail of a mineral called serendibite, which Uncle Vince told me about. He said it is only found near Crane Mountain and in Sri Lanka! I know someone who may have collected the last specimens from around here.

Evelyn Greene, North Creek NY
(1 of 21 Nieces/Nephews)

62 A relative on the Schaefer side of the family.

Wildflowers, Barns—and Serendipity

When Uncle Vince died, my husband Perry and I were in Seattle at a Peace Corps reunion. I had seen Vince in the hospital just before we left and brought him a stalk of "Pearly Everlasting." This wildflower grows in the Adirondacks, but finding a clump of it growing by the sidewalk near Ellis Hospital was unusual. It was the kind of thing Vince would have noticed. It was hard to come back from Seattle and simply find him gone, and know we would not see him again. It left a big ache. He hadn't gone, though; he was just waiting for us to find him again.

One Sunday afternoon in September, Perry and I took a ride out of Delmar southwest into the countryside to look for a place to walk and pick dried grasses for the house. Driving through the hamlet of Feura Bush, we approached a fork in the road. "Which way?" asked Perry. "Oh…left," I said, attracted by what looked like a road less traveled. It met my expectations, narrowing and getting wilder. After about a mile, we crossed a small bridge and passed a magnificent old Dutch colonial manor house, flanked by huge, twisted locust trees, with outbuildings and a big outdoor fireplace in the rear.

As we slowed down to look, we noticed, in a field beyond the house, a Dutch barn just like the model Uncle Vince had once constructed. On the fence by the road was a sign that read "Barn Visitors Welcome." "Perry, stop!" I said. "Uncle Vince had a friend who relocated and restored a Dutch barn. Could this be the one? What was the man's name? Touhey, I think…Carl Touhey."

We approached the barn, feeling a little like trespassers, but--"Barn Visitors Welcome"! "The Jan Wemp Barn" read the New York State historical marker by the path. We entered the open door and wandered through. Much work had been done to make interested visitors welcome, because on each of the main posts were hand lettered placards explaining the structural elements of the barn, how they were crafted and assembled, and how the whole structure was used. The uniqueness of the Dutch barn as a design, the understanding of physics that the original builders had brought to bear, and the beauty and functionality of the structure came through in these brief, lucid explanations.

The Wemp Barn was clearly a tribute, by a group of devoted historians, to the artistry and skill of a past generation, but it also seemed like a gift, to us and to future generations. With a sense of recognition and gratitude we read, at the bottom of each hand lettered sign, the initials, "VJS."

RoseAnne Fogarty and Perry Smith
Delmar, New York

Sue Remembers

Dad had asked me to go on the shakedown cruise for the eclipse in 1973, which left New York in February. Mom had been told by her doctor not to go because of her respiratory tendencies to pneumonia when climate changes were experienced. Even though I wondered about the validity of such limitations on her fun, she said she'd happily be the fill in for me (at home) if I'd go with Dad. NO PROBLEM!!

Dad and I boarded a bus in Schenectady bound for New York and we were on our way.

We chatted, watched scenery and read. Looking up from my book at one point partway to the city, I saw Dad had a notebook on his lap and was looking, then writing, looking again and writing columns of numbers and checking the second hand on his watch with regularity. Curious, I asked what he was doing. From our seat on the right side of the bus, we could observe the driver, who was smoking. What held Dad's fascination was the drivers' puffing and inhaling the smoke of the cigarette. Dad recorded each puff taken, and when a new cigarette was lit from the previous one, he made a note of the new start. The smoker was totally unaware that he was the subject of this scientific curiosity that we, who knew VJS, saw manifested throughout his long, much-enjoyed life.

While on this cruise, my job was to be scientific assistant when Dad sampled the air on the open sea, and at the ports of call in the Caribbean. I also had to figure out how to darken the windows in some of the salons on board so that slides and movies could be shown in the daytime during the eclipse cruise in June of that year.

Dad always bought papaya on an island and kept it in his room. Early each morning he'd knock on the wall of his room and say, "Sue, are you up?" Some quick dressing happened after the "Yes," and we'd share a piece of the fruit and head up to the top deck to do some early readings of the particulate matter in the air as the sun came up.

The privilege of being executrix of the Schaefer estate provided some interesting discoveries. One being that Dad proposed to Mom at one of their favorite haunts in the Albany area, Ghost Fire Bend, on a creek in Guilderland, New York.

When I was a teenager and in high school, one of my fun outings happened once a month when Dad would let me go with him to the meeting of the Van Epps-Hartley Historical Association at the Shrine of Kateri Tekakwitha in Fonda, New York. There were two requirements; that my homework was done, and that I bake some kind of cake or cookies to take to the meeting.

At the meeting, discussions concerned the archeological digs progressing at the sites of the Iroquois long houses discovered on the hills above the shrine. The remnants of the poles that held the buildings' roof were there in the ground as well as the cook fire charcoal spots. Much was theorized about life in the days of those native peoples from evidence found at the digs. Plans for future work were outlined, and locations of recent finds were plotted on permanent graphs.

But the real fun of it all was being with my Dad who always made me feel valued. He gave as much credence to my ideas as he did to other people in his fascinating group of friends.

After exams one year at Potsdam College, the Dean called me to his office because my mark on one exam was low. He said he was calling my Dad to let him know, then he wanted me to talk to Dad.

When I got on the phone, Dad asked me how school was going. He said it must be disappointing to receive a low grade. Was I doing my best? I replied, "Yes, I am." He then said, "All we ask is that you do your best." He hoped to see me soon and signed off. The Dean was surprised to see me so soon. When I related what Dad had said, the Dean looked disappointed!

Dad had a massive heart attack on July 15, 1993. He was in the hospital for ten days, lucid, listening, making comments about our conversations, sleeping, before he gently died. During the time he was resting and we were getting used to his slow journey to his reward, we made sure his friends had a chance to see him, discuss ideas and in so many ways, say goodbye.

One day, while Kathie, Jim and I were visiting, Dad said to get the nurse with the blue jacket and ask her to come into the room. When she came in he "laid down the law," that, "at no time are my children to be kept out of my room. They are to have access day and night." She humbly agreed.

One day during that time, I began to realize what a task I had ahead of me as executrix of an estate of a genius who loved books, made notes on all sorts of paper, and kept letters, photos and had collections of all kinds of scientific paraphernalia in his lab. Visiting Dad that day, I jokingly remarked, "Dad, it is going to take five years to take care of all your estate business!" He replied, with a twinkle in his eyes, "That is a conservative estimate!" (He is right!!)

When I was small, we spent lots of time at Camp Cragorehol in the Adirondacks. Many days, Dad would go fishing early in the day, and we would not see him until almost supper time. We could see him coming up the path with his easy gait holding a creel made of birch bark sewn together with twigs. We kids would run to him and ask, "How many did you catch?" His usual reply was a casual, "Oh, a couple." Upon opening the creel and looking in, we would see a bed of fresh green ferns, cradling beautiful, cleaned, Rainbows, Browns, and Eastern Brookies ready for the cooking pan. The count??? EIGHT!!

Susan Schaefer Sullivan
Sudbury, Massachusetts

Kathie's Dad

So much has been written about Dad– about his youth, his early working experiences, projects and discoveries he made after becoming an assistant to Dr. Langmuir, and so many other facets of his life…and there were many!!

As a family, we had a closeness that wasn't touchy-feely, but we all loved each other in a very special way, and knew it without having to dissect our relationships! Of course, Dad's word was 'it,' but he never 'lorded' it over us, instead coming at decisions from a 'let's figure this out together' way.

He supported each of us individually, and I can't tell you the trust and confidence he instilled in me, the middle kid who never wanted to be out in front or taking risks!! Yet when I was struggling with uncertainty as a sophomore at Potsdam State Teacher's College in the late '50s/early '60s, and after attempts to discover how to proceed with my life, he and Mom helped me find a school in Boston and a place to live (at great time and expense, I know!!) so I could learn to be a dental assistant. Then what to do? Heck…I couldn't stay near home! I had to go clear to Montana to live! How? Dad and Mom bought me a little blue Volkswagen Beetle. They sent me off, without a list of "don't do's"…trusting me to head to Montana and find a job, which I didn't have when I left home! They really trusted that they'd brought me up right, apparently, and also had a huge faith in God that He'd protect me! (He did!!)

But the thing I'd like to share about Dad is something that perhaps many don't know and few would expect. When Mom was in her mid-to-late 80s, she became more and more incapacitated by osteoporosis. Moving was painful and so walking and standing– as in doing her normal homemaker thing– was just not possible. She tried, but found it too painful. Having been a nurse before marrying, she was the consummate 'server' type of personality. So it was very difficult to be in the position of having to have Dad make dinners, do the laundry, change bed linens, grocery shop, etc. He took on those tasks seemingly with gladness…made friends with the butcher at the grocery and came home with fabulous meats for cooking over the fire in the basement, doing the other household tasks without complaint. I don't think he knew how to complain! He was just a joyful, happy person at whatever he was doing.

During Mom's final five years or so, Dad got up early every morning, warmed water in a Dutch oven on the stove, soaked a towel in the warm water, squeezed it out and put it on Mom's back to warm her up and make her feel good! With Mom's input, he made bread, cookies, even wild grape jelly! He fixed up a desk chair (the kind with rollers) so Mom could easily move from the bed to her chair in the living room, or get to the bathroom easily. He even set up a porta-potty chair right near the bed toward the end of her life, to make her life easier. Mom is the one who complained to me occasionally– not about how much she hurt, but about the fact that Dad had to do 'her stuff' around the house! But his selfless, kind care of Mom will always remain as a special quality of this great Dad of mine! He was a special man in many, many ways!!

Kathie Miller
Sun City West, Arizona

The First Arrowhead—and the Bear 'Facts'

It's hard to pick one or two memories of Vince…there are so many!

One very vivid to me comes from long ago, when our mother asked Vince to take our brother, Carl, and me (ages about eight and ten) for a walk while she prepared Sunday dinner. We went out Campbell Avenue from Arthur Street where we lived at the time, to the old racetrack, and walked around the track. All of a sudden, Vince stopped and picked something up from the ground, gave a yell, and started running for home– with Carl and me trying to keep up. When we finally reached home, he showed Mom his first arrowhead. It was the start of his collection, which I'm sure numbered hundreds.

Another story appears in the Camp Cragorehol record book, and was written by our granddaughter, Maggie Flad recently. Right before we came to Camp, I heard that Great Uncle Vince had died. I wasn't able to come to the funeral, but Ethan [brother] went. The first thing I thought of when I heard the news, was one summer when Rowan and I invited some friends up to stay at camp. There was some kind of party or reception going on, and we kept sneaking food back to the tent on the platform. We were planning to have a feast that night. We thought no one had seen us, but I guess Uncle Vince did, because he came to us and told us not to be afraid of sleeping in the woods, and that the bears wouldn't want us unless we had food in the tent. Because the bears would smell it and come after us. Well, we were all so scared, that we ate all the food right away, and stayed up almost all night waiting and listening for the bears.

Gertrude (Schaefer) Fogarty
[Sister of Vince]

Short and to the Point

What a great world this would be if everyone contributed even a small fraction of what Vince did!

Peggy (Schaefer Mearns) Allen
(sister)
Albany, New York

A Letter from Vince to His Daughter Sue

February 1989

Dear Sue,

Sorry I missed talking to you this evening. I thought you would get a kick out of a series of problems I had during the past week or so. It started with a fairly heavy snowfall which plugged our driveway.

I got out the Bob Cat Snow Blower, which I have had for many years, and, despite minor repairs, was doing a fine job. As I was approaching the road with my first run, the chain driving the auger came apart. This meant a trip to the hardware store (Wallace Armor). I decided to try getting out with the Escort, and its front wheel drive. Managed to get out of the garage, but soon got bogged down since the snow was so deep, it served to slow down forward progress because it was dragging underneath. As I turned to go up the driveway, it was obvious that I couldn't make it, especially since the plow had piled a ridge at the edge of the road. Then I discovered I had a completely flat tire! It would have been a real mess to change the tire where I was, but I hesitated to try to retrace my route into the garage. Then I remembered I had a can of inflatable goo that I hadn't used. Getting it, I applied it to the tire and it inflated and is still OK! So I decided I had to shovel the driveway, which I proceeded to do.[63] I gave it a rough job– just enough to get out, which I did. I got two links, returned home and soon had the chain repaired. Then I found the chain was too loose, and I had to spend more time shimming up the gear which helped to tighten it to drive the augur. Within another hour or so, I had the driveway and paths cleared.

Then we had another storm– this time nearly a foot. About noon, the next day, I figured the storm was about over, so started to get the blower out. I found that one of the tires of the blower was flat, and in serious trouble! I took the wheel off and got chains off and finally, after struggling for a while, got the tube out and found the valve was ripped! So – back to the shovel. This was the first time I shoveled the driveway two times in a row! I finally got out of the driveway and stopped at Randy's to get a new tube. Found they didn't have one– Randy suggested All Seasons. Went there, and eventually got a tube. Took it back to Randy's to have the tube put into the tire. After quite a long wait, he came out and said the tube was too big! Back to All Seasons and eventually got a smaller one. Back to Randy's and after another half hour, got the inflated wheel as good as new! $3.00! Thence back to Woestyne, replaced wheel and changed oil in blower and finally got underway.

Started clearing paths and suddenly the blower made a loud noise and stopped– a broken piston rod! So– nothing to do but to buy a new snow blower!

Went back to All Seasons– looked at new models– and bought a 24" 7hp Ariens. However -- there was only one on the floor and I was promised an assembled unit by evening. I forgot to mention that to get the blower, I had to reactivate my truck which meant changing into snow tires! Did so at about the time that Lo asked me if I'd take her to Peggy Wilkins in Charlton. Did so, under blizzardy conditions. On the way back, we stopped at All Seasons and picked up new snow blower. Returning home, I put the truck in the garage with plans to try it on Sunday. Thence to 5 pm Mass.

Thus it looked like at least part of my problems were solved. However, I still had to fix the garage door! Both springs had broken recently so I had to get new ones and figure how to install them. The weather had been so cold that I couldn't feature handling screws, clamps, pulleys and wire with bare hands! However, the last few days have been milder, so I got one spring in yesterday and the second one today. I then had to get one of the tracks straightened and then finally got it working.

How's that for problems– successfully solved– eventually!

---

63 Editor's note: VJS was 83 years old at this time!

The new blower is as good– possibly better than my old "Bob Cat" and about half of the weight.

Mom seems to be having a good time with her knitting and weaving. Hope you are having fun!

As ever,

Dad

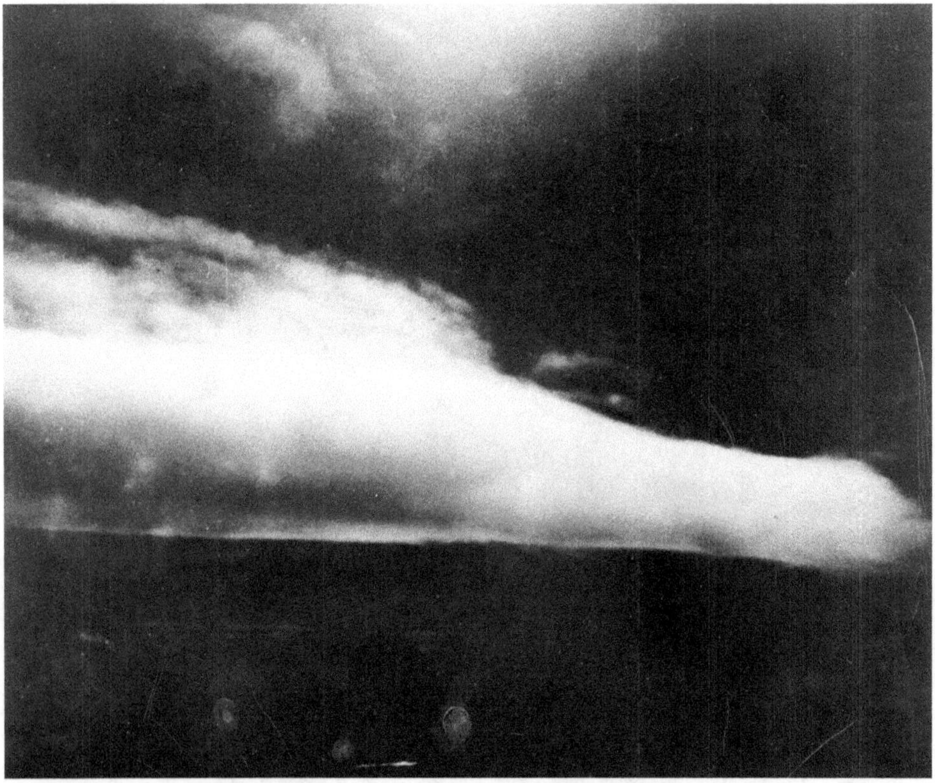

First cloud from which man created snow on November 13, 1946. Courtesy of Schenectady Museum.

Appendix A
Project Cirrus Photos

The following series of photographs of Project Cirrus are courtesy of the Schenectady Museum (now the Museum of Innovation and Science).

Cloud Seeding Flight Team, 1947. Vince Schaefer is tallest standing under bomber canopy.

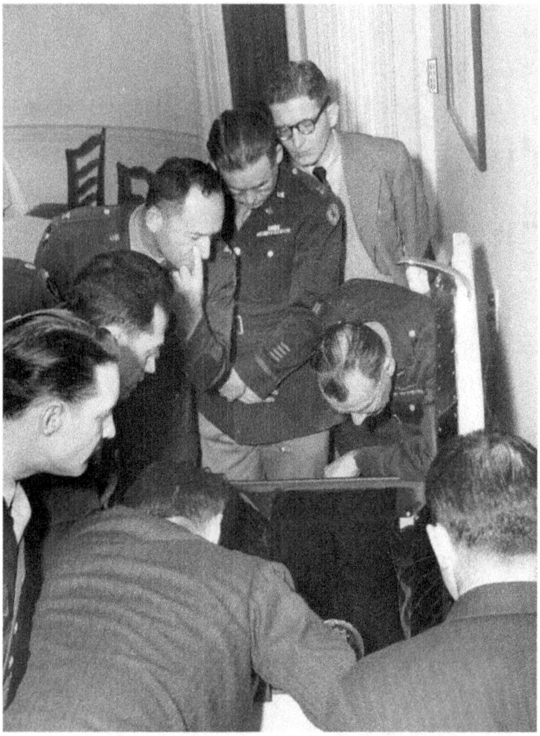

Army Signal Corps visit in March, 1947.

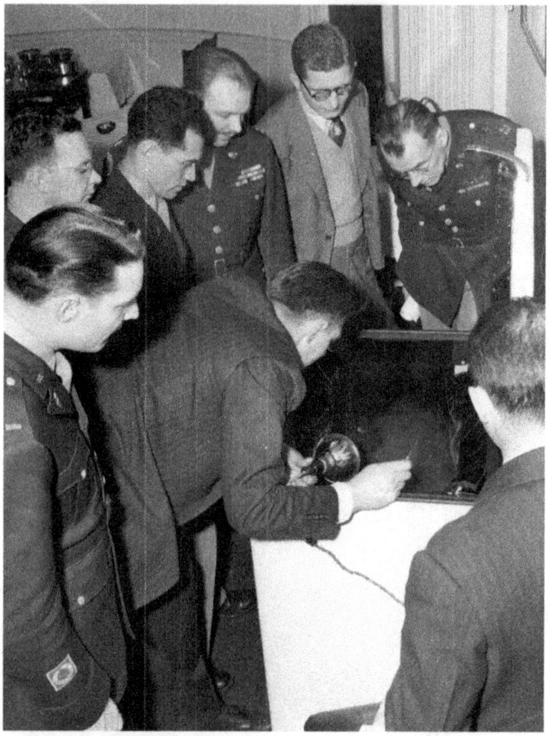

Army Signal Corps visit in March, 1947.

Dr. Irving Langmuir talking to an unidentified man.

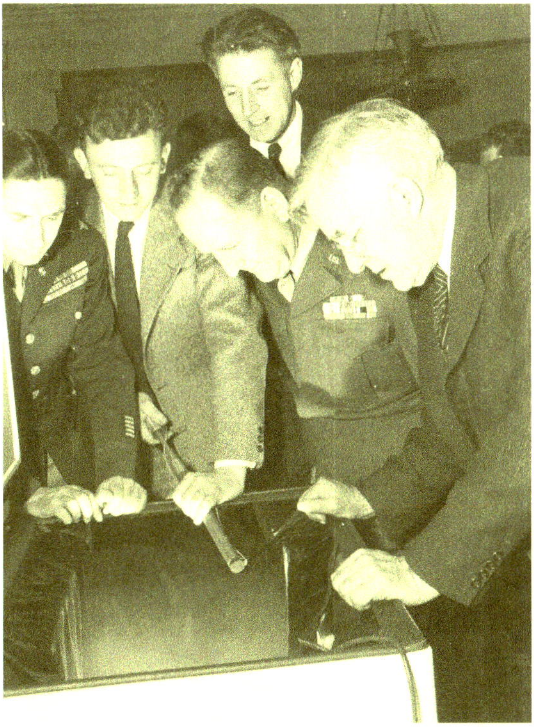

Personnel examining ice box experiment shown by Vince Schaefer, Bernie Vonnegut and Irving Langmuir. Vonnegut is using a pop gun to create clouds in the chamber. These photos were taken March 13, 1947.

Bernie Vonnegut using a silver iodide generator operating at night, which generates a shower of sparks. The microscopically small particles of silver iodide get carried by air currents into clouds over a wide area. This photo appeared in the April 1950 issue of Popular Science *in an article titled, "Can We Make It Rain?"*

Crewman carrying bags of dry ice or silver iodide to this B-17 Bomber used for Project Cirrus. This photo also appeared in the April 1950 issue of Popular Science.

Sugar Plantation Co. makes first manmade rainbow in rainmaking experiment.

Falling shower of rain, sugar plantation.

Segment of loop of greek letter in cloud.

Vince recording cloud movement.

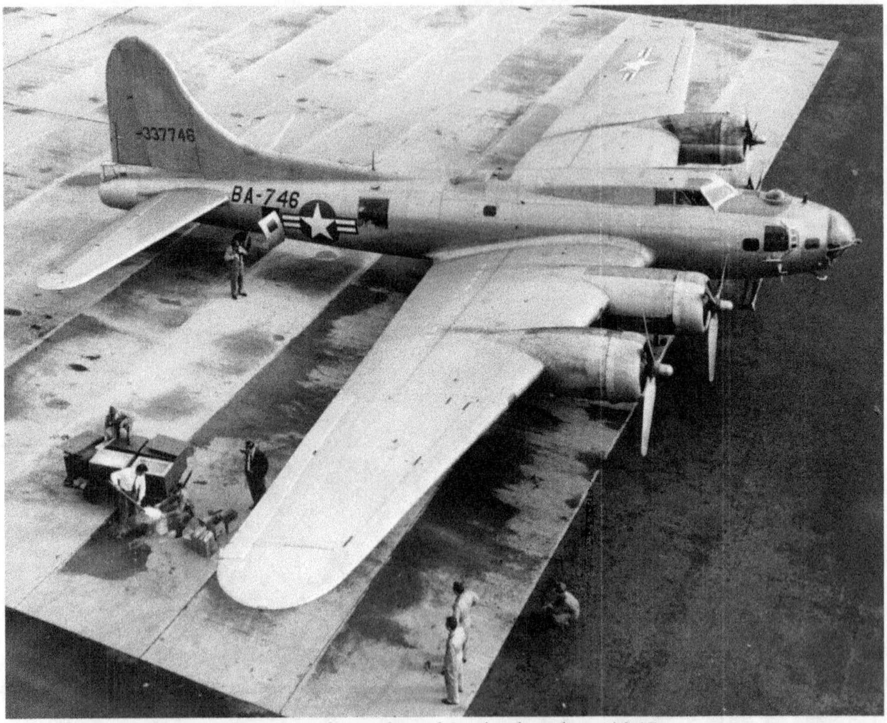

Top view of aircraft used in cloud seeding, 1947.

Vince putting on flight suit.

Vince and crew looking over flight plan.

Loading dry ice.

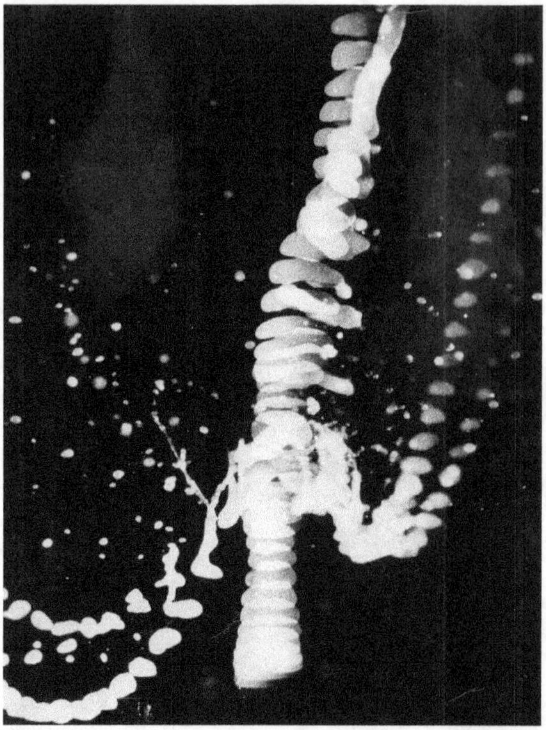

Water drop in shock breakup experiment on March 2, 1949.

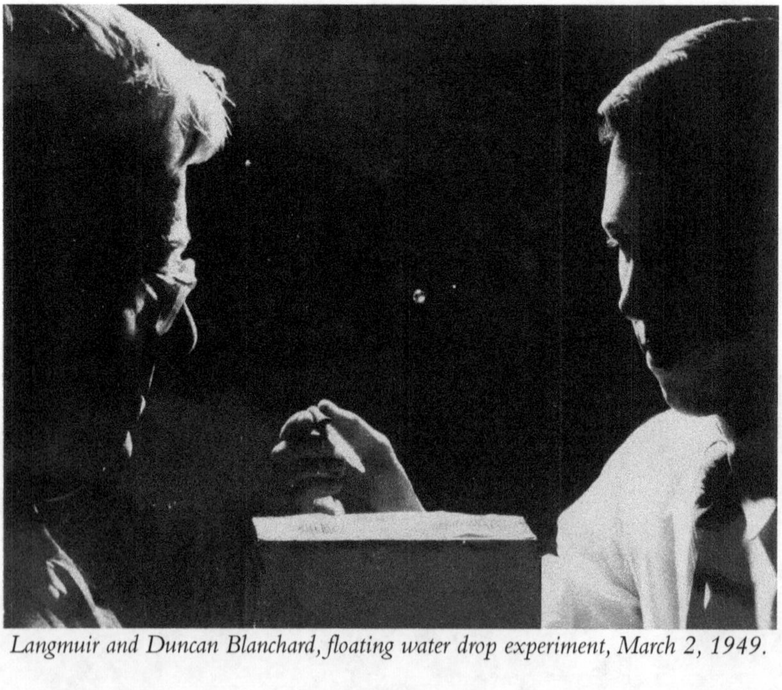

Langmuir and Duncan Blanchard, floating water drop experiment, March 2, 1949.

Seeding of cloud in shape of gamma.

Loading dry ice into plane.

Math calculations before flight.

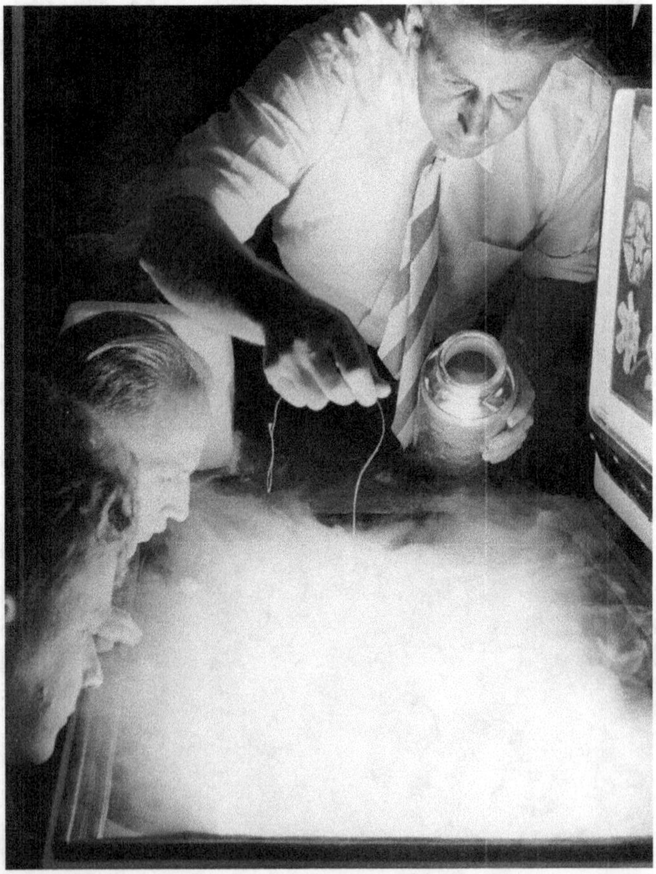

Heat wave snowstorm created by Vince.

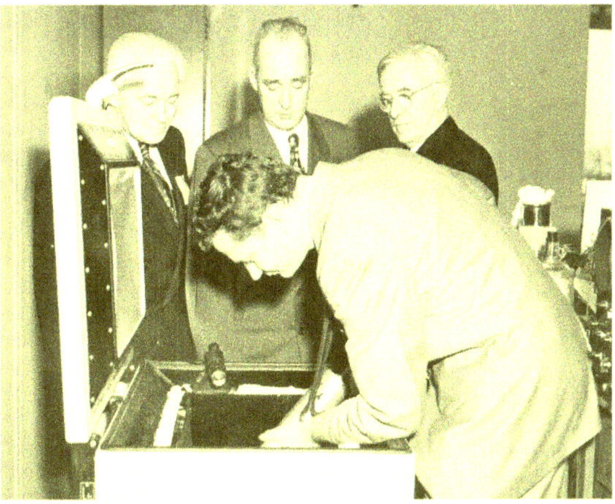

Top left: Vince showing the snow making experiment. Top right: Bernie Vonnegut shows snow making with a pop gun, February, 1950. Center left: Vince showing map of rainfall in New Mexico, February 1950. Bottom left: Bernie Vonnegut and Vince Schaefer with preserved snowflakes, January 11, 1951. Bottom right: Snow crystal recorder capturing the shape, size, and frequency of falling crystals.

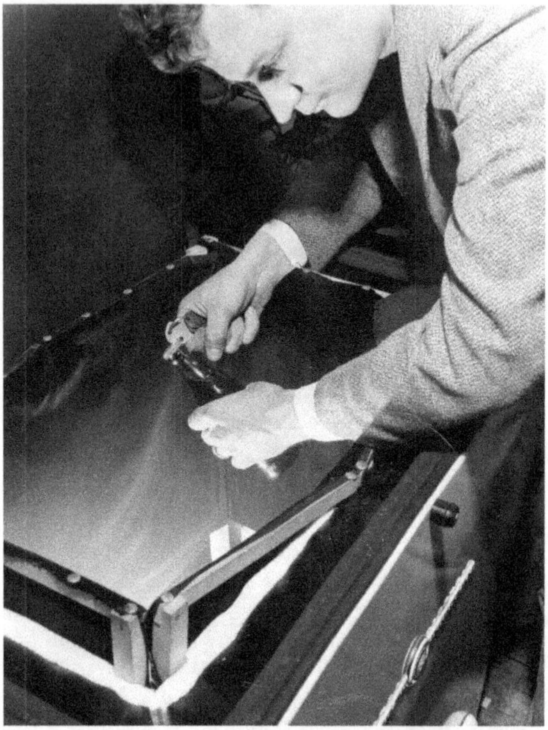

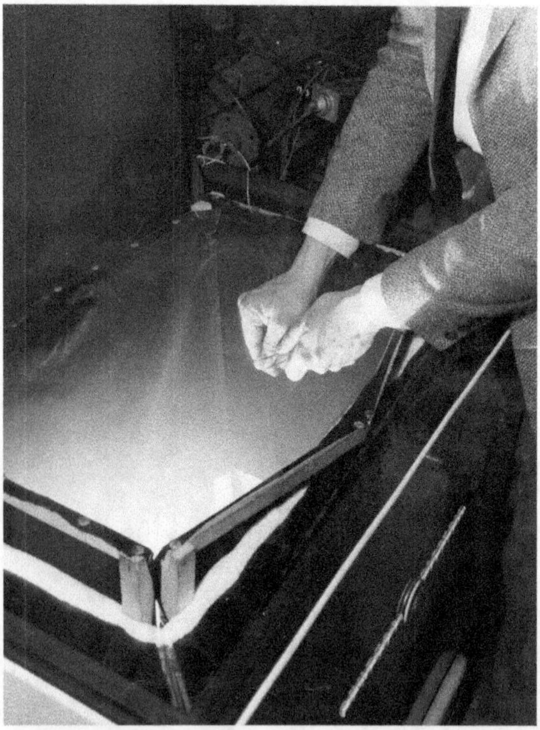

Top left: Bernie Vonnegut uncapping a soda bottle in a cloud, showing that it produces a snowstorm (February 14, 1952). Top right: Bernie's popgun turns a cloud to snow. Left: Bernie shows that a tiny balloon produces a blizzard in a cold chamber.

Opposite page: The "Rain by Fire" press release issued by GE on October 28, 1948. (See photo on page 371.)

From General News Bureau (H)
GENERAL ELECTRIC COMPANY
Schenectady 5, New York

For Release after 7 a.m. EST,
October 28, 1948.

RAIN BY FIRE. Dr. Bernard Vonnegut, weather scientist of the General Electric Company's Research Laboratory, uses fire as an agent to dispense tiny silver iodide particles into the atmosphere. Fiercely-burning charcoal impregnated with a silver iodide solution emits thousands of sparks, each of which produces millions of silver iodide particles.

In the sky, the particles serve as nuclei upon which super-cooled or below-freezing water droplets in a cloud crystallize into snow. The snow then may turn to rain, dependent upon temperature and humidity of atmosphere near the ground.

According to Dr. Vonnegut, the charcoal-burning generator dispenses one hundred million million particles of silver iodide per second, which theoretically would be sufficient to seed a cubic mile of atmosphere at a rate of one particle per cubic inch.

Laboratory tested thus far, this generator and others developed are expected to undergo extensive experimentation in actual weather conditions by the U. S. Army Signal Corps and the Office of Naval Research under the weather research program known as "Project Cirrus."

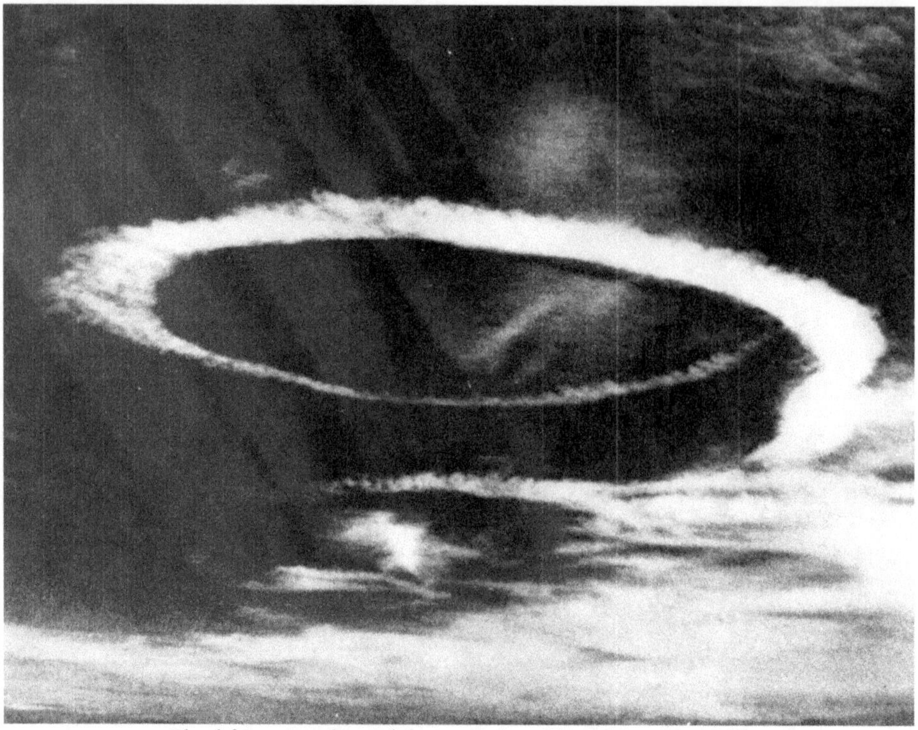

Cloud formation observed during Project Cirrus, Nov. 20, 1952.

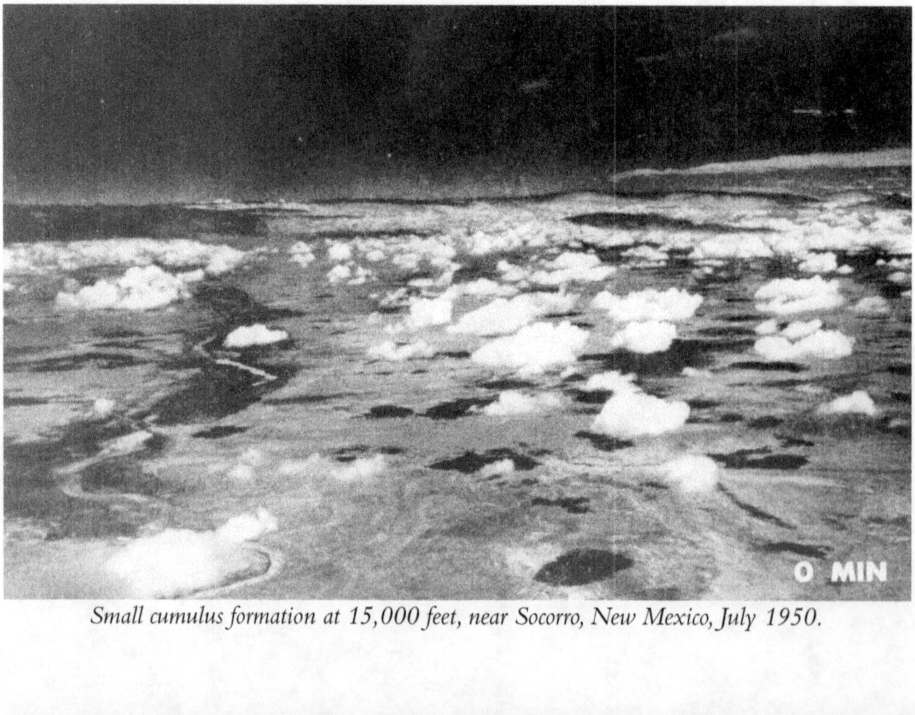

Small cumulus formation at 15,000 feet, near Socorro, New Mexico, July 1950.

Cloud formation 20 minutes after seeding (November 20, 1952).

Figure "4" formed in stratus cloud deck, November 20, 1952.

Gamma figure formed in stratus cloud deck.

Snowing in a cleared area of a cloud after seeding.

Appendix B
Project Cirrus Hurricane (H-1) Seeding in 1948

The following series of photographs of Project Cirrus are courtesy of the CECOM LCMC Staff Historian's Office, Aberdeen Proving Ground, Maryland. These were taken in 1948, when Project Cirrus first began cloud seeding of hurricanes.

Members of Project Cirrus H-1 flight team. Vince is in the bottom photo, fourth from left.

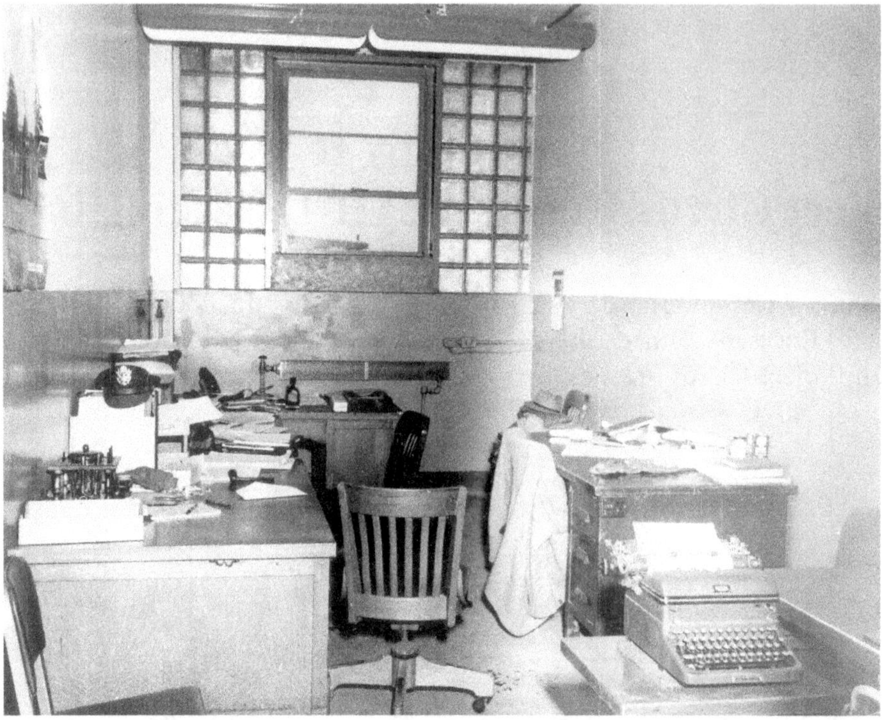

Top: Project Cirrus office at Schenectady County Airport, in the General Electric Hanger. The room was nine by twenty feet. Bottom left: Meeting of the minds at the the GE Hanger. Bottom right: Irving Langmuir taking photographs for the mission.

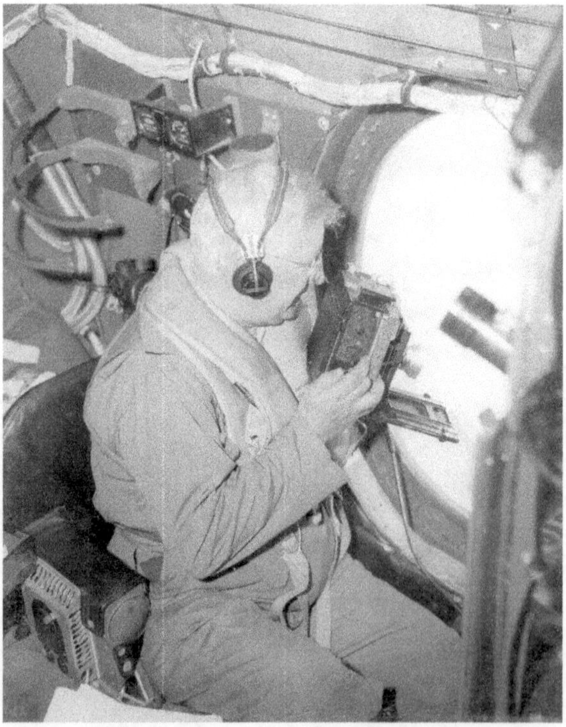

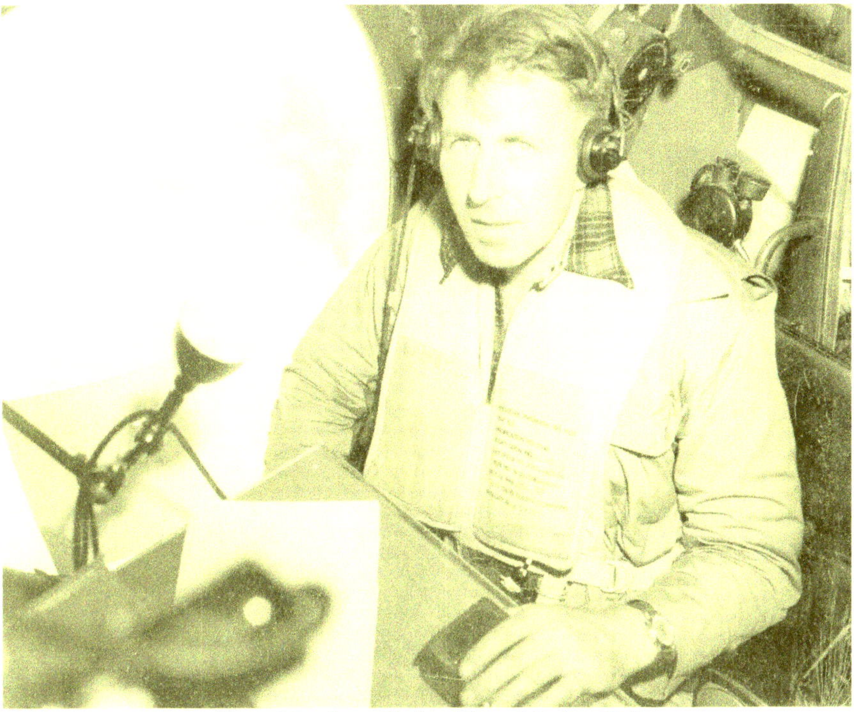

Vince Schaefer on Project Cirrus Mission H-1.

An unidentified member of the Project Cirrus H-1 team.

Unidentified members of the Project Cirrus H-1 team.

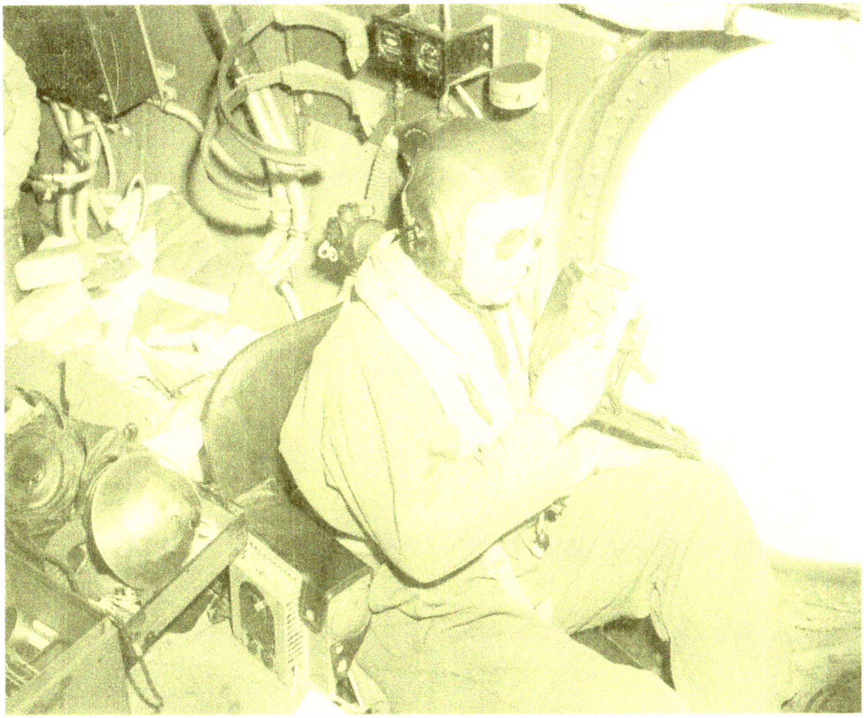

Unidentified members of the Project Cirrus H-1 team.

Unidentified members of the Project Cirrus H-1 team.

The Project Cirrus flight team prepping equipment fora mission.

A Project Cirrus mission comes to a halt.

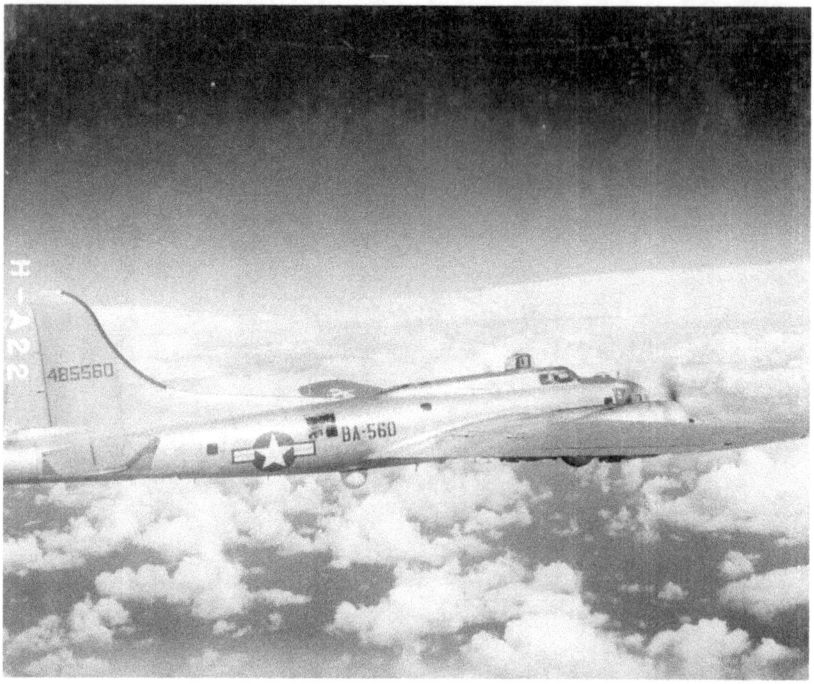

Boeing B-17s used for the Project Cirrus flight missions.

H-A 47

H-A 49

Index

GENERAL NEWS BUREAU

(RH)

GENERAL ELECTRIC

SCHENECTADY 5, N. Y.

For release in morning papers
of Thursday, November 14, 1946

ARTIFICIAL SNOWFALL CREATED FOR FIRST TIME

IN GENERAL ELECTRIC RESEARCH LABORATORY

SCHENECTADY, N. Y., Nov. 13 - Man-made snow, every bit as real as that which makes for a "white Christmas," has been produced for the first time by Vincent J. Schaefer, scientist of the General Electric Research Laboratory, it was announced here today.

Although the snow was created in a laboratory cold chamber, Schaefer believes that the technique used will work just as well out-of-doors to make actual potential snow clouds crystallize and shed their snow when and where man wants it. In pursuit of this possibility, he plans to conduct experiments from an airplane in natural clouds during the next few months.

-more-

First page from the GE press release announcing Vince Schaefer's first artificial snow creation on November 13, 1946. The release was embargoed until the following day. Courtesy of Schenectady Museum.

www.ingramcontent.com/pod-product-compliance
Lightning Source LLC
Chambersburg PA
CBHW082058230426
43670CB00017B/2882